Elements – Continents

NOVA ACTA LEOPOLDINA
Abhandlungen der Deutschen Akademie der Naturforscher Leopoldina

Im Auftrage des Präsidiums herausgegeben von
HARALD ZUR HAUSEN
Vizepräsident der Akademie

| NEUE FOLGE | NUMMER 360 | BAND 98 |

Elements – Continents
Approaches to Determinants of Environmental History and their Reifications

Leopoldina Workshop

Deutsche Akademie der Naturforscher Leopoldina
in Collaboration with
DFG Research Training Group
"Interdisciplinary Environmental History"

Göttingen, Germany
November 14 to 15, 2007

Editors:

Bernd HERRMANN (Göttingen)
Member of the Academy

Christine DAHLKE (Göttingen)

With 63 Figures and 6 Tables

Deutsche Akademie der Naturforscher Leopoldina –
Nationale Akademie der Wissenschaften, Halle (Saale) 2009
Wissenschaftliche Verlagsgesellschaft mbH Stuttgart

Redaktion: Dr. Michael KAASCH und Dr. Joachim KAASCH

**Die Schriftenreihe Nova Acta Leopoldina erscheint bei der Wissenschaftlichen Verlagsgesellschaft mbH, Stuttgart, Birkenwaldstraße 44, 70191 Stuttgart, Bundesrepublik Deutschland.
Jedes Heft ist einzeln käuflich!**

Die Schriftenreihe wird gefördert durch das Bundesministerium für Bildung und Forschung sowie das Kultusministerium des Landes Sachsen-Anhalt.

Cover image:
Map of the world by Henricus HONDIUS from 1630, exhibiting the antique quaternary pattern of a nature in balance and harmony still as an influential episteme. The then four continents of the map are accompanied by four famous geographers in the corners (CAESAR, PTOLEMY, MERCATOR and HONDIUS' father JODOCUS), furthermore by the four elements (fire, air, water and soil). A European emperor (representing the Caucasians) receives delegates from the Middle East, Africa and America (centre bottom), an allusion to the four human races known at that time. Map taken from the Atlas Major of 1665, courtesy of Bijsondere Collecties, Universiteitsbibliotheek Amsterdam – UvA.

Bibliografische Information der Deutschen Nationalbibliothek
Die Deutsche Nationalbibliothek verzeichnet diese Publikation in der Deutschen Nationalbibliografie; detaillierte bibliografische Daten sind im Internet über http//dnb.ddb.de abrufbar.

Die Abkürzung ML hinter dem Namen der Autoren steht für Mitglied der Deutschen Akademie der Naturforscher Leopoldina.

Alle Rechte, auch die des auszugsweisen Nachdruckes, der fotomechanischen Wiedergabe und der Übersetzung, vorbehalten.

Die Wiedergabe von Gebrauchsnamen, Handelsnamen, Warenbezeichnungen und dgl. in diesem Heft berechtigt nicht zu der Annahme, dass solche Namen ohne Weiteres von jedermann benutzt werden dürfen. Vielmehr handelt es sich häufig um gesetzlich geschützte eingetragene Warenzeichen, auch wenn sie nicht eigens als solche gekennzeichnet sind.

© 2009 Deutsche Akademie der Naturforscher Leopoldina e. V. – Nationale Akademie der Wissenschaften
06019 Halle (Saale), Postfach 11 05 43, Tel. + 49 345 4723934
Hausadresse: 06108 Halle (Saale), Emil-Abderhalden-Straße 37
Herausgeber: Prof. Dr. Dr. h. c. mult. Harald ZUR HAUSEN, Vizepräsident der Akademie
Printed in Germany 2009
Gesamtherstellung: Druck-Zuck GmbH Halle (Saale)
ISBN: 978-3-8047-2604-8
ISSN: 0369-5034
Gedruckt auf chlorfrei gebleichtem Papier.

Contents

LEMMERMÖHLE, Doris: Welcome Address 7

SCHULZ-HARDT, Stefan: Greetings 9

HERRMANN, Bernd, and DAHLKE, Christine: Introduction 11

Elements

ALLGÖWER, Britta: Fire – Benefit or Nuisance? Landscape and Fire Management in the Swiss National Park 19

VAN DAM, Petra J. E. M.: Water, Steam, Ice. Environmental Perspectives on Historical Transitions of Water in Northwestern Europe 29

LUCHT, Wolfgang: Air: A Planetary Hybrid 45

BORK, Hans-Rudolf, DAHLKE, Christine, DREIBRODT, Stefan, and KRANZ, Annegret: Soil and Human Impact 63

TILZER, Max M. VON: The Fifth Element: On the Emergence and Proliferation of Life on Earth 79

Continents

SIEFERLE, Rolf Peter: Europa: Umwelthistorische Determinanten 111

BEINART, William: Ecological Imperialism, Plants Transfers, and African Environmental History 133

ELVIN, Mark: Nature, Technology and Organization in Late-Imperial China 143

BARGATZKY, Thomas: The Iconic Quality of Land in Australia and Oceania 159

McNEILL, John: Environmental History in the Americas: The Two Great Invasions 183

Research Training Program „Interdisciplinary Environmental History"

HERRMANN. Bernd, and DAHLKE, Christine: Preface 201

KLAMMT, Anne, and STEINERT, Martin: "Slavs, Waters and GIS" – Methods and Base Data to Search for Watercourses and Floodplains in a Meso Scale Study 205

POTSCHKA, Jens: Water and Waters in the Latin Medieval Sources – an Evaluation of the Settlement Area of the Slavs by a Semantic Analysis 219

WILGEROTH, Cai-Olaf: City – Forest – Man. Environmental-Historical Studies of a Fundamental Urban Relation: Goslar and Hildesheim between Medieval Desertification Period and Thirty Years' War 223

HÜNNIGER, Dominik: Cattle Plague in Early Modern Germany: Environment and Economy/Knowledge and Power in a Time of Crisis 235

CORTEKAR, Jörg, and MARGGRAF, Rainer: Cameralistic and Utilitarian Conceptions of Happiness and their Implications in Respect of Today's Environmental Crisis 241

HÖLZL, Richard: Contested Forests – Environmental Crimes between Science and Rural Society: Bavaria 1780–1860 249

MUTZ, Mathias: Nature's Product? An Environmental History of the German Pulp and Paper Industry 259

SPICALE, Jessica, and BÜRGER-ARNDT, Renate: 200 Years of Flora Development in the Natural Landscape Unit "Göttinger Wald" 265

HENNIG, Anna-Sarah: Urban Environments and their Perception Reflected in 18th and 19th Century Medical Topographies 273

WINDELEN, Steffi: Mice, Maggots, Moles: On the Discussion of Vermin in the 18th Century 279

STÜHRING, Carsten: Perception and Control of Cattle Epidemics in the Electorate of Bavaria in the 18th Century 281

ARMENAT, Manuela: Schwarze Elster in the Flow of Time – Landscape Change and History Supported by Hydraulic Engineering 283

BADER, Axel: Silva Nervus Belli 287

KREYE, Lars: Replanting the World. Colonial Forestry in the German "Kaiserreich" 1884–1918 295

MASIUS, Patrick: Flooded. Social Perspectives on Natural Disaster in 19th Century Germany 299

ZWINGELBERG, Tanja: The Influence of Medical Topographies on Urban Development and Residents' Health in Urban Environments in the 18th/19th Century 301

SCHWARZER, Markus: Conceptions and Ways of Dealing with Post-Mining Landscapes. A Cultural Analysis of Planning Discourses 303

Welcome Address

Doris Lemmermöhle (Göttingen)
Vice-President of the Georg August University of Göttingen

Ladies and gentlemen,
Members of the Research Training Group,
Colleagues,

It is my pleasure to welcome you on behalf of the Georg August University of Göttingen to the opening of the Workshop "Elements – Continents".

This workshop marks the official start of the second cohort of doctoral students of the DFG Research Training Group "Interdisciplinary Environmental History" at our university. I congratulate the new graduates on their personal success at having been admitted to this research group, which is breaking new ground with regard to its organization and its subject. Hopefully, this workshop will have a long-lasting positive impact on your own projects.

I would like to express thanks to the *Deutsche Forschungsgemeinschaft* (DFG) for supplying the material backup. Thanks also go to those colleagues who made the proposal for having enriched the fields of research in Göttingen. I acknowledge with special thanks that the German Academy of Sciences Leopoldina, which is meeting for the first time the challenge of such an event, has chosen this research training group in Göttingen for this purpose. My special thanks go to the representative of the Leopoldina who is here today, Professor Berg, for the dedication and commitment which has finally made this event possible in this form.

I would like to welcome especially the speakers of this workshop, some of whom have travelled a long way to come to Göttingen. I hope that both the town and the university will be good hosts so that you will have pleasant and inspiring memories of this event and of your stay.

The topic of this workshop picks up an important discussion of fundamental parameters in environmental history, which is presently being challenged by neo-deterministic explanatory approaches. I hope that in the scientific discourse you will succeed in finding positions where a consensus can be reached. I learned that the Leopoldina will also see to the publication of the workshop. So I am awaiting with great interest the results of this conference.

Coming to close I would like to remind you that the University of Göttingen is closely connected with the subject of this conference not only through this research training group. Through its collection of the South Seas, Göttingen is constantly linked with the history of discoveries in one of the continents, Oceania. And after all, the element Cadmium was discovered here, one of those elements which present the biggest problems for today's industrialized world. In Göttingen lived and taught scientists like Haller, Grisebach, Bergmann and Firbas: well-known biologists, who laid the foundation-stone for today's understanding of the geographical and biological equipment of the continents.

Doris Lemmermöhle

In this respect, Göttingen is, in a scientific-deterministic sense, by all means the place for a conference of this kind, since the weight of our historical disciplines still reinforces this determinism. I wish you success, inspiring discussions, and a fruitful conference.

>Prof. Dr. Doris LEMMERMÖHLE
>Vice-President
>Georg-August-Universität Göttingen
>Pädagogisches Seminar
>Waldweg 26
>37073 Göttingen
>Germany
>Phone: +49 551 399446
>E-Mail: dlemmer2@uni-goettingen.de

Greetings

Stefan SCHULZ-HARDT (Göttingen)

Dean of the Faculty of Biology of the Georg August University of Göttingen

Vice-President LEMMERMÖHLE,
Speakers and members of the Research Training Group "Interdisciplinary Environmental History",
Members of the German Academy of Sciences Leopoldina,
Ladies and gentlemen,

Today, two groups of people are in the focus of our attention: The first one is the new group of doctoral students of the Research Training Group "Interdisciplinary Environmental History", which will be officially admitted to the academic community by this workshop. We wish them a lot of pleasure in their scientific work and much success for their projects.

The other group consists of those scientists who make this workshop possible by their unhesitant participation and who, through their lectures, remind us of the roots of environmental history in terms of the history of ideas. They are all renowned and important representatives of international environmental history. I would like to thank them for coming to Göttingen to discuss their insights and the results of their research together with the next scientific generation.

As the Dean of the Faculty of Biology it is a pleasure for me to see this event as yet another sign that our long-lasting engagement for environmental history will also be fruitful in the future. Of course, it is a special pleasure for me to point out that it was the Faculty of Biology in the first place which offered the scope for the development of activities in environmental history to the university. At the same time, I would like to emphasize that it is the funding by the *Deutsche Forschungsgemeinschaft* that makes this Research Training Group possible, and therefore I would like to thank the *Deutsche Forschungsgemeinschaft*, today represented here by Dr. PURSCHE. Of course, we do hope that the work which has already been done in this Research Training Group will be acknowledged and lead to a second grant period.

My special thanks go to the German Academy of Sciences Leopoldina. It is for the first time ever that the Leopoldina organizes an event like this in cooperation with young scientists. We have noticed with gratitude but also with pride that you, the representatives of the academy, have chosen a graduate school in Göttingen and specifically one from my own faculty. We hope that this example will set standards for further collaboration between the University and the Academy.

Finally, let me add a few words from the perspective of my own discipline, which is Social Psychology. The talks and the discussions in the next two days, as well as the projects in this Research Training Group, are certainly not what is called "mainstream". In terms of Social Psychology, what you are doing here can be regarded as "minority influence". One interest-

Stefan Schulz-Hardt

ing thing about minority influence is that, if minorities act consistently over time, they foster creativity and innovation. In one of the seminal experiments in Social Psychology it has been shown that people who are exposed to a minority advocating a different position to their own one, come up with better and more creative solutions to a given problem – even if the position advocated by the minority was completely wrong! Therefore, you should not feel under pressure to generate ultimate truths within this workshop or within your projects in the next year – the sole fact that you advocate an alternative to the mainstream should be beneficial for your academic discipline.

In this spirit I wish you all a fruitful conference, much pleasure in the exchange of ideas, and sustainable inspiration for your scientific works.

> Prof. Dr. Stefan SCHULZ-HARDT
> Dean
> Faculty of Biology
> Department of Economic and Social Psychology
> Institute of Psychology
> Georg August University of Göttingen
> Goßlerstraße 14
> 37073 Göttingen
> Germany
> Phone: +49 551 3913561
> Fax: +49 551 3913570
> E-Mail: schulz-hardt@psych.uni-goettingen.de

Introduction

Bernd HERRMANN ML and Christine DAHLKE (Göttingen)

With 1 Figure

On November 14th and 15th, 2007, the German National Academy of Sciences Leopoldina held a workshop at the Georg August University Göttingen in collaboration with the Research Training Program 1024 (*Graduiertenkolleg*) "Interdisciplinary Environmental History" of the German Science Foundation (DFG). The subjects of this workshop offered a certain but programmatic ambiguity: "Elements and Continents".

Why is it that "Elements and Continents" can be worth of being discussed prominently in environmental history in the days of climatic change, global dimming, flooding Central America, impending volcano eruptions etc.? In fact, hazards and catastrophes were core subjects of environmental history at its beginning. Evenly the beginning of German environmental history is marked by the famous analysis of the large earthquake and landslide of medieval Villach, which was published by Arno BORST in 1981. But environmental history has turned into broader scopes and views of what is usually addressed as the two-realm-area of "Man and Environment" in history during follow-up studies. The interest in events later to become issues of environmental history has never been the interest in events as such but is has always been the interest in structures behind those events. This is not talking about structures in terms of natural principles or laws that govern the natural. It is about structural components of the human action in coping with challenges of their natural and culturally transformed environment. It is mentalities and human perception that form the character of human responses. Those responses necessarily depend on times and places, depend on cultures, on attitudes, and whatever features else. Thus they might be judged as arbitrary, or random, or superstitious, enlightened or sophisticated. But they are the result of a specific and structured concept of and towards the natural. More than that, they are "rational" in terms of stringent patters of actions within a given society. If they were merely consistent, any historical study would be useless.

Historical studies turn the past into experience by recollection. If we are interested in a continuing and sustainable environment, we ought to understand how humans interact with their environment. There is no understanding of this interaction if we do not understand the structures that shape and influence human perceptions and human concepts of the environment. There is a clear difference between a reflected action and a reflex action, as the reflex is usually considered of being free of reflection. However, as any human action, even reflexes are influenced. They are influenced for example by the tacit knowledge that underlies a given society. We are not taught superstition, we live it. We might be educated in style, but mostly, we live it. We are not introduced to matters of course, we live them. All these form powerful influences on our reflexes as well as on our reflections. But, if history was only a sequence of enlightened positions and rational arguments, any attempt of historicization and historic analysis would be unnecessary. Instead, historic analyses refer to factors and conditions that influenced aims of actions and thoughts of individuals beyond the obvious and beyond the rational.

Bernd Herrmann and Christine Dahlke

During the past two decades Environmental History, especially in the German speaking countries, has slowly changed from an "uncommon ground" towards a "common ground" in historic sciences and the humanities. Mostly because of solid, sometimes popular work, and in a few cases even through world sellers like John MCNEILL's *Something New Under the Sun*. The daily business in environmental history has identified so many interesting fields of research and dedicated its work to subjects so various, that quite a few contributions have already left far behind basic questions of epistemic significance and of shaping the field. In fact, many papers not even refer to epistemological issues. It appears as if there was a common feeling for the underlying structures. Scholars seem more attracted by their actual problem and take the crude assumption for granted that environmental history is what environmental historians do. But we (BH and CD) were brought up in our sciences that one should never explain a phenomenon through the phenomenon itself. Therefore, we appreciate William BEINART's presence at this conference, since he was one of the first scholars to attempt a definition for environmental history. Definitions however provide an inclusive epistemological framework.

Given this situation, what would be the guideline for graduate students to orientate, to see and understand mainstream epistems in environmental history rather than developing a vague feeling for them?

In the field of history of science there is an approved strategy to get down to the essence of a field by a set of operations. Most common is "the rethinking-approach ...". So, rethinking environmental history does mean to newly discover or invent its basics, while the full knowledge that has been brought up since its beginning is in the back and available. If one prefers, this is sort of an "*against the grain*"- or an "*a rebours*"-strategy to locate *ex negativo* the basic ideas.

The major prerequisite for environmental history is space, where humans can exist. And space is made of and filled with matter. There are different spaces on earth suitable for human living. These spaces differ not by matter but by arrangements of matter, shaping the unique challenges of those spaces for human responses. Therefore we have identified matter and space as a baseline for environmental history. Humans entered those spaces only long after the spaces had been formed. However, once in those places the coevolution of the two systems "man" and "environment" started. Coevolution is the process where human opinions come into play, opinions provided by experience and ontological assignments. Neither "man" nor the "environment" does exist in sheer objective and absolute way. Jakob VON UEXKÜLL discovered the specific and species depending "environment" (which is admittedly far from his ideas, today). Ernst CASSIRER transformed this idea into his definition of man as "animal symbolicum", demonstrating that the only approach where humans can experience the environment is through symbolic forms.

So, whoever deals with environmental history has to be aware not only of the natural principles that govern matter and space, but has to consider predominantly human opinions about matter and space. If there is any doubt about that, just recall human attitudes of using the environment. A society practicing a natural religion surely has different opinions on matter, space and the bio-components than a technically improved and enlightened society.

It is remarkable that opinions about the matter that influenced human ideas about "nature" and the "natural", not only in European societies but in many other cultures, basically refer to the elements "fire, water, soil, and air". It is not necessary to go into details of presocratic philosophy and its influence on harmonistic concepts (cf. BÖHME and BÖHME 1996). The ancient ideas were transformed into corresponding scientific concepts as the four directions, associated with deities, colours, landscapes and human compositions. What became the ancient humoric system of GALEN survived in a very practical way and coined concepts of science during the transformation process from magic to experimental science, to enlightened and rational positions.

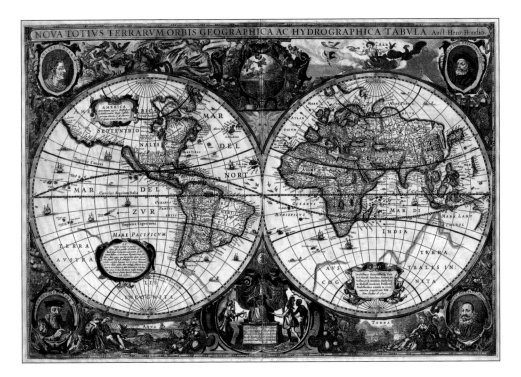

Fig. 1 This Map by Henricus HONDIUS from the early 17th century exhibits the founding influence of the "four elements" and related ideas based on the number four for the emergence of science and for the reception of the natural world. At that time only four continents were known. They were corresponding to four seasons, four winds and four large rivers. (Atlas Major of 1665 [2005], p. 44; by courtesy of: Bijzondere Collecties, Universiteitsbibliotheek Amsterdam – UvA.)

This and related reasoning finally brought up the idea of this conference: "Elements – Continents – Approaches to determinants of environmental history and their reifications". Two main fields have determined the relations between man and nature in the course of history. Our intention was to encourage rethinking by linking the features of the two fields with modern environmental history approaches.

The first main realm refers to those elementary qualities which have been relevant since the beginning of history because of their material quality and, at least since antiquity, because of their metaphysical quality. These "elements" are fire, water, air and soil. They mediate directly between man and nature used and exploited by him.

Stephen PYNE pioneered the environmental history of fire. Fire is the element that has shaped a whole continent (Australia) in a specific way and accompanies humans in history in so many and culturally important ways. Britta ALLGÖWER's (Zürich, CH) approach demonstrates the impact of fire on natural and cultural issues in an ideal type of a microhistoric study.

The overview on water by Petra van Dam (Amsterdam, NL) has much to do with her Dutch homeland that was in some ways water born. It was not only the purveyor of what became the Netherlands but helped also to defend and protect the country. It provided an infrastructure and helped the Dutch early exploiting their bogs as fossil energy source. Since water comes in different conditions of aggregation, her paper also focuses on aspects of steam and ice. As a surprising outcome Petra van Dam points to a desideratum in research since knowledge on fresh water supply during the frost season has not yet been investigated in detail.

"Air" probably is the most investigated element on earth at present, since it carries climate, microorganisms, pollution, and energy in terms of storms that scare people. We are understanding the role of "air" within the climate context, the most important issue for humans within forthcoming years. Thus the contribution of Wolfgang Lucht (Potsdam, D) is the timely approach to link insights from science with humanities. Climate change turned out to provide very strong push and pull factors for the development of civilizations during human history (e.g. Issar and Zohar 2004).

All life depends on water, but most organisms cannot survive without being linked with soil, be it directly or indirectly. In fact, there is more life (biomass) in soils than on top of its surface. However, humans walk on the ground and tie their cultures to it. Obviously this is done more intensively than we would expect. For example, only for a short time we have known that "untouched pristine rainforests" of the Amazon have a long lasting history of anthropogenic soils (Woods et al. 2008), but the anthropology of soil (Herrmann 2006) has still to be written. Hans-Rudolf Bork (Kiel, D) and colleagues outline the principles of scientific understanding of soil.

By introducing a fifth element along with the four, the Göttingen workshop presented a special variant: the diversity of organisms. This corresponds to a minor extent to a philosophical understanding, where "life" cannot only be seen as a mere additional quality of material components. Here it rather corresponds to a rhetorical gestus, by which the workshop focused on the truly "elementary" importance of biodiversity. By giving his abbreviated version of life on earth Max von Tilzer (Konstanz, D) illuminates the many facets and the importance of biota for whatever approach to environmental history. His contribution was given as the public evening lecture.

The second main realm refers to the material equipment of large human living areas for which the term "continent" stands. The natural determinants fauna, flora, microorganisms and qualities of soil determine the spectrum of possibilities of anthropogenic land use but also set limits for human activities. We have looked for scholars that will either provide outlines of environmental history for given continents, or who are specifically concentrating on aspects that are underrepresented in environmental history approaches. We appreciated that this idea could successfully be realised.

Concentrating on its "special course" Rolf-Peter Sieferle (St. Gallen, CH) opened the second section with a concise environmental history of Europe. William Beinart (Oxford, GB) made accessible how mistaken Afrika's contribution is to resources that are nowadays used globally by humans. Since "Asia" is far too big to be dealt with in a short conference contribution, Mark Elvin (Canberra, AUS/ London, GB) kindly concentrated on the hottest spot of human history in Asia, namely China. The experience of Thomas Bargatzky (Bayreuth, D) as ethnologist doing field work in Australia and Oceania reminds us that ideas of "continents", "land", and "soil" are all concepts of the mind. He pointed out how deeply they interfere with human approaches of nature, the natural and the self-understanding.

Introduction

As one could assume, there is no "special message" from the workshop, the outcome is in the papers that could be jointly published through the help of the Academy. It was important at least to us, that the workshop helped to recall that environmental history owes a lot to previous ideas, thinkers and scientists and how lively and even fruitful it can be to change directions of viewing.

The workshop would not have been possible without the help and assistance of many people. First, our thanks go to those scholars who made the workshop possible by their lectures. All of them are renowned colleagues from the environmental history community. Most of them turned into friends over the years, which brings us to a very personal "Thank you" for their support and cooperative friendship in realizing this event.

We thank the German National Academy Leopoldina for its support that made participation possible for most speakers from abroad. The understanding and help of the secretary general Jutta SCHNITZER-UNGEFUG is greatly appreciated. We thank also Joachim and Michael KAASCH of the academy office for their meticulous and patient work with the publication.

Our thanks go to Katharina BOUFADEN for office support, who managed many impossible things during the days of the strike of German locomotive drivers. Everybody got to the venue in time and nobody got lost after the workshop. Our PhD-students helped a lot with organization and catering. Finally, we have to thank the German Science Foundation for its support and engagement to promote the Research Training Group 1024 and the Georg August University in Göttingen, our home university.

References

BEINART, W., and COATES, P.: Environment and History. The Taming of Nature in the USA and South Africa. London, New York 1995
BÖHME, G., und BÖHME, H.: Feuer, Wasser, Luft, Erde. Eine Kulturgeschichte der Elemente. München 1996
BORST, A.: Das Erdbeben von Villach. Historische Zeitschrift *233*, 529–569 (1981)
HERRMANN, B.: Man is made of mud. "Soil", bio-logical facts and fiction, and environmental history. Die Bodenkultur *57*, 215–230 (2006)
HONDIUS, H.: World Map. In: Bibliographischer Nachdruck Österreichische Nationalbibliotek. Atlas Major of 1665. J. BLAEU and texts by P. VAN DER KROGT. Köln: Taschen 2005
ISSAR, A., and ZOHAR, M.: Climate Change – Environment and Civilisation in the Middle East. Heidelberg 2004
MCNEILL, J.: Something New under the Sun. An Environmental History of the Twentieth-Century World. New York, London 2000
PYNE, S.: Fire. A Brief History. Seattle 2001
WOODS, W., TEIXEIRA, W., LEHMANN, J., STEINER, C., WINKLER PRINS, A., and REBELLATO, L. (Eds.): Terra Preta Nova: Wim Sombroek's Vision. Berlin 2008

Prof. Dr. Bernd HERRMANN
Georg-August-Universität Göttingen
Johann-Friedrich-Blumenbach-Institut
für Zoologie und Anthropologie
Abteilung Historische Anthropologie
und Humanökologie
Bürgerstraße 50
37073 Göttingen
Germany
Phone: +49 551 393642
Fax: +49 551 393645
E-Mail: bherrma@gwdg.de

Dr. Christine DAHLKE
Georg-August-Universität Göttingen
DFG-Graduiertenkolleg Interdisziplinäre Umweltgeschichte
Bürgerstraße 50
37073 Göttingen
Germany
Phone: +49 551 393890
Fax: +49 551 393645
E-Mail: cdahlke@gwdg.de

Elements

Fire – Benefit or Nuisance?
Landscape and Fire Management in the Swiss National Park

Britta ALLGÖWER (Davos/Zürich)

With 4 Figures

Abstract

Fire is a driving factor in landscape and forest evolution. However, as the Alps have been subject to intense land use practices over many centuries, fire has been excluded from the range of natural disturbance activities and regimes. It is only recently that fire has come into focus again as a natural disturbance agent in Alpine ecosystems. Due to changing economic and demographic conditions many forested areas are not maintained any longer, but left to themselves. Consequently, fuels are building up. This process is especially visible in the forests of the *Swiss National Park* (SNP), Engadine Valley, where Swiss Federal law protects all natural processes occurring within the borders of the SNP since it was founded in 1914. Natural wildland fires are counted among them and should not be extinct. Strict nature conservation represents the top goal of the SNP (IUCN category 1) and does not allow any mitigation measures being undertaken unless the park is put at risk in its very existence. Nevertheless, for societal reasons all fires are put out at present regardless whether they are of natural or human origin. Almost 100 years of strict nature protection have triggered fuels to build up in the boreal type forests of the SNP reaching the point where natural fire cycles could come into play again. The core questions are whether or not (natural) fires play a role in the forest ecosystem of the Swiss National Park and to which extend?

Such questions stand at the origin of intense forest fire research studies in the Swiss National Park and its surroundings. With field-based fuel investigations and high-resolution Remote Sensing (i.e. LIDAR and Imaging Spectroscopy) we get a very good picture of the present forest and fuel structures, allowing us to predict potential fire behavior. On the other hand pollen and charcoal analysis show us that fire has been an important and regular disturbance factor in the SNP area, shaping vegetation succession long before men became a dominant factor in this remote landscape. All these elements are input to the fire management policy of the Swiss National Park which is particularly challenging as it needs to respect and allow the natural processes occurring in the area, together with the demands of the local population.

Zusammenfassung

Feuer ist eine treibende Kraft in der Landschafts- und Waldentwicklung. Durch die Jahrhunderte dauernde intensive menschliche Nutzung der Alpen ging Feuer als ökologischer Faktor jedoch verloren. Erst in jüngerer Zeit rückte Feuer als natürlicher Faktor in den Alpinen Ökosystemen wieder ins Blickfeld. Aufgrund sich ändernder ökonomischer und demographischer Verhältnisse werden Gebirgswälder heute nicht mehr intensiv genutzt und sind vermehrt sich selbst überlassen. In der Folge sammelte und sammelt sich Brandgut an. Dieser Prozess ist besonders sichtbar im Schweizerischen Nationalpark, der 1914 gegründet wurde und wo seither sämtliche natürliche Prozesse unter gesetzlichem Schutz stehen. Als Park der IUCN-Kategorie 1 stehen im Schweizerischen Nationalpark Natur- und Prozessschutz an oberster Stelle. Natürlich verursachte Brände gehören auch zu den natürlichen Prozessen und sollten nach Möglichkeit nicht gelöscht werden. Aus gesellschaftspolitischen Gründen werden sämtliche Brände jedoch – ob durch die Natur oder den Menschen verursacht – gelöscht. Beinahe 100 Jahre strikter Naturschutz und ausbleibende Nutzung haben in den Borealen Waldtypen des Schweizerischen Nationalparks eine Brandgutsituation entstehen lassen, bei welcher natürliche Feuerzyklen wieder eine Rolle spielen könnten. Die zentrale Frage ist, welche Rolle natürlich verursachte Feuer in dieser Gegend gespielt haben könnten und was dies für die zukünftige Landschaftsentwicklung bedeutet. Und was bedeutet dies für die Verwaltung eines Schutzgebietes, das natürliche Prozesse von Gesetzes wegen zulassen – ja schützen muss?

Britta Allgöwer

Solche Fragen stehen im Zentrum intensiver Untersuchungen zur Waldbrandthematik im Schweizerischen Nationalpark und dessen Umgebung. Mit Hilfe hochauflösender Fernerkundungsmethoden (im speziellen Laserscanning und Bildspektrometrie) werden einerseits die Brandgutsituation und das potentielle Waldbrandverhalten abgeschätzt. Andererseits beleuchten Kohle- und Pollenanalysen die Langzeitvegetations- und Feuergeschichte und zeigen, dass Feuer die untersuchte Landschaft prägten, lange bevor der Mensch darin aktiv wurde. Alle diese Daten fließen in die Bildung von Zukunftsszenarien ein, welche dazu dienen, Feuermanagementkonzepte für den Schweizerischen Nationalpark zu entwickeln. Dies ist besonders anspruchsvoll, da letztere sowohl die Interessen der Natur wie der lokalen Bevölkerung berücksichtigen müssen.

1. Introduction and Motivation

1.1 Fire – Companion of Mankind and Nature in 'Good and Evil'

Mankind has a long relationship with fire, driven by both fascination and fear at the same time. Used wisely, fire is a life spending and live saving tool. Human wellbeing by and large depends on the availability of sufficient energy sources. The manufacturing of most tools requires a heat source. Without fire, the evolution of mankind would not have been possible. Tribute to this fact is the implementation and praise of the goddess *Vesta* in the ancient Roman culture, or *Hestia* in the Greek culture as being the keeper of the sacred flame, essential for a fortunate home. Gone out of control or used in evil intention, fire is a devastating, deadly force. Countless are the victims and examples of destruction where fire has been used as a weapon to destroy the enemy and his life supporting infrastructure. Profound and comprising books on how fire and mankind did co-evolve were written by the well-known fire historian Stephen J. PYNE, i.e., *World Fire* (PYNE 1995) and *Vestal Fire* (PYNE 1997).

Fire also has two faces in nature. Regularly occurring, low intensity (natural) fires in e.g. *Pinus Ponderosa* forests in North America have a cleansing, life sustaining and renewing effect. Rigorous suppression of fires in such ecosystems may lead to heavy fuel build-ups and finally severe, nearly uncontrollable fire behavior once such forests are ignited, be it by lightning or by (human) negligence. It took several decades to understand these mechanisms and to refrain from fire extinction by all means. 'Smokey the bear' and 'all fires must be avoided' dominated American fire management policies. For many decades, nature conservation principles did not foresee the beneficial and live renewing force of fire. The same was true for large predators. American national parks were founded for recreational purposes first of all and not for understanding and investigating natural processes. Hence fire and large predators were extinct whenever encountered. Out of this, unbalanced wild ungulate populations and catastrophic fires resulted. In Europe, and especially in the Mediterranean, a widely distributed wildland urban interface (WUI) area developed. Being dominated by highly flammable vegetation, the Mediterranean WUI counts among the most fire prone areas of this planet. However, even the most restrictive fire mitigation measures do not seem to diminish the fire problem in the Mediterranean. On the contrary, the number of fires and the extent of area burned seem to increase. Hence, a modern and spirited tribute to the wise use of fire is the fact that the European Commission is funding a large European research project – the *Fire Paradox*[1] project – that studies what it takes to safely apply the techniques of prescribed burning within the different

---

[1] *Fire Paradox* – An Innovative Approach of Integrated Wildland Fire Management Regulating the Wildfire Problem by the Wise Use of Fire: Solving the Fire Paradox. European Research Project FP6-018505. http://www.fireparadox.org. The author of this contribution is also a project member.

European countries that are severely affected by wildland fires, but partly do not allow to use fire as a fuel reduction measure.

1.2 Fire – Essential Agent in Alpine Forest Ecosystems

Fire often is a driving factor in landscape and especially forest ecology. However, as the Alps have been subject to intense land use practices over many centuries, fire has been excluded from the range of natural disturbance activities and regimes. It is only recently that fire has come into focus again as a natural disturbance agent in Alpine ecosystems (e.g., CARCAILLET 1998, CARCAILLET and BRUN 2000, STÄHLI et al. 2006) as fuels are building up due to decreasing forest management measures. Changing economic and demographic conditions have triggered substantial alterations in traditional land and forest management practices, leading to an increase of biomass. While forest and alpine pasture resources were overused in the late 19th century, the second half of the 20th century was characterized by the partial retreat of human land use in the Alps. In many areas, forests and shrubs started growing back. Moreover, logging in rugged terrain is costly and avoided wherever possible. Consequently, we may be faced with forests that build up substantial fuels. Furthermore, in 1876 the first Federal forest police law (*Forstpolizeigesetz*) was implemented in order to protect Swiss forests from being overused. Today, Alpine landscapes are rather polarized: on the one hand we find intensely used areas with a sophisticated network of tourism infrastructure facilities and on the other hand whole regions are being left behind. In light of this development it may well be that natural fire cycles can come into play again. However, most landscape and management policies do not anticipate this possibility and are not prepared to meet this challenge.

1.3 Fire in the Swiss National Park

Located in the *Engadine* Valley, the *Swiss National Park* (SNP) was founded in 1914. Ever since, the forests are not managed anymore. *Swiss Federal Law* protects all natural processes occurring within the borders of the *Swiss National Park*. Natural wildland fires are counted among them and should not be extinct. Strict nature conservation represents the top goal of the SNP (IUCN category 1) and does not allow any mitigation measures being undertaken unless the park is put at risk in its very existence. Nevertheless, for societal reasons all fires are put out at present regardless whether they are of natural or human origin.

The leasing contracts between the *Swiss National Park* and the land owning communities specify that the SNP is liable for damages emerging from within its borders. Therefore, all fires are extinct at present regardless whether they are of natural or human origin. After almost 100 years of strict nature protection – meaning no management measures whatsoever – the question arises whether or not fuels are building up in the boreal type forests of the SNP reaching the point where natural fire cycles could come into play again and what to do from a managerial point of view?

The goal of this long-term study is to develop fire management strategies that integrate ecological and societal demands alike and thus have a chance to be accepted by both the local and cantonal authorities and the population.

2. Material and Methods

Fire history and vegetation conditions are part of the puzzle required for the elaboration of a Wildland Fire Management Plan that meets the requirements of a nature conservation area such as the Swiss National Park.

2.1 Study Area (Swiss National Park)

The *Swiss National Park* is situated in the Southeast of Switzerland in the midst of the Engadine Valley, one of the top tourist destinations of the Canton of Grison. Rugged topography, Dolomite limestone and little precipitation (900 mm per year) create harsh environmental conditions. Forests cover one third of the Swiss National Park. Before the foundation of the SNP in 1914 these forests were heavily logged and livestock grazing was going on (PAROLINI 1999).

2.2 Fuel Assessment

Specific SNP fuel models were established in former investigations (ALLGÖWER et al. 1998) and introduced to GIS-based fire propagation modeling (SCHÖNING 1996/2000, BACHMANN et al. 1997).

In 2002 a high resolution remote sensing campaign took place where the *Ofenpass* area was investigated with a LIDAR sensor (Falcon II by Toposys, Ravensburg, Germany) as well as two Imaging Spectroscopy sensors (DAIS and ROSIS by DLR, Germany). Both flight and ground truth campaigns are described in detail by MORSDORF et al. (2004) and KÖTZ et al. (2004).

2.3 Long-term Fire History

Little is known on the long-term fire history of the *Swiss National Park*. Unfortunately, natural archives such as mires are very rare in that area. Nevertheless, it was possible to core one mire south of Il Fuorn (central Ofenpass) and to perform pollen, plant macrofossils, microscopic and macroscopic charcoal analysis. A detailed description of our Holocene fire history investigations can be found in STÄHLI et al. (2006).

3. Results

3.1 Fuel Models and Fuel Build-up

Former studies produced three fuel models (A, B, and C) for the Swiss National Park (ALLGÖWER et al. 1998). Species names follow the nomenclature of the *Flora Helvetica* by LAU-

---

Fig. 1 (A) Model A "Mixed conifers": *Lárix decidua, Pinus sylvestris* L., *Pinus montana* ssp. *arborea, Picea abies, Pinus cembra*. Typical for N, NW und NE locations. Understorey: Ericacea. Gramineae and Cyperaceae; middle to little fuel loads (Swiss National Park). (B) Model B "Mountain pine": *Pinus montana* ssp. *arborea, P. cembra* aufkommend. Typical for S, SW and SE locations. Understorey: *Eríca cárnea* L. 'Teppiche'; middle to heavy fuel loads (i.e. Eastern Ofenpass, SNP). (C) Model C "Legföhren": *Pinus montana* ssp. *prostrata*. Typical for disturbed areas like avalanche paths. Understorey: *Eríca cárnea* L. 'carpets' or eventually gravel; high fuel loads (SNP)

Fire – Benefit or Nuisance? Landscape and Fire Management in the Swiss National Park

BER and WAGNER (1996). The following has also been presented in this condensed form by ALLGÖWER et al. (2005) in the proceedings contribution at the alparc 2005 conference, held at Kaprun (Austria).

Model A (Fig. 1A) is an assemblage of "mixed conifers", namely *Larix decidua*, *Pinus sylvestris* L., *Pinus mugo* ssp. *uncinata*, *Picea abies*, and *Pinus cembra*. This vegetation set-up is typical for N, NW and NE oriented slopes. The understorey is formed by *Erica carnea* L., *Rhododendron hirsutum* L., and *Rhododendron ferrugineum* L. as well as various *Gramineae* and *Cyperaceae* associations. Fuel loads are low to medium. Fire behavior is expected to be of low to medium intensity. However, some stands could easily develop intense fire behavior as they are rich in vertical fuels consisting of lichens and low reaching branches.

Model B (Fig. 1B) consists of "Mountain Pine" *Pinus mugo* ssp. *uncinata* mainly and is typical for the S, SW and SE facing slopes. The understorey is formed by dense *Erica carnea* L. 'carpets' that would maintain a low intensity surface fire. Depending on the amount of fine dead fuels and the distribution of vertical fuels fire behavior is expected to vary from medium to severe. In general surface fires are expected to develop into crown fires easily wherever vertical fuels, especially low reaching branches allow for torching.

In Model C (Fig. 1C) the "dwarfed Mountain Pine" (German *Legföhre*) *Pinus mugo* ssp. *mugo* is the dominant tree species. Fuel loads are high and fire behavior is expected to be severe. This *Pinus* variation forms a brush like, very dense vegetation cover that grows to an average height of 2 to 3, seldom 5 m; it is typical for avalanche shoots and other disturbance areas. The understorey often contains dense *Erica carnea* L. 'carpets' but may also consist of gravel which then reduces fire potential considerably.

The LIDAR investigation (Fig. 2) provided very good data on the vertical and horizontal distribution of the fuels as single tree geometry can be derived, in particular tree height, crown base height, crown diameter and from that fractional cover. The main outputs of the DAIS campaign (Fig. 3) are Leaf Area Index LAI (m^2/m^2), crown water content (g/m^2), equivalent water thickness (g/cm^2), and live fuel moisture content (%). All results are described in detail in MORSDORF et al. 2004 and KÖTZ et al. 2004.

3.2 Long-term Fire History

Our pollen and charcoal analysis (Fig. 4) shows that fire has been an important and regular disturbance factor in the SNP area, shaping vegetation succession long before men became a dominant factor in this remote landscape. The Il Fuorn core allows us to study approximately 8000 years of landscape and vegetation history. Thereafter we can distinguish two phases: (*i*) 6000 BC to 0 and (*ii*) 0 and 2002 AC. Contradictory to all expectations the first phase is characterized by a rather high fire frequency – one event per approximately every 250 to 300 years, whereas fire cycles become almost twice as long during the second phase. During the investigated time period *Pinus mugo* ssp. *uncinata* has always been the dominant tree species and seemed to nicely intertwine with *Picea abies* which was present in higher concentrations than today. Herbs and *Graminacea* pollen do not vary significantly between the two phases; *Cerealia* pollen cannot be found at any time at the coring site of Il Fuorn. This indicates that the Ofenpass was settled only late and for logging purposes mainly. This may also explain why fire cycles increased despite human presence. Logging and livestock activities kept fuel loads low and hence decreased fire ignition probability.

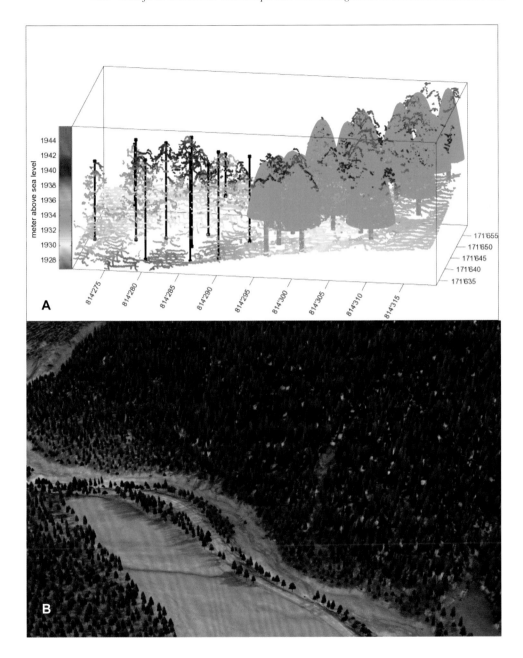

Fig. 2 (*A*) 3D visualization of LIDAR data: Geometry of single trees. (*B*) Calculated 3D forest stand scene, based on the true x.y positions of the trees, derived from LIDAR data

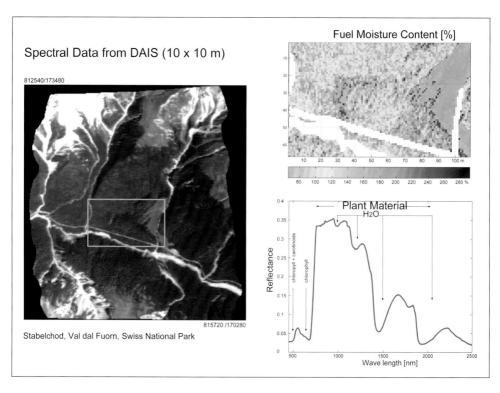

Fig. 3 Derivation of biophysical fuel parameters from hyperspectral data

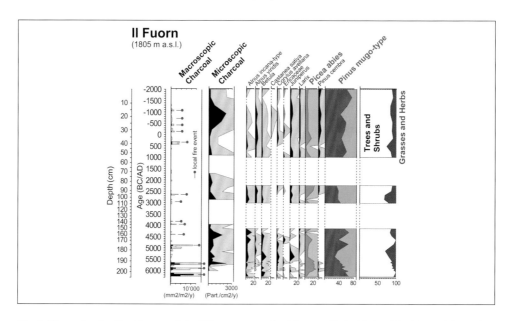

Fig. 4 Charcoal and pollen diagram of the Il Fuorn area on the Ofenpass (Swiss National Park). Special attention require the curves of *Pinus mugo* ssp. *uncinata* together with *Picea abies* as well as herbs and Graminaceae.

4. Discussion and Outlook

Fire and Mountain Pine stands seem to go along fine – they even appear to promote each other! This comes as no surprise really as it is a well-known fact that many pine species and fire have a close relationship and depend on each other. Hence, given the fact that fuels are building up in the Swiss National Park it is maybe only a question of time until a natural fire regime will establish itself again. Future fire management that fully respects the park's goals and legislation needs to respect ecological and societal requirements alike and will have to balance them carefully. If the SNP succeeds to handle this challenge successfully, he will become a pioneer in dealing appropriately with (natural) fires in the Alps in the future.

Based on the described results, previous analysis of the present fire situation (LANGHART et al. 1998) as well as fire risk analysis (BÄRTSCH 1998) wildland fire management strategies are now being developed that meet the requirements of the SNP and the surrounding areas. With that a visionary but challenging path for the successful and well-received 're-introduction' of a natural and moderate fire regime to the landscapes of the Swiss National Park is set.

Acknowledgments

The presented work is the result of a whole network of successful collaborations. Without the help of a large number of persons this work would not have been possible. Special thanks go to Markus STÄHLI and Michael BUR who both investigated fire history during their Master thesis at the University of Zürich (Department of Geography). Benjamin KÖTZ and Felix MORSDORF dedicated their entire work to the investigation of the fuel situation in the Swiss National Park during their PhD thesis at the University of Zurich (Department of Geography). Walter FINSINGER (University of Utrecht, Laboratory of Palaeobotany and Palynology) and Willy TINNER (University of Bern, Institut für Pflanzenwissenschaften) supported us in all paleobotanical tasks and questions. We are grateful to the financial support of the European research project SPREAD (EU contract Nr. EVG1-2001-00027) enabling us to perform the high resolution remote sensing campaigns in the SNP, Prof. Robert WEIBEL (University of Zurich, Department of Geography) for financing the charcoal dating analysis, and finally the Swiss National Park for providing a superb environment and '1:1 free range laboratory' for the observation of natural processes. Last but not least, it was Paul GLEASON (†) (Colorado State University, CO, USA), famous fire fighter and prescribed burning specialist of the US Forest Service, who put us on track to investigate vegetation and fire history in order to truly understand the landscape of the Swiss National Park.

References

ALLGÖWER, B., HARVEY, S., and RÜEGSEGGER, M.: Fuel models for Switzerland: Description, spatial pattern, index for crowning and torching. In: VIEGAS, D. X. (Ed.): Proceedings of the 3rd International Conference on Forest Fire Research/14th Conference on Fire and Forest Meteorology, Luso, Portugal, November, 16–20, 1998. Vol. 2, 2605–2620 (1998)

ALLGÖWER, B., STÄHLI, M., BUR, M., KÖTZ, B., MORSDORF, F., FINSINGER., W., and TINNER., W.: Long-term fire history and remote sensing based fuel assessment: Key elements for landscape management in the Swiss National Park. In: KASERER, S., BAUCH, K., *Hohe Tauern National Park and ALPARC* (Eds.): 2005, September 15–17, Kaprun, Austria. Network of Alpine Protected Areas & Hohe Tauern National Park Council, 11–14 (2005)

BÄRTSCH, A: Konzeption eines Waldbrandmanagements für die Regionen Engadina Bassa, Val Müstair und den Schweizerischen Nationalpark. Diplomarbeit, Geographisches Institut der Universität Zürich. Abt. Geographische Informationsverarbeitung/Kartographie. Zürich 1998

CARCAILLET, C.: A spatially precise study of Holocene fire history, climate and human impact within the Maurienne valley, North French Alps. Journal of Ecology *86*, 384–396 (1998)

CARCAILLET C., and BRUN J.-J.: Changes in landscape structure in the northwestern Alps over the last 7000 years: lessons from soil charcoal. Journal of Vegetation Science *11*, 705–714 (2000)

KÖTZ, B., SCHAEPMAN, M., MORSDORF, F., BOWYER, P., ITTEN, K, and ALLGÖWER, B.: Radiative transfer modeling within a heterogeneous canopy for estimation of forest fire fuel properties. Remote Sensing of Environment 92/3, 332–344 (2004)

LAUBER, K., and WAGNER, G.: Flora Helvetica. Flora der Schweiz. Bern, Stuttgart, Wien: Paul Haupt 1996

LANGHART, R., BACHMANN, A., and ALLGÖWER, B.: Spatial and temporal patterns of fire occurrence (Canton of Grison, Switzerland). In: VIEGAS, D. X. (Ed.): Proceedings of the 3rd International Conference on Forest Fire Research/14th Conference on Fire and Forest Meteorology, Luso, Portugal, November, 16–20, 1998; pp. 2279–2292 (1998)

MORSDORF, F., MEIER, E., KÖTZ, B., NÜESCH, D., ITTEN, K., DOBBERTIN, M., and ALLGÖWER, B.: LIDAR based geometric reconstruction of boreal type forest stands at single tree level for forest and wildland fire management. Remote Sensing of Environment 92/3, 353–362 (2004)

PYNE, S. J.: World Fire. The Culture of Fire on Earth. Seattle, London: University of Washington Press 1995 (paper back edition 1997)

PYNE, S. J.: Vestal Fire. An Environmental History, Told through Fire, of Europe and Europe's Encounter with the World. Seattle, London: University of Washington Press. Paper back edition 1997

PAROLINI, J. D.: Zur Geschichte der Waldnutzung im Gebiet des heutigen Schweizerischen Nationalparks. Diss. ETH Nr. 11187. ETH Zürich 1995

STÄHLI, M., FINSINGER, W., TINNER, W. and ALLGÖWER, B.: Wildfire history and fire ecology of the Swiss National Park (Central Alps): New evidence from charcoal, pollen and plant macrofossils. The Holocene: 16/6, 805–817 (2006)

        Korrespondenzadresse
        Dr. Britta ALLGÖWER
        Wissensstadt Davos – Science City Davos
        Rathaus
        Berglistutz 1
        CH-7270 Davos Platz
        Switzerland
        Phone: +41 81 4143313
        E-Mail: britta.allgoewer@wissensstadt.ch

        Dr. Britta ALLGÖWER
        GIS Swiss National Park (GIS-SNP)/
        Geographic Information System (GIS) Division
        Department of Geography
        University of Zurich
        Switzerland
        Phone: +41 44 6355253
        Fax: +41 44 6356848
        E-Mail: britta.allgoewer@geo.uzh.ch

Water, Steam, Ice.
Environmental Perspectives on Historical Transitions of Water in Northwestern Europe

Petra J. E. M. van Dam (Amsterdam)

With 2 Figures

Abstract

This contribution deals with the element of water in its different states. Viewed from the angle of environmental history, insights into the transformations of the aggregate states and their impact on the countryside and the everyday lives of people are analyzed in various historical epochs, with the Netherlands as well as various other European regions serving as examples. Here, issues regarding the supply and safeguarding of drinking, service and sanitary water as well as water in the form of ice as a medium of winter sports are addressed.

Zusammenfassung

Der Beitrag beschäftigt sich mit dem Element Wasser in allen seinen Zustandsformen. Unter dem Gesichtspunkt der Umweltgeschichte werden die Kenntnisse über die Umwandlungen der Aggregatzustände und ihre Auswirkungen auf die Landschaft und das alltägliche Leben der Menschen am Beispiel der Niederlande, aber auch anderer europäischer Gebiete in verschiedenen historischen Epochen analysiert. Dazu werden Fragen zur Bereitstellung und Sicherung von Trink- und Brauchwasser ebenso betrachtet wie Wasser in Form von Eis als Medium des Wintersports.

1. Introduction

Worldwide we currently witness a growing consciousness about the meaning of water. We are becoming increasingly aware that bodies of water communicate, both the visible bodies of water on the surface of the earth, and the invisible ones underneath. Also, we are learning that good water (water that is good for us) is scarce, and that we are extremely dependent on water. In fact, water is the basis of all life processes on earth; for instance, 66% of the human body consists of water.[1]

In this article I reflect upon water as an object of environmental history. As a starting point I take the transitions between water and the other classical elements. This has an ancient tradition, as Hartmut Böhme (2002, p. 239) explains in the proceedings of the Leopoldina Academy, in a volume devoted entirely to the theme of water. Plato and Aristotle ordered the four elements from heavy to light, as rings around the earth: earth, water, air, and fire. A dynamism characterizes the four elements. Fire is light and strives upwards, all that is heavy will sink and fall back to earth. In between are water and air. Both are a middling kind of

---

[1] www.umweltbundesamt.de/wasser-e/themen/drinking-water/index.htm.

element, since water that contains a lot of cold earth will flow down or become ice even, and water with lots of fire (read energy) will rise as vapor or steam. Thus, ice and steam may be considered as transitions between water and earth on the one hand and water and fire on the other.

Relevant general questions from an environmental-historical perspective are: How does our experience of these states of water change over time? How does our control over the transition from one state to the other change over time? And, to what extent does nature determine human history or *vice versa*? This article does not pretend to give a survey of existing trends in water history. The subject is far too comprehensive to do that within the limits of an article. Rather, my aim is to come up with potential new research questions, starting from the concept of the transitions of water.

Ice connects us to experiences of frozen landscapes, frozen surface water, and frozen drinking water. Ice connects us to weather extremes and how these affect human life. In this way, it is easily associated with one of the mainstreams of environmental history, the history of catastrophes. But I am at least as interested in how people dealt with common environmental extremes in daily life. How did people cope with severe winters? How did these winters influence people's relationship with wetlands and affect their access to drinking water, how damaging was frost to drainage systems? Steam connects us to narratives about steam as a source of power. It leads to other chapters in environmental history, such as the history of energy, and how changing energy regimes and changing water technologies interact over time in particular.

Regarding the selection of sources, I take examples from Northwestern Europe and in particular from the wetland region *par excellence*, Holland.[2]

2. Ice and Landscape: Environmental Determinism

Between the Middle Ages and the nineteenth century, Europe had a period with severe, long winters. The characteristics of the Little Ice Age are a much debated topic. Here, I go into the consequences of frozen landscapes. Figure 1 shows how ice enables people to use wetlands in a different way in winter. To some extent ice could have economic functions in the premodern period, as water becomes treadable when frozen. In winter peasants cut the reeds or other plants along the shores, using the ice as a working platform. Particularly in boggy areas, where little firm land existed anywhere, this was the most efficient way to maintain the waterside. Thus, bogs turned into extensive reed plantations; it was the regular way of harvesting. Another economic use of frozen water was the measuring of the water table. Such measuring, in particular the relative tables of several connected canals and lakes, was needed in preparation of new drainage projects, in order to estimate the capacities of new drainage canals and sluices. Measuring water tables also formed part of the data collection for making

---

[2] In the medieval period Holland was the county of the Counts of Holland. During the period of the Dutch Republic (1581–1795) Holland was the main province in terms of political and economic weight. Now Holland is the western region of the Kingdom of the Netherlands, and consists of two provinces, North Holland and South Holland (*Noord-Holland* and *Zuid-Holland*). Relevant introductory chapters for water history on a world scale in McNeill 2000. English language introductions on the history of water in the Netherlands are Van de Ven 2004 (drainage, water authorities) and Huisman 2004 (drinking water, groundwater). For Holland a Dutch language comprehensive edited volume covers many aspects of the history of water, Beukers 2007.

new survey maps, at the demand of public water authorities or private owners.[3] It goes without saying that frozen landscapes were the setting for specialized winter outdoor recreation, too, varying from playing children on sledges to serious winter sports such as long distance skating competitions. Among the first testimonies of winter sports are (excavated) medieval ice-skates made of animal bones, a material that survives pretty well in the soil conditions of wetlands.[4]

Fig. 1 A winterlandscape painted by Aert van Neer (1603–1677) shows economic and recreational uses of ice: reed cutting and ice-skating. Source: Rijksmuseum Twenthe, Enschede. Photograph by R. Klein Gotink.

All these ice-related activities show how people commonly adapted to severe winter conditions, and actually made good use of them, by developing specialized seasonal activities. In the premodern period such activities were fairly harmless, in my view. They did not affect the landscape or the organisms underneath the ice cover in any substantial way. This makes an interesting contrast with modern winter sports. Glacier skiing can be quite a polluting activity, affecting the drinking-water resources for very large areas. Large numbers of tourists on glaciers, and the accompanying snow preparation vehicles, cause serious environmental damage by polluting water sources, not to mention the disastrous effects of the surrounding over-used skiing pistes on local vegetation covers.

---

3 A very large collection of specialized water maps: www.rijnland.net/archief_en/archieven/rijnlands_beeldbank.
4 Willemsen 1998, p. 109.

How disruptive could ice be for daily life? The first example concerns spring flooding of major rivers. In lowlands ridden with large rivers, ice on the river could be a severe danger. Particularly in the eighteenth century, in the last phase of the Little Ice Age when many severe winters occurred, ice formation on the big rivers was frequent.[5] When thaw set in, local ice and ice from higher up the river started to float and formed ice mountains wherever obstacles occurred. These obstacles could be high grounds in the river or human-made groins (breakwaters), aimed at deflecting the current or erected for land reclamation, or both. Although such dams were not entirely water-tight, the water level rose and the water streamed over the dikes. As a result the dikes softened, collapsed and the water would flood the land. Broken dikes meant broken roads, disrupted connections and damage to trade.[6]

For the countryside river flooding was very harmful. It certainly meant substantial damage for countryside inhabitants, if only because they had to repair the dikes again. Also, extensive flooding of agricultural land would shorten the growing and grazing season. Yet, human and animal casualties were most probably relatively low, particularly when compared to similar events in China for instance: 275 human casualties and some thousands of cattle were high figures in 1809.[7] The peculiarity of this kind of river flooding as opposed to flooding caused by heavy rain in upstream regions, for example, was that people would be warned in advance, so they could flee. Ice mountains are highly visible, and cracking ice is a very noisy natural phenomenon. Peasants would flee in time, often simply to the upper floor of their farm. In the Dutch river area not only farms were frequently situated on (man-made) elevations, also sheds.[8] The so-called flood shed (*vloedschuur*) stood on top of a hill or was surrounded by a hill on three sides. The hill served as a wide ramp leading up to the upper floor, where the cattle was kept during the flood. The roof of the farm contained a special small door. This gave the residents camping on the upper floor access to their rowing boat, which would transport them to the flood shed in order to take care of the cattle.

An extensive case study of dike breaches in the river area was carried out by A. DRIESSEN (1997). This study concerns all river floodings in the region between the rivers Maas and Waal, from Nijmegen (German: *Nimwegen*) to Loevestein, during the period 1750–1820 (Waal is a Dutch name for the Rhine). This seventy-year period comprised fifteen years with one or more floodings, some thirty incidences in total. All but two dike breaches were caused by ice dams.[9]

DRIESSEN emphasizes that ice formation was not the only cause of river floodings. Human agency also has to be taken into account, in particular the changes wrought in the river bed. In the eighteenth century big waterworks were carried out at the German-Dutch border, changing the water loads of the several Rhine branches, the rivers Waal, Nederrijn and IJssel. Large tracts of the rivers no longer carried sufficient water, so they silted up and reeds and trees grew in the middle of the river.

The rivers were bordered by two lines of dikes, the low summer dikes and the high winter dikes. A tendency existed to keep the summer dikes very high in order to protect the lands behind the dikes for as long as possible. However, the lands between the summer and winter

---

5 VAN HEEZIK 2007, p. 44, 47.
6 VAN DE VEN 2004, p. 347.
7 VAN DE VEN 2004, p. 347. 1 million deaths after the breach of the Yellow River in 1117, see M. ELVINS contribution to this volume.
8 HARTEN 1993, pp. 44–45.
9 DRIESSEN 1997, pp. 15, 28–30, 35.

dikes belong to the streambed of the river and are meant to receive surplus (spring)water. As a result of the (too) high summer dikes, the water level of the river rose too much, and the channel was too narrow, which again contributed to the formation of ice mountains. In addition, the high water levels, which could persist for over a month, led to an increase of seepage water (under the dikes), which caused the dike to erode from within and from the land side.

A second example of how ice disrupts a society concerns the freezing of water defense systems. In a very watery countryside such as Holland people learned to use their natural environment in specialized defense systems, the so-called 'water-line'. The basic idea was that a designated part of land was flooded on purpose so that the foreign army would get stuck in the mud.

At the beginning of the Eighty-Year Revolt against the Spanish Habsburgs, several towns were besieged by the Spanish, such as Brielle (1572), Alkmaar (1573) and Leiden (1574–1575). The defense tactics that spontaneously developed in the context of a kind of guerrilla warfare comprised inundating the surrounding countryside. To this effect, drainage sluices were opened and dikes were breached to let sea and river water in. The water did not always flow very easily, because in this period the countryside was still elevated above sea level quite a bit (now it is below sea level due to centuries of man-induced land subsiding[10]). In the case of Leiden, the success of these tactics depended on having the wind at the right angle to blow the water into the canals, and on the occurrence of springtide. (Due to moon influence, the common sea tide is higher once a month.) So after months of waiting, and while the population of the besieged town was near starvation, the land finally flooded sufficiently to drive the Spanish out of their campsites and to allow the Dutch armed forces to approach by boat.

A most famous tapestry by Joost Jansz LANCKAERT made after a design by Hans LIEFERINCK (1587–1588), now in the Lakenhal Museum in Leiden, shows the fleet sailing over the inundated pastures from Rotterdam to Leiden. This design was used again in a stained-glass window in the St. Jans Church in the town of Gouda.[11] The boats were rowing boats or flat-bottomed sailing boats with stabilizing leeboards on the sides, a type of boat in common use all over the western and northern Netherlands, where relatively shallow lakes and canals predominated. The Dutch armed forces, led by WILLIAM Prince of Orange, who had thus far been the representative of the Spanish emperor in the Netherlands, had detailed landscape knowledge about which parts of land were flooded and how deeply. Thus, they were able to penetrate the myriad of dikes and roads defended by special armed constructions (entrenchments) without getting trapped. Apart from dikes and roads the countryside offered few obstructions. The predecessor of barbed-wire fences, vegetative hedges, must have stood out sufficiently, but were probably rare, the common parcel boundary in the western Netherlands being a water-filled ditch.[12]

The improvised use of water as a means of defence was successful in military terms, but also had disadvantages. The dependence on natural forces such as the tides and winds made the defence vulnerable. Severe damage to the drainage and agricultural systems was accepted for once, but not something to be repeated. After the victory at Leiden it took the Regional Water Authorities of Rijnland and many private landowners until 1580 before dikes were

---

10 VAN DAM 2000.
11 GROENVELD 2001, pp. 25, 34.
12 KLINKERT 2007, pp. 458–462.

repaired, sluices and windmills rebuilt, and the canals freed of shipwrecks. Agricultural production was slow to pick up again. In 1578 large areas were still inundated. Many peasants had no harvests for six years in a row, and the large landowners had little income, since land rents were cancelled or substantially reduced.[13]

From 1575 onwards, the water defense system became professionalized. Rather than flooding the entire countryside, the so-called water-line formed the basis of the new system. Firstly around the major cities in Holland, and later in the east of Holland at the border with Utrecht and in the south in the province of Brabant, corridors were built which could be flooded at will, using designated territories and specialized hydraulic equipment.[14]

Thus, guerrilla tactics were replaced by a regular water defense system, and the tide and wind were eliminated as obstructive forces. However, one unaccountable natural element remained: frost. Ice formation in severe winters made the military use of water vulnerable. For a long time the Republic survived in spite of this perpetual risk. A famous test of the new defense system occurred in the disastrous year of 1672. Having signed the peace treaty with Spain in 1648, the Republic of the United Netherlands had new enemies to face. France, England and the German bishops of Münster and Cologne invaded the country with an exceptionally large force of 100,000 men. The water-line proved difficult to realize. It still took a long time to flood the whole water-line. The building of temporary dams to prevent salt water from flooding the countryside delayed the process, and also peasants secretly tapped off water. On top of internal troubles, the winter was severe. In December the surface water froze solidly. Fortunately, this also caused the enemies to retreat, to the neighboring conquered province of Utrecht. For the rest of the winter small attacks by Dutch troops on skates on a minor foreign camp formed the main military action. Meanwhile, thousands of laborers worked to keep the water moving and to weaken the ice in the water-line with the aid of icebreakers and windmills. Using simple utensils, they freed the sluices of ice by hand. Attempts by the French troops to use similar water tactics for counterattacks failed, partly due to a lack of local knowledge. Whatever water they let in was tapped off by the Dutch army. In the end, the water-line was successful to the extent that it drastically slowed down the advancing troops, particularly when the battling season started again and the ice thawed in spring. Since the enemy troops were also hindered by a lack of funding, they gave up in the autumn of 1673.[15]

So, the winter of 1672 proved to be a narrow escape, but in 1794 the ice turned out to be too strong. A first attack was launched by the French in 1792–1793 but failed. A water-line south of Holland was realized, inundating large regions of the province of Brabant. Yet it was not really tested, because the French retreated after the victory over the Austrian army at Fleurus in June 1794. In the summer of 1794 the French took an easterly route and large regions were inundated again, both to the south and the east of Holland. However, the winter of 1794 was extremely cold. The frost turned the river Merwede (Rhine) from a major defense moat into a great access road into Holland. The French army invaded the towns of Dordrecht and Rotterdam across the ice.

The low temperatures helped the French to conquer the Republic, but this chapter in history is not ruled entirely by environmental determinism. The French also received a lot of aid from Dutch Patriots who adhered to the ideals of the French Revolution. The Patriots

---

13 GROENVELD 2001, pp. 24–25.
14 KLINKERT 2007, p. 464.
15 KLINKERT 2007, pp. 465–473, 466: map of inundations 1672.

welcomed support from outside in order to reform the Republic, which still bore institutional traits of medieval origin. Many towns were in the hands of revolutionary committees of Patriots and hailed the French troops with tricolour flags and revolutionary posters and pamphlets. Contemporaries saw the events as a new Dutch Revolution, inspired by the same principles as the Dutch Revolt of the sixteenth century, the liberation of tyranny (of the Orange party this time) in other words, and this was celebrated by numerous victory parades, thanksgiving ceremonies, theatrical performances, and banquets. By the Treaty of The Hague of 1795 the Republic was renamed the Batavian Republic. It would serve as a kind of satellite to France and remained so until the international system was turned upside down again in 1813.[16]

3. Ice and Drinking Water: Environmental Buffers

Another form of disruption involving ice is the freezing of water supply systems. In medieval cities both private wells and centrally-maintained networks of pipes (and gutters) existed, in particular in new cities in Southern Germany founded by the Hohenstaufen and Zahringer. Evidence about the waterworks of Nuremberg in the late Middle Ages is very detailed, as fifteenth-century documents of a master of the waterworks have been preserved. The town had at least seventeen aqueducts for the supply of water, which carried water from the nearby chalk grounds to the city.[17]

In Great Britain, too, such systems existed, but the oldest evidence mostly concerns piped water systems of religious institutions. In the later Middle Ages cities integrated such systems in their public piped water system, as examples from Paris, Dublin, London and several English towns show.[18] In winter, however, such piped systems were frought with troubles. The author of the sixteenth-century manuscript called the 'Rites of Durham' presents a sunny image of monks of the Cathedral Priory washing at the laver fountain and drying their hands on clean towels before entering the refectory. But the Durham account rolls paint a far grimmer picture. During the Great Freeze of 1495–96 eighteen men were hired by Durham Cathedral Priory to help to de-ice the aqueduct. Also, the accounts mention payments to women water-carriers hauling water up the steep hill from the River Wear, because the abbey's pipes were once again fractured or frozen. Archaeological excavations have confirmed the details and suggest that the pipe trenches were too shallow to protect the pipes against frost damage.[19]

How did piped water function in regions with long and harsh winters? What action could a city take when the piped water system froze? There were several alternatives. The obvious choice was using other water sources. Often this implied reverting one or more steps in technological advance. To this end, many cities kept a great number of deep private and public wells, alongside the piped systems, well into the nineteenth century. For instance, in the Middle Ages Bremen had eleven public wells, governed by source communities (*Brunnengemeinschaften*), but Nuremberg had many more: ninety five to be precise (besides the seventeen aqueducts). In the same period, for a population of around 35,000 inhabitants, Brussels had some forty common and public places providing water (fountains, sources), not

---

16 KLINKERT 2007, pp. 473–477; ISRAEL 1995, pp. 1119–1121.
17 GREWE 2000, pp. 145–150.
18 MAGNUSSON 2001, p. 33.
19 MAGNUSSON 2001, p. X, 117.

counting the wells belonging to private houses, which were often shared properties.[20] Paris was a huge city in the nineteenth century and had twenty five to thirty thousand wells all over the city in 1830.[21]

Another option was to collect rainwater. Many houses in Amsterdam had cisterns filled by rainwater coming from the roofs of buildings. At the end of the nineteenth century the municipal authorities ordered the construction of thirty such cisterns. This was not necessarily good water. As investigations in the nineteenth century showed, this water was often heavily polluted with lead.[22]

Yet another alternative was using surface water: lakes, rivers and canals could contain large amounts of good water in spite of local pollution from cities. In many places the self-cleaning capacity of surface water was quite sufficient.[23] In this respect it is important to note that in many cities professional water carriers were active, even when an extensive piping system was present, like in London. In late-medieval London small, family-based companies made a living of loading water from the Thames at a designated place in the harbor and transporting it downtown in privately-owned carts. In 1349, Johanna, the widow of the London waterlader Geoffrey PENTHOGG, left five horses and two carts, apart from a house and a garden.[24]

Can we conceptualize this exploitation of surface water as an 'environmental buffer'? Mark ELVIN introduced this concept in his recent survey of the environmental history of China, *The Retreat of the Elephants* (2004, p. 30). He shows that in cases where agricultural systems failed (due to mismanagement, sedimentation of irrigation works, bad weather etc.), the Chinese people could fall back on non-cultivated patches of forest that would serve as reserves of food. Similarly, we could consider large bodies of surface water such as lakes, rivers and canals outside towns as buffers of drinking water to be used in cases when more advanced water provisioning systems such as pipes or wells fell dry due to frost or drought.

Inevitably, the availability of rivers and canals as drinking water buffers was hampered by really severe winters. Then frost acted like desertification. Frost diminished the amount of water available. It meant an increase in the costs of drinking water, in whatever terms these costs were expressed (labor, money, deaths). This connects to the concept of environmental justice. Natural sources should be available to all, but were not.

Battling the river ice as a means to safeguard the water supply was available to societies with a sufficient level of technical sophistication and, in particular, a sufficient reserve of capital. Amsterdam is a good example of a very rich and technologically advanced city in premodern Europe. By 1550 the water in the city canals became too salty, due to large-scale changes in the balance between land and the North Sea. Private entrepreneurs started a water-shipping service to large bodies of lake and river water outside Amsterdam. Daily, the water ships left the city in the morning and returned with their loads of fresh water in the afternoon. In 1781 this train comprised 43 brewer ships and 114 private water-company ships. In 1786 the town authorities assumed responsibility over the fresh-water supply in the city and founded a Fresh Water Society (*Verschwater Sociëteit*) to exploit water ships. By 1825, 40 water ships were in operation, and 240 small ships (*leggers*), equipped with pumps, distributed the water along

---

[20] DELIGNE 2002, pp. 111, 116.
[21] BARLES 2002, before p. 139. Such double systems remained effective well into the twentieth century, like in Rabbat, Marocco, see FRITSCHY 2007, p. 160.
[22] HUISMAN and BUITER 2007, p. 394.
[23] GUILLERME 1988.
[24] MAGNUSSON 2001, p. 138; KEENE 2001, p. 169.

the canals. This was sufficient to supply the Amsterdam populace with a ratio of 4 l of water per person a day.[25]

In 1651 the brewers founded a company to exploit an icebreaker (Fig. 2).[26] The ships were drawn by horses. In normal winters the water ships needed two horses each, and the icebreaker was drawn by eight horses. In severe winters such as 1755 or 1777, the water ships needed twelve horses and the icebreaker thirty six horses. In extremely severe winters, however, the icebreaker was not strong enough. In those years the water route from Amsterdam to the river Weesp was cut open by hand. Notorious years in this respect were 1687–1690, 1740, 1763 and 1783. In these years the ice was divided into parcels and put out to entrepreneurs, often professional water companies. The costs for such undertakings were enormous. In 1740 the price for one such ice-breaking project that lasted several days, amounted to a total of 2800 guilders, in 1763 to 7522 guilders. Operating the icebreaker cost between 400 and 800 guilders per journey, depending on the (huge) number of horses and the army of laborers working on the ship and along the shores of the rivers in order to keep it moving. As the brewing industry declined and the number of brewers decreased at the end of the eighteenth century, the brewers became less and less willing to carry the costs for keeping open the canal connecting the town to the fresh water resources. Since the private water companies profited from their work, the brewers began to see the ice costs as an indirect tax. In 1786, after a few years of negotiating, the icebreaker was taken over by the town. The city maintained the icebreaker until 1860, i.e. the advent of a piped water system.

The price of drinking water sold in Amsterdam reflected the weather conditions. In very cold years, the water price would rise to 3 stivers per bucket (1757, 1763), which is the equivalent of about an hour's wage of a contemporary laborer.[27] In such years the poor would resort to melting ice from the canals and get sick, since from the sixteenth century onwards, the water in the canals downtown was not only too salty, but also heavily polluted. During weather extremes poor people were at a disadvantage in procuring natural resources such as water. From very early on, water in towns was commercialized; water became a commodity, a product subject to market conditions.

4. Water and Fire: International Water Techniques

The other transition of water I would like to touch on in this article is the transition between water and fire, or vapor. In environmental history the wider meaning of fire as energy is most relevant. In order to become vapor, water needs energy. Also, vapor can be a medium for transporting energy, in particular in the age of steam engines. But water can also provide energy by just staying water, chiefly energy as derived from gravitation forces. In order to create a long-time perspective, I take examples of two successive power regimes: the early-modern period, when water-driven mills spread, and the nineteenth century, when steam-driven pumps were introduced.

As mentioned above, in Northwestern Europe in the Middle Ages gravity-powered piped systems existed in certain privileged cases, such as southern German towns and London. In the

---

25 GROEN 1978, pp. 14–27; FRITSCHY 2007, p. 146.
26 VAN EEGHEN 1954.
27 VAN EEGHEN 1954, p. 70; DE VRIES and VAN DER WOUDE 1997, p. 610, table 12.1.

Petra J. E. M. van Dam

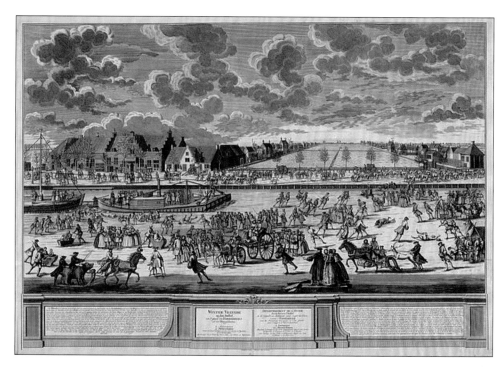

Fig. 2 When entering the city of Amsterdam the icebreaker pulled by dozens of horses draw a lot of attention. Behind the icebreaker a train of some 150 water ships followed transporting drinking water to the city from the river Vecht, the nearest source of sweet surface water, receiving part of its water from the sandy Utrechtse Heuvelrug. Engraving by Tieleman van der Horst, ca. 1730–1736. Source: Collection Atlas Dreesman, Muncipal Archive of Amsterdam.

fifteenth and sixteenth centuries water-driven pumps were built. The basic idea was that a watermill driven by a river provided energy for pumps, which lifted up the water. So this differs from the concept of the type of watermill with scoops that had been in use since Roman Times. The lifted water might come from a source or from the same river that provided the energy for the pumps. Important contributions to the development of this new technology came from at least two entirely different disciplines, the pumping of mines and the irrigation of private gardens.

Gardens are intriguing objects of study for environmental historians, since they convey a lot of information about the perception of nature. With the aid of water, sixteenth-century garden owners engineered nature in order to reproduce a landscape that could serve as an adequate backdrop for a 'civilized' way of life. Travelers from Northern Europe visited Italy and France and admired the gardens and their waterworks. For instance in St. Germain-en-Laye they found grottoes featuring surprising special effects with water, with the sea god Neptune on a sea horse, plus nightingales and cuckoos that began to sing and call out whenever a faucet was turned on. This kind of water culture was highly appreciated: it imitated or rather outdid nature. In contrast, when travelers crossed the Alps they commented that the waterfalls in the Alps were frightening (*effroyables*). 'Wild water' in unstructured landscapes was ugly, uncivilized and unuseful.[28]

---

28 De Jong 2000, pp. 15–16.

The new love for water engineering in gardens led to great innovations, which ultimately contributed to better drinking water systems for citydwellers. An exceptional example of a garden in Italian-French style was the early sixteenth-century Coudenberg Park in Brussels. The park was the property of the archdukes ALBERT and ISABELLE. Around 1500 or so, the park comprised several palaces, fields lined with trees, decorative gardens, a vineyard, a labyrinth, and artificial caves with running water.[29] In 1602 an irrigation system with pumps was installed.[30] The watermill was powered by the river Maelbeek. The mill powered the pumps, which lifted the clear water of the source called Broebelaer. The water arrived at the pumps via an underground water tube. The water served the gardens and the waterfalls of the caves, but also the palaces and some important mansions outside the park. Thus, the project combined the provision of irrigation and drinking water.[31]

It is known who were involved in the Brussels project. An Italian and a French engineer designed the system. Even some drawings of pumps by the Frenchman, the famous Salomon DE CLAUS, have been preserved. But the execution of the project was entirely in the hands of German and Belgian engineers, originating from mining regions in Southern Germany and Liège. So, Italian and French design was fused with German and Belgian engineering, and the whole project was made possible by the sponsorship of the highest European nobility; a truly international project.

At the end of the Middle Ages, elsewhere in Europe, too, watermills were founded to pump up water to the city, but as in Brussels, the drinking water was intended first and foremost for specific privileged groups. Often, pumps were installed in cities situated in flat landscapes, such as the northern German plain, where gravity as a source of energy was not available. Commercial users, brewers especially, often took full control over the machinery. Good examples are Lübeck, Hanover, Bremen and Wroclaw.[32] In many cases we see that the common waterwheel with scoops or buckets was succeeded by pumping stations, but the older system remained in use until the industrial period.

Lübeck is situated at the confluence of two rivers, the Wakenitz and the Trave. In 1294 a common waterwheel was in function there, and in order to make it work, a dam was built in the river Trave. It was connected with wooden pipes to some 200 houses situated above the river. This construction was financed by the brewers, and it was named after them: 'Brewing water fountain'. In the early fifteenth century the Brewing water fountain probably passed into the possession of the town council. Towards the end of the century the construction was replaced by a pumping station and passed into the brewers' possession. In Hanover in 1535, a water work with six pumps was founded. It could draw up 8000 barrels a day, each barrel probably containing 200 l. The water was transported to the city by at least three water wagons drawn by teams of horses. In Breslau in 1538, an extraordinary pumping station of fifteen meters high was built, extracting water with 160 wooden tubes to a height of nine meters. It pumped up some 500 l per minute. Five conduits conveyed all this water to the town. In the city of Bremen an association was formed to build a waterwheel at the bridge on the river Weser in 1394. This waterwheel powered a rotating chain with pitchers and lifted the water into a collecting basin. The list of founding members, among whom some city councilors,

---

29 DELIGNE 2002, pp. 201–205.
30 LOMBAERDE 1991, p. 163.
31 DELIGNE 2002, p. 201.
32 GREWE 2000, pp. 151–156.

and the particulars of the statutes of the association reveal that only well-to-do citizen could join it. The entrance fees were very high. In 1790, 490 taps were connected to the basin. The whole network remained intact for centuries. The waterwheel was demolished in 1822, followed up by a horse-driven windlass for some time and replaced by a steam engine in 1873.

In Great Britain, too, pumping stations were introduced from 1500 onwards. In 1581 Peter Morris, of Dutch or Flemish origin, built a water engine within the first arch of London Bridge. It supplied Thames water directly to individual houses in the eastern part of the city by means of wood and lead pipes. The water engine was privately-owned and operated, with customers paying fees for the service. Possibly, London was one of the most advanced regions with respect to providing drinking water on a commercial basis. Private ventures in water supply systems became more common in the seventeenth century as shareholders obtained patents and invested in profit-seeking schemes to lift and convey water. The Derby engineer George Sorocold built a new, improved water engine at London Bridge in 1701. He was also responsible for the design and installation of water engines in many provincial towns in the late seventeenth and eighteenth centuries.[33]

In the Netherlands, like on the German and English coastal planes, no fast-flowing river existed. Yet lots of wind was available. Since 1408 this great source of power was being used by windmills to drain the countryside. In the sixteenth century this invention was put to a new use. Towns suffering from polluted water due to high population growth and industrial expansion started to employ windmills to speed up the flow in their inner-city canals. For the poor these canals were still a source of drinking water, even though attempts at shipping water were made in some towns, as described above. In Amsterdam in 1649, the New Canal (*Nieuwe Vaart*) was dug as a direct connection between the town centre and the deep waters of the Zuider Zee. At both ends of the city canal two pairs of large and strong watermills were built (*stadsvuilwatermolens*) to stimulate the flow from the Zuider Zee. A mud barge and horse-driven, floating mudmills passed along the canal continually to keep it at the right depth.[34] Thus, in the Netherlands the natural environment did not allow for water technologies based on gravitation (energy transported by water) to improve healthy water provisioning, but gave rise instead to water technologies based on wind power (water moved by energy).[35]

Three centuries later, as part of the new energy regime of fossil energy, the steam engine was introduced. In England the first steam engines for drinking-water supply were already in existence at the end of the eighteenth century.[36] Some historical patterns seem to have persisted, particularly the link with mining and gardens. However, this time the new technology came from England, where the steam engine had been developed for mining operations in the first place.

In the 1770s the city of Amsterdam contracted the English engineer William Blakey to demonstrate his engines, which it wished to employ for drinking water systems. Blakey had put his engines to practice in French aristocratic gardens. He was unlucky, for his engine exploded when one of the Amsterdam burgomasters visited him. However, the Amsterdam patrician and banker John Hope did not despair. Courageously, he had a steam engine built on his estate Green Valley, situated at the border of the coastal sand dunes, in 1781. The pump

---

33 Magnusson 2001, p. 169.
34 Prak and Frijhoff 2005, p. 85.
35 De Bont 2000, p. 83; Smit 2001, p. 57–69; van Tielhof and van Dam 2006, p. 160; van Dam 2008, forthcoming.
36 Robbins 1946, pp. 191–192.

was used to lift the great quality dune water to his gardens. This was especially important in summer, because when the canals fell dry, they produced a terrible stench. At the beginning of the nineteenth century Amsterdam solved its problems with steam engines and also started to pump water from the dunes to the town.[37]

5. Conclusion

I have investigated how the classical transitions between water and earth (ice) and between water and fire (vapour) can inspire us to formulate new research questions for environmental history. Referring to my general questions in the introduction, it is clear that for some states of water our experience does not change that much over time. We still enjoy winter sports.

In certain respects our control over transitions of water has increased. We have overcome the problems posed by severe winters. Our piped water systems no longer freeze and we no longer need icebreakers to maintain our defense systems. This improved control leads to more daily comfort and security. Nevertheless, examples in the history of ice have also shown that nature by itself is never a sufficient explanation for historical developments. One could break through defense systems relying on water-lines in severe winters, or not. Human agency also influenced river beds to such an extent that the risk of flooding as a result of ice formation increased. Human agency interacted with weather conditions and determined the outcome, not nature as such.

Extreme situations reveal a lot about the characteristics of a society. The topic ice has led us to reflect upon such concepts as environmental justice and environmental buffers. While our control over transitions of water increased, our vulnerability to failing technological systems increased as well. It is worth pondering on the fact that major centers of population had complementary water provisioning systems at their disposal until well into the nineteenth century, so that in times of crisis they could rely on a technically less advanced but at least secure water system. It is revealing, too, that in many cases only selected groups profited from the technological advance in powered water systems, in particular religious houses, commercial water users such as brewers, and elite town-dwellers. Some social groups remained excluded from the improvements in water provisioning systems. Poor people simply could not afford the service fees charged by water companies, let alone the rising costs of water in times of scarcity. Which water buffer could they fall back on? Such unjust consequences of weather extremes deserve further research.

Whatever transition of water we observe, ice or steam, in densely-populated Northwestern Europe water became a commodity very early on. The benchmarks in the environmental history of water transitions seem very tightly connected to those of economic, social and technological history.

Acknowledgement

The language corrections by Caroline F. MEIJER, Utrecht, are gratefully acknowledged.

---

[37] ROBERTS 2004, pp. 259–263: It was only after such experiences that the local drainage units of Holland, called polders, started to experiment and replace the traditional windmills by steam-driven pumps. The first one was installed in the Blijdorppolder near the city of Rotterdam in 1787; FRITSCHY 2007, p. 151.

Petra J. E. M. van Dam

References

Barles, S.: L'invention des eaux usées: l'assainissement de Paris, de la fin de l'Ancien Régime à la seconde guerre mondiale. In: Bernhardt C., et Massard-Guilbaud, G. (Eds.): Le démon moderne. La pollution dans les sociétés urbaines et industrielles d' Europe. The modern demon. Pollution in urban and industrial European societies (= Collection 'Histoires croisées'), pp. 129–156. Clermont-Ferrand 2002

Beukers, E. (Ed.): Hollanders en het water. Twintig eeuwen strijd en profijt. 2 Vol. Hilversum 2007

Böhme, H.: Kulturgeschichte des Wassers im Rahmen der Vier-Elementenlehre. In: Parthier, B. (Ed.): Wasser – essentielle Ressource und Lebensraum. Vorträge anlässlich der Jahresversammlung vom 6. bis 9. April 2001 zu Halle (Saale). Nova Acta Leopoldina NF Bd. *85*, Nr. 323, 227–250 (2002)

Bont, C. de: Delft's Water. Tweeduizend jaar bewoning en waterhuishouding in en rond Delft. Delft/Zutphen 2000

Dam, P. J. E. M. van: Sinking peat bogs: Environmental change in Holland, 1350–1550. Environmental History *5*/4, 32–45 (2000)

Dam, P. J. E. M. van: Frühmoderne Städte und Umwelt in den Niederlanden. In: Schott, D., und Toyka-Seid, M. (Eds.): Die europäische Stadt und ihre Welt, S. 83–104. Darmstadt 2008

Deligne, C.: Bruxelles et sa rivière. Genèse d'un territoire urbain (12ᵉ–18ᵉ siécle) (= Studies in european urban history 1) (1100–1800). Turnhout 2002

Driessen, A.: Watersnood tussen Maas en Waal: overstromingsrampen in het rivierengebied tussen 1780 en 1810. Zutphen 1997

Eeghen, I. H. van: De ijsbreker. Jaarboek van het Genootschap Amstelodanum *46*, 61–75 (1954)

Elvin, M.: The Retreat of the Elephants. An Environmental History of China. New Haven, London 2004

Fritschy, W.: L' eau potable á Amsterdam et á Rabat (1850–1950): deux expériences historiques. In: Kaddouri, A., et Obdeijn, H. (Eds.): L'Eau entre moulin et noria. Actes du Colloque Marrakech 14–16 Novembere 2005, Rabat (2007)

Grewe, K.: Water technology in Medieval Germany. In: Squatriti, P. (Ed.): Working with Water in Medieval Europe. Technology and Resource-use. (= Technology and Change in History Vol. *3*); pp. 129–160. Leiden 2000

Groen, J. A.: Een cent per emmer. Het Amsterdamse drinkwater door de eeuwen heen. Amsterdam 1978

Groenveld, S.: 'Van vyanden und vrienden bedroevet' De gevolgen van het beleg van Leiden voor de omgeving van de stad. Leiden 2001

Guillermé, A. E.: The Age of Water. The Urban Environment in the North of France, 300–1800. College Station, Texas 1983 (translation 1988)

Harten, J. D. H.: Vloedschuur: veiligheid voor het vee bij overstromingen. In: Barends, S., Renes, H., Stol, T., Triest, H. van, Vries, R. de, and Woudenberg, F. van (Eds.): Over hagelkruisen, banpalen en pestbosjes; historische landschapselementen in Nederland; pp. 44–45. Utrecht 2ⁿᵈ print (1993)

Heezik, A. A. S. van: Strijd om de rivieren: 200 jaar rivierenbeleid in Nederland. The Hague 2007

Huisman, P.: Water in the Netherlands. Managing Checks and Balances. Utrecht 2004

Huisman, P., and Buiter, H.: Het zoete nat. Zorg om drinkwater en omgang met afvalwater in Holland. In: Beukers, E. (Ed.): Hollanders en het water. Twintig eeuwen strijd en profijt. 2 Vol., pp. 383–438. Hilversum 2007

Israel, J.: The Dutch Republic. Its Rise, Greatness, and Fall 1477–1806. Oxford 1995

Jong, E. de: Nature and Art: Dutch Garden and Landscape Architecture 1650–1740. Philadelphia 2000 (Translation of: Natuur en Kunst: Nederlandse tuin en landschaparchitectuur 1650–1740. Bussum 1993

Keene, D.: Issues of water in medieval London to c. 1300. Urban History *28*/2, 161–179 (2001)

Klinkert, W.: Water in oorlog. De rol van het water in de militaire geschiedenis van Holland na 1550. In: Beukers, E. (Ed.): Hollanders en het water. Twintig eeuwen strijd en profijt. 2 Vol., pp. 1–504. Hilversum 2007

Lombaerde, P.: Pietro Sardi, Georg Müller, Salomon de Claus und die Wasserkunst des Coudenberg-Gartens in Brüssel. Die Gartenkunst *3*, 159–171 (1991)

Magnusson, R. J.: Water Technology in the Middle Ages. Cities, Monasteries, and Waterworks after the Roman Empire. (= The John Hopkins Studies in the History of Technology) Baltimore, London 2001

McNeill, J.: Something New under the Sun. An Environmental History of the Twentieth Century. (= Global Century Series). London etc. 2000

Prak, M., et Frijhoff, W. (Eds.): Geschiedenis van Amsterdam II-I. Amsterdam 2005

Robbins, F. W.: The Story of Water Supply. London 1946

Roberts, L.: An Arcadian apparatus: the introduction of the steam engine into the Dutch landscape. Technology and Culture *45*, 251–277 (2004)

Smit, C.: Leiden met een luchtje. Straten, water, groen en afval in een Hollandse stad, 1200–2000. Leiden 2001

Tielhof, M. van, and Dam, P. J. E. M. van: Waterstaat in stedenland. Het hoogheemraadschap van Rijnland voor 1857. Utrecht 2006

Technology and Culture *3*, 43. Special issue: Water Technology in the Netherlands (2002)
VRIES, J. DE, and WOUDE, A. VAN DER: The First Modern Economy. Success, Failure and Perseverance of the Dutch Economy, 1500–1815. Cambridge 1997
WILLEMSEN, A.: Kinder delijt: middeleeuws speelgoed in de Nederlanden. Nijmegen 1998

        Prof. Dr. Petra J. E. M. VAN DAM
        Vrije Universiteit
        Faculteit der Letteren
        De Boelelaan 1105
        1081 HV Amsterdam
        The Netherlands
        Phone: +31 20 5986434
        E-Mail: pjem.van.dam@let.vu.nl

Natur und Migration

Vorträge anlässlich der Jahresversammlung vom 5. bis 7. Oktober 2007
zu Halle (Saale)

Nova Acta Leopoldina N. F., Bd. 97, Nr. 358
Herausgegeben von Harald ZUR HAUSEN (Heidelberg)
(2008, 225 Seiten, 81 Abbildungen, 2 Tabellen, 29,95 Euro,
ISBN: 978-3-8047-2500-3)

„Natur und Migration" – assoziiert sehr verschiedenartige Phänomene, die sich durch Wanderungsprozesse auszeichnen. In diesem Band wurden besonders interessante Gebiete ausgewählt, u. a. Migration und Seuchen, Reisen und Epidemien in einer globalisierten Welt, der Vogelzug, aber auch die Migration geologischer Fluide, die Elektronenmigration in Halbleitern, die Migration als treibende Kraft in der Organogenese, die Biophysik der Zellbewegungen, die Migration von Tumorzellen, Migration als Phänomen in der Neurobiologie oder die Migration wissenschaftlicher Ideen. Besondere Akzente setzen die Themen „Diversität als neues Paradigma für Integration?" und „Vorspiel der Globalisierung. Die Emigration deutscher Wissenschaftler 1933 bis 1945".
Die Beiträge sind von herausragenden Experten der jeweiligen Gebiete, u. a. durch die Leopoldina-Mitglieder Markus AFFOLTER, Lorraine DASTON, Wolfgang FRÜHWALD, Michael FROTSCHER, Jörg HACKER, Hans KEPPLER und Otmar WIESTLER, in anspruchsvoller, aber durchaus gut verständlicher Form verfasst.

Wissenschaftliche Verlagsgesellschaft mbH Stuttgart

Air: A Planetary Hybrid

Wolfgang Lucht (Potsdam)

With 3 Figures

Abstract

Air is the most elusive of the classical elements. Today it is at the centre of the global change challenge as humankind realizes the extent of anthropogenic climate change and the magnitude of the impacts it will have on sociocultural structures. Air is a planetary hybrid: originating as a joint product of geosphere and biosphere co-evolving through Earth's history, it has recently become a co-construct of geosphere, biosphere and the newly emerged anthroposphere. Avoiding climate change will require not just scientific earth system analysis through global observations and modeling, but also cultural narratives of the elements that resonate with scientific findings, anchoring them in the sociocultural practices that determine the dynamics of the anthroposphere. Geophysiology is a form that such knowledge might take, taking the whole of the planetary body and its elements into a narrative of reflective adjustments.

Zusammenfassung

Luft ist das am wenigsten greifbare der klassischen Elemente. Heute steht sie im Zentrum der Herausforderung des globalen Wandels, indem die Menschheit das Ausmaß des anthropogenen Klimawandels und die Größe der Folgen erkennt, welche sie auf soziokulturelle Strukturen haben wird. Luft ist ein planetarer Hybrid: entstanden als ein gemeinsames Produkt von Geosphäre und Biosphäre in ihrer Ko-Evolution im Verlaufe der Erdgeschichte, ist sie inzwischen zu einem Ko-Konstrukt von Geosphäre, Biosphäre und der kürzlich entstandenen Anthroposphäre geworden. Die Vermeidung des Klimawandels wird nicht nur einer wissenschaftlichen Erdsystemanalyse durch globale Beobachtung und Modellierung bedürfen, sondern auch kulturvolle Erzählungen von den Elementen erfordern, die in Resonanz mit den wissenschaftlichen Befunden stehen und diese in den soziokulturellen Verfahrensweisen verankern, welche die Dynamik der Anthroposphäre bestimmen. Geophysiologie ist dabei eine Form, welche derartiges Wissen annehmen könnte, welche das Ganze des planetaren Körpers und seiner Elemente in eine Erzählung der reflexiven Anpassung aufnimmt.

1. Air

Air is the most elusive of the classical elements. It is high as the sky, invisible and shapeless, a breeze or a storm. It is nobody's property: a borderless movement, local and continent-wide at the same time, a well-mixed global common, a planetary layer of haze.

In the experience of the individual, air is a resistance in the wind that remains elusive to the grasp, associated with movement and flow. It is the physical element of life, felt in the depth of breath. It is integral part of creation as the divine transfers breath into dead matter.

Seen culturally, air is the godly element, more so than intransparent water, which reflects the depth of the organic soul more than that of the divine, or the firm solidity of earth, from which the organic body is made and to which it returns. Air shares a fleeting spirit with the modifying, consuming power of fire, but unlike fire it has a physical body that links it into the real world. Air carries birds in the sky as well as the luminance of light.

But air does not just provide the spacious clarity of life and spirit while being the physical resistance of wind and breath. It may also carry disease and death, the destruction of storms, the emanations of bogs, a cold wind. And it carries the haze that clouds perceptions. From the viewpoint of an elevated position it obscures the wider landscape. It is the haze that lingers in the air which limits our perceptions of things distant, veiling human knowledge of the true relations of the world (GRAZCYK 2004).

In all of this, air is the last element to have yielded to technology. Humans had left their ecological niche and conquered the land to become a global species, had learned to use and maintain fire, and were using water as passageways for travel long before air was conquered for the speed and lightness of flight. Though wind has long been used for sailing, only very recently has air also been liquefied, chemically altered, compressed and thrust through engines to perform work as an element of power.

Today, air is mostly seen from a scientific point of view. The explosion of scientific knowledge in just a few generations, amounting to not more than some two centuries, has led to the scientific reconstruction of origins and evolutions, revealing the properties of air in deep time and across the continents of one world, in terms of the languages of chemistry, physics, biology and geology. And none too late: it is one of the foremost concerns of our time that the air is changing (WEART 2003). The term by which this is being addressed is: climate change.

But climate change is not an isolated phenomenon, it is a symptom of the much more fundamental problem of our times, of global change. Evolving on the currents of global air circulation patterns, climate change makes apparent air the element as being in the middle of human interactions with the planetary world, and at the centre of the challenge faced by current humankind, achieving a transition to sustainability.

2. Climate Change and Global Change

Global change is as unique and difficult a challenge to the current generation as the largest challenges have been that previous generations had to meet in trying to ensure continuity for their respective societies when living in times of rapid change. It is equally fraught with uncertainties, controversies and the dangers of complacency.

One element of this challenge is that, due to the sum of human activities, the Earth's climate system is gradually entering into a no-analogue state that is dangerous (*IPCC WG1* 2007). No-analogue here means that in the last several million years, the Earth's atmosphere has not been in a radiative and chemical state comparable to the one now being created. The consequences will be fundamental and severe, changing the world we know to one we do not yet know (*IPCC WG2* 2007).

As climate change arrives, soon more strongly than in its current beginnings, it will begin to be a reality that will not suddenly disappear again. Already it is being detected in the analysis of observations made by global networks of measuring stations, and scientific confidence in the projections of large change coming made by computer models in electronic simulations is high.

A second element of the global change challenge is that, due to the concurrent strong increase in human population numbers, and even more due to a historically unprecedented

Air: A Planetary Hybrid

Fig. 1 Air is the most elusive of the classical four elements. It has the reality of a body and the spiritual dimension of breath. The Earth's airy atmosphere is a hybrid co-creation of the Earth's geological, biological and, most recently, human processes. Digital photograph taken by astronauts on the International Space Station in Earth orbit on July 20, 2006. Image credit: NASA (photo ISS013-E-54329).

wealth in energy and material consumption of a sizeable fraction of that population, brought about by the processes of industrialization, Earth's surface with its biosphere is independently also being transformed into a no-analogue state (HABERL et al. 2007). As far as we know, never before in Earth history has one single species altered, as recently humans have, as many ecosystems as extensively and converted as much land to suit its particular needs as to fundamentally change the whole complexion of the global biosphere itself.

This ongoing land use transformation is intricately linked to climate change: both are a direct consequence of industrialization, made possible by the enormous energy provided by fossil fuels and allowing a lifestyle of large-scale technologically and scientifically based economies. The magnitude of land use change largely reflects the industrial magnitude of the material input streams required for the metabolic reproduction of these industrial societies, while climate change is among the consequences of the corresponding waste output flows of transformed materials produced in the process (HABERL and FISCHER-KOWALSKI 2007). The change occurring in the element of air composing the Earth's atmosphere goes hand-in-hand with the enormous volume of inputs harvested by modern societies.

The urgency of the sustainability challenge on the international table lies in the observations that neither climate change nor the land use change process are anywhere close to being complete: in fact the largest changes will take place in the next several decades, within the lifespan of the current generations. As a consequence, the Earth as a system will operate in a fundamentally new regime, a comprehensively no-analogue state before the end of this century.

When this happens, nobody can assume that Earth system processes such as air and ocean flow patterns or spatial balances in ecological processes will continue in a shifted, adjusted, but stable manner in all respects. As any physicist knows, complex systems are capable of switching into different states of internal organization, for example by re-ordering their spatial structures. Evidence of climate change in earlier periods of Earth history shows that the Earth system is indeed also capable of very rapid changes under forcing (BRAUER et al. 2008). This past system variability, documented in the record of past climates and environmental conditions, serves as a dire and important warning that the Earth system can and will fluctuate in important parameters when affected by changes in driving variables such as the radiation balance or surface reflectivity (HANSEN et al. 2007). A strong reaction of the Earth's planetary system to the current large anthropogenic forcing is therefore to be expected (ALLEY et al. 2003). Anything else would be a surprise. Precisely because climate has always changed is it dangerous to be causing such a large driver of change as global land use change, greenhouse gas emissions and atmospheric pollution are.

An additional element of the global change challenge to current humankind lies in the still unfulfilled state of one of the oldest projects of the modern human mind: to rid, for religious, ethical and humanitarian reasons, the world of poverty and hunger, and to create a minimum of global justice (*United Nations* 2000). This requires to consciously break out of the many detrimental cycles currently operating in the world where societal interactions with the environment cause substantial degradation of living conditions (PETSCHEL-HELD et al. 1999). Today, it is a fraction of the world's population that uses most of the world's resources for their own benefit. Many of the international structures governing global exchanges are characterized by historically skewed relations. Such structures are not stable in the long run and their development into more equitable practices is therefore an integral part of the sustainability challenge.

These elements of the global change challenge of this century have to be considered under the increasingly visible severe constraints imposed by a finite planet that is by now governed by global-scale interactions between human societies and their environments. It is clear, though still not the basis of many ongoing debates of the problem, that sustainability will not simply emerge as an optimization of current practices (KATES et al. 2001). To collectively find, understand and achieve more sustainable paths into the future, where change will be limited to the adaptive capacities of the affected societal and ecological systems and the chosen extents of tolerable change, will require deep transformations in societal structures and self-understanding. These will have to be fueled as much by advanced scientific analysis as by deeply reflective cultural introspection. The world is finding itself in a new state not only environmentally, but also, though this is frequently not realized, culturally.

A hybrid scientific-cultural narrative has to emerge explaining where humans and their societies stand individually and collectively in a world of strong change. That narrative will have to fully resonate with the growing body of empirical and theoretical knowledge being accumulated about the world, while remaining closely in touch with the socially meaningful universe represented by the depth of meaning of images of nature, for example the classical elements (HUMBOLDT 1807). The challenge requires, both scientifically and culturally, the deconstruction and reconstruction of contemporary cosmologies of the world and the human position in it, a renewed construction of origins and futures, of transformations and continuities (SACHS 1999). Only then, if it is possible, can visions of alternatives, of a changed world, and of paths that lead out of the increasingly severe and problematic current circumstances materialize.

I propose that a hybrid, jointly scientific and cultural narrative of this type is indeed not just practically necessary, but that it is closer to the workings of the real world than either a purely scientific analysis that neglects the deep symbolic drivers of human action, or a purely cultural reflection that is largely oblivious of the very real changes it causes in the physical environment into which it is embedded. Paths into a sustainable future will require changes in the socio-ecological input/output patterning of current human societies and therefore new developments in both scientific analysis and in relevant cultural production.

Earth system analysis is the required science. A distributed, collective earth system consciousness rooted in history and science is the socio-cultural phenomenon that is now necessary for achieving a transition of humanity out of the problem zone in which it finds itself.

3. Atmosphere and Biosphere

Air is an element that spans all continents. It is a global phenomenon. It is composed of nitrogen, with a fifth of oxygen and important small fractions of carbon dioxide, methane and other gases.

Air forms the outer layer of the planet in form of the atmosphere. This layer is very thin compared to the planetary body with its diameter of 12,722 km. Traditionally, the upper boundary of the atmosphere is taken to be at 100 km above the ground. The highest clouds, formed of particles of ice, form in the mesosphere at a height of 82 km. This height is not more than a mere wisp surrounding the planet. If one takes a globe and measures on it the distance between two locations 100 km apart, it will be a tiny distance away from the globe when tilted upright to show the maximal height of the atmosphere. Even spacecraft in low-earth orbit such as the international space station are not further removed from the planet's surface than around 350 km. Polar-orbiting satellites are stationed at heights of 650–850 km, still merely skimming the surface of the globe. Earth's atmosphere truly is only a very thin layer of outgassing hugging the surface of the planetary mass.

Below the atmosphere is a thin layer of life, even more minute when compared to the size of the planetary body and the forces that move its continents and oceans. Its thickness is mostly not more than some 50 m, some of which is above ground and some below ground. Life extends beyond that layer both into larger depths and greater heights, but only in sparse traces through microbes in deep earth layers and some birds, insects and spores in the higher air.

However, the chemistry of that thin layer is firmly under the control of life and it controls many of the chemical processes of the earth's surface and in the atmosphere by exchanging considerable volumes of bulk and trace chemicals with the atmosphere and the soils, changing the composition of both (VERNADSKY 1926). All organic matter of the world's soils is a product of life, producing in its decay the water-holding soils upon which vegetation and thereby the whole trophic sequence of life depends as well as returning gases to the atmosphere that were taken from it for growth. The weathering rate of rocks is greatly increased by the biological activity acting on them, introducing an important and strong chemical feedback on the planet's radiative state and therefore temperature. Globally, vegetation transpires as much water from the ground into the atmosphere as runs off in all of the world's rivers. Some even suggest that the topographical form of landscapes may be detectably shaped by the processes of life (DIETRICH and PERRON 2006).

It is for this reason that air, that the atmosphere is not a purely geochemical phenomenon. It is a biogeochemical phenomenon, a co-construct of life and planetary geology, because the biosphere is chemically exceedingly active, and the atmosphere so thin that the volumes of chemical exchange are large enough to cause alteration of both atmosphere and biosphere. The carbon contained in life (Ajtay et al. 1979, Saugier et al. 2001), of which the largest fraction is locked in vegetation (99%), amounts to roughly the same amount as the carbon stored in the atmosphere (both on the order of 800 GtC). The oxygen contained in the atmosphere is a product of life, produced in photosynthesis, consumed in the decay of life, and set free when organic material is buried without decay, by sedimentation.

4. Origins

What can be said about the position of air in the Earth system? It is useful to look at origins.

Viewed scientifically, Earth as a planet is an evolving chemical system that has self-organized into dynamic spatial patterns from global-scale flow systems to the encapsulated self-replicating membrane systems of organisms. It is materially largely closed but energetically open, fed by solar energy. Air, the atmosphere, is originally a formation of the earth's early geochemical history, an emanation of the crust that formed the planet and an accumulation of material from impacts. In the next several billion years, however, air increasingly became a joint product of geosphere and biosphere (Lenton et al. 2004a). Life quickly emerged on early Earth after the end of the last bombardments, around 3.85 billion years ago, and began to so strongly alter the atmosphere as to make its properties incomprehensible without consideration of the continuous metabolic processes of organic life.

Evidence for methanotrophic early life forms could indicate a comparatively methane-rich nitrogen atmosphere on early Earth. This would have produced a greenhouse warming effect that would have been crucial for life at a time when solar luminosity is thought to have been a quarter or more less than it is today. Not before 2.7 billion years ago did oxygenic photosynthesis appear, possibly as a product of nutrient-rich, light-flooded environments on stabilized continental shelves. But it required another immensely long period, five hundred million years, before the still relatively weak photosynthetic activity of life had fed sufficient amounts of oxygen into global chemical budgets and enough organic carbon had been sedimented in the oceans, removing it from the global budgets, to convert the atmosphere from a reducing state to the oxidizing state it has ever since maintained. This great global oxidation event of 2.2 billion years ago marks the largest and most fundamental change in Earth's chemical makeup. Though its causes have not been conclusively resolved, it was possibly life that altered forever the nature of air by tipping the system into an oxidizing state, thereby changing the subsequent geological and biological evolution of the planet profoundly (Goldblatt et al. 2006).

Oxygenic photosynthesis provided a new source of energy to life that fueled it on a different level. Possibly interlinked with such break-through changes as the evolution of eukaryotic cells, the basis of all higher life, atmospheric oxygen began its measured episodical rise to present levels (Canfield 2005, Falkowski 2006). At around 1 billion years ago, small animals and land plants, operating at an unprecedented degree of internal differentiation, are thought to have made their first appearances. But it was only after the breakthrough of multicellular life in the Cambrian explosion of 543 million years ago that large, complex terrestrial plants with roots evolved, some 400 million years ago, which greatly increased global photosynthetic production and, combined with continued oceanic carbon sedimentation, led to a second strong increase in atmospheric oxygen. Atmospheric oxygen rose to current levels or slightly above, allowing the oxygen-hungry metabolism of large land animals to succeed, and creating the planetary atmosphere of nitrogen mixed with oxygen that we know today.

The atmospheric concentration of carbon dioxide has in turn been on the decline in the last 400 million years. The precise causes and the contributions of life to this gradual drawdown are under debate, with interactions between enhanced weathering, tectonic activity, the biosphere and sedimentation rates likely important. But the gradual cooling of the planet that resulted from this decrease allowed Earth to remain in the habitable temperature zone when rising solar insolation from a slowly aging Sun acted to increasingly warm it (ZACHOS et al. 2001). By some 20 million years ago, carbon dioxide levels had dropped to a level that allowed the formation of polar ice caps. Around 2.6 million years ago, the current phase of ice ages began, characterized by extensive ice sheets on land, interrupted only by relatively short but warm interglacials. Once more, details of the co-regulation of temperature and carbon dioxide during these latest glaciation cycles remain unclear, but there is no doubt that the triggers are periodic changes in solar radiation due to changes in the Earth's orbit around the Sun. These cause relatively small changes in the latitudinal and seasonal distribution of solar radiation, the effects of which are then strongly amplified by feedbacks within the Earth system to cause glaciations, feedbacks to which life as a dynamic planetary factor reacting to climate change probably contributed important factors.

Seeing how life since its origins has co-evolved in close interaction with its planetary environment and is intimately linked with the global geochemical cycles that co-govern its energetic state (LENTON 1998), some have gone as far as suggesting that life itself co-creates the environment in which it exists, including the atmosphere, such that it is able to exist (LOVELOCK 1979). In other words, life has acted, it is suggested, through feedbacks on the chemical cycles of the planet in such a way as to always have maintained temperatures and chemical conditions within the habitable corridor of conditions. If this were the case, and currently it is far from having been convincingly proven, the undoubtedly uninterrupted continuation of life on Earth and the associated atmospheric evolution would be an incredible occurrence, a grand exception to the rule that atmospheric evolution will not normally maintain a planet within a small temperature zone for billions of years. Earth, of all places that produced life, would then be the one place where circumstances combined in a miraculous way to allow today's observer-biased present where humans consider whether their evolution is the rule or the exception. Alternatively, planet Earth as an evolving, self-organized chemical system that quasi automatically encompasses biological dynamics could be an example of a very normal planetary program that is set in motion if earth-like conditions that allow a rich carbon chemistry prevail. In that case, the existence of life, of an atmosphere co-formed by life, and perhaps even of higher life and human-like life forms, might not be exceptional. Though, perhaps, such conditions may be rare to begin with (CONWAY MORRIS 2005).

Could life on extraterrestrial planets then be detected based on its metabolic chemical signature in the environment, most notably the atmosphere? This was indeed first suggested in the 1960ies (LOVELOCK and GRIFFIN 1969). Even a cursory look at Earth's atmosphere would reveal the presence of a strong biological component as the chemical composition of its atmosphere of air is clearly not that of a dead planet governed only by the laws of physics and chemistry. One would observe the simultaneous presence of materials that, if not constantly replenished by biological metabolic processes, would not co-exist for long, such as of oxygen and methane. Accordingly, it should be sufficient for a spacecraft to sample a distant planetary atmosphere to infer the presence or absence of life. Good optical observations of light reflected by an atmosphere would potentially even allow the inference of life without even having to travel there, or at least without having to land, by means of spectral analysis. Planets with life carry a strong signature of life in their air. This most poignantly demonstrates how the element air is a product also of life, if present.

5. From the Holocene into the Anthropocene

The latest geological period is the Holocene, which began when the last ice age ended around 10,000 years ago and the current interglacial commenced. The Holocene differs in character from previous interglacials. Earth's orbit displays at present an unusually low eccentricity; in another 20 thousand years it will be more circular than it has been at any time in the past 350 thousand years. While preceding interglacials have lasted ten to twenty thousand years, computer models project that, as a consequence, the Holocene with its unusually stable solar input, will last for another 50 thousand years (BERGER and LOUTRE 2002). Only then will a new ice age set in.

Due to this extraordinary climatic stability of the Holocene, environmental change in the last ten thousand years has largely been limited, both in magnitude and extent, with moderate regional fluctuations. A major exception to this is the formation of the Saharan desert around 4–6 thousand years ago, which seems to have been caused by a flip in the bistable climate-vegetation system of northern Africa. Otherwise there was little change in the Holocene up to the present: slow shifts in various biogeochemical pools, gradual changes in vegetation composition, but nothing with strong or rapid global dynamics.

It was in this interglacial Holocene that humans first started to become noticeable as a global force. This happened after humans evolved without much disruption of their environment for the previous 3 million years. As a genus, humans had previously lived through several glacial cycles, very slowly developing their technology (AMBROSE 2001), expanding into continental Europe mainly during the intervening interglacial periods. At the height of the last ice age, Neanderthals lived in southern Europe. In the later stages of the last glaciation, *Homo sapiens sapiens*, the only extant species of humans, migrated from Africa and the Near East into Europe, bringing with it new abilities, expressed in refined composite tools, art, jewellery and spiritual world views (MELLARS 2006).

In the Holocene, human agriculture appeared and developed into a global-scale land transformation, causing extensive deforestation of a hitherto widely forested warm period world. For the first time human population numbers started to exceed a few scattered million, followed more recently by the staggering rise to currently 6.7 billion individuals alive. The Holocene saw the first major development of villages, towns and cities, of global travel becoming a reality linking distant regions, of industrialization, technology and science emerging with profound impacts on the environment (TAKÁCS-SANTA 2004, COSTANZA et al. 2007).

This transition from hunter-gatherers to agriculturalists and then, in some parts of the world, to inhabitants of industrial cities, greatly increased the material consumption that humans required per capita and per area. The machines and procedures created that allow humans to more widely transcend the physical limits of their bodies and abilities force them to also maintain considerable additional metabolic budgets. The materials used originate in the environment, where their production or extraction transforms the environment, for example through extensive land use. After the materials have been used, they are the cause of waste, exhaust and pollution streams returning into the environment, changing its biogeochemical balances. It is the volume of these additional material flows, a consequence of the social and technological transformations of human societies in the Holocene, that have turned humans into a major driving force of the Earth system in an otherwise stable interglacial. Of this climate change and the land use transformation are the leading examples. Currently, the volume of materials consumed per capita and per area is on another strong rise that began in the middle of the 20th century, when unprecedented growth was realized in many relevant statistics.

Throughout its long history, the atmosphere of Earth had been a product of geosphere-biosphere co-evolution, a co-construct of the physical and chemical environment of the planet and its biological processes (LENTON et al. 2004b). Now, in the developing Holocene, the material exchanges of the newly emerged anthroposphere are changing the nature of air. It is now a hybrid of natural and cultural processes. The co-evolutionary double strand of biosphere-geosphere co-evolution has widened into a three-sided relationship between geosphere, biosphere and anthroposphere (Fig. 2).

This introduces a new quality into planetary development. Today, even the most basic measurements in the environment, such as measurements of air temperature, of river sediment loading, of the chemical composition of air or the species composition of ecosystems

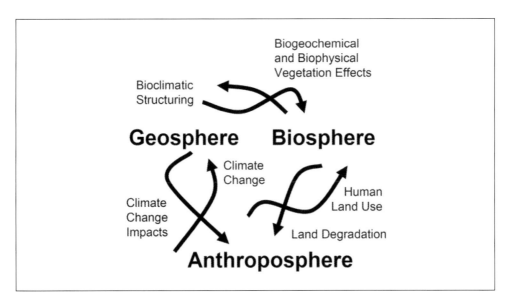

Fig. 2 The co-evolution of geosphere and biosphere that has charactized the past nearly 4 billion years of Earth system history has today, in the Anthropocene, been transformed into a three-sided relationship involving the anthroposphere as a major part. In fact, in the current relatively stable interglacial, most of the change originates in the dynamics of the anthroposphere. The increasingly problematic nature of the changes caused by human activity in geosphere and biosphere and their feedbacks onto the anthroposphere require that societies now begin taking into account an understanding of the joint dynamics of the whole of the Earth system when charting paths into the future (LUCHT 2007).

can no longer be understood without reference to human activities. The dynamics of the Earth system are shifting away from those of the Holocene, towards the dual no-analogue states of a climate warming beyond its normal interglacial warmth, and an spatially extensive transformation of the biosphere to intensive agro-industrial production systems. Humans are acting as a planetary force. It is for this reason that, with good arguments, a new planetary period has been proposed, to have begun in the recent past, the Anthropocene (CRUTZEN and STOERMER 2000).

Earth system dynamics in the Anthropocene are driven by the dynamics of the anthroposphere: by social, cultural, economic and political dynamics. These are founded in the workings of societies, realized within the framework of environmental conditions but mostly without substantial reference to their longer-term dynamics. The laws and rules governing societies arise from the properties of human social interactions, cognitive capabilities and cultural adaptations to the challenges of societal reproduction but do not substantially make reference to the dynamics of the environments into which they are embedded. Until the present, the material basis of human societies has not been a central aspect of mainstream theories of economy and sociology.

This was perhaps a sufficient state of reflection as long as the environment was not substantially altered by human activities on a global scale. Regionally, though, the material basis of societal processes and the repercussions of the change caused by human use of resources had to be considered much earlier. For example, recent evidence shows that some very early societies already had learned to incorporate into their behavior longer-term considerations of their interactions with the environment. Some late hunter-gatherer societies before the Neolithic transition seem to already have managed wild-living herds of the animals they hunted by being selective in the gender and age of the animals they took, with the aim of ensuring viable reproduction of the herds they depended upon. Pre-neolithic societies seem to regionally have encouraged the growth of wild occurrences of edible plants so as to ensure a continued harvest (MITHEN 2003), which might have led to the domestication of plants. Nomadic herders in arid environments are known to have adapted the intensity and timing of their grazing to the reproductive capacities of sparse grasslands to ensure their availability for grazing in the future.

These are examples of societies that have consciously taken into account the impacts of their activities on the natural systems they use and have adapted the pattern of their use such that their longer-term viability is safeguarded. Now, in the Anthropocene, such considerations have to reach a global scale. The viability of the whole of the global system is in question in the face of current overuse through industrialization and a large human population. Due to the limitations of a finite world with finite resources and reproductive capacities, the repercussions of change in the environment on the human social organizations that depend on this environment have now to be taken into active consideration when attempting to manage global societies into the future. This requires a new type of reflective awareness of the planetary system. Achieving this global view is at the core of the global change challenge (RASKIN et al. 2002).

The current paradigms of sustained, shared and ever-increasing socioeconomic global growth will falter as the magnitude of feedbacks from change in the natural systems on human sociocultural reproduction mounts. Awareness will have to rapidly increase that human activities are now not just causing change in the global environment and the availability of its natural resources, but that these changes, by being rapid, global and large, are beginning to produce feedback impacts on societies that threaten their sociocultural dynamics.

Air as element is currently at the center of this ongoing realization of humankind of its impact on the planet. The world's societies did not arrive at this point intentionally. It was not known that the tremendous opportunities provided by the energy from fossil fuels would alter the whole of the earth's atmosphere so profoundly as to endanger the well-being of the environment and the societies living in it. This was discovered largely in the 20th century and not fully realized until its second half. And only in the first years of the 21st century has that insight reached a general public and have countermeasures been seriously discussed. In this debate, climate change is now the spearhead topic that has led to substantial, though not yet sufficient, political action at the international level.

Controlling human emissions into the atmosphere is a good starting point for more general discussions of sustainability because the emission sources are generally well known and the atmosphere largely well-mixed, causing a general and global problem. All important political questions can be developed and explored in tackling the problems of air and climate change, such as: Who is responsible, who is affected by the impacts, who pays the costs of mitigation and who pays the damages, how can states organize to tackle a global problem, what economic instruments are suitable for management of the problem, which legal constructs have to be devised, and how is influence on the negotiations and agreements shared and organized?

In all of this, however, emissions into the atmosphere are just the tip of the iceberg, representing just a part, albeit an important part, of the material output of human societies into the environment, essentially of waste from socioeconomic production processes. The climate change debate does not address the more general large pressure on the environment of socioeconomic demands for material input reflected in land use change. Sustainability is a challenge far larger than the challenge of climate change, but climate change and the alteration by humans of air are the breakthrough topic that is leading humankind into a realization of the danger of the currently developing global situation.

6. Earth System Analysis

How can sustainable paths into the future be found? The Anthropocene is characterized by a co-evolution of societies and the environment, where geosphere and biosphere are folded into one environmental dimension (SCHELLNHUBER 1999). To mathematically describe attainable trajectories in this co-evolutionary joint space of environment and societies would require mathematical theories of the dynamics of both dimensions. However, such theories are currently not available, if they even exist. Theories of the geosphere are advanced and available; here the main problems are dealing with complex networks of interactions and setting the values of relevant parameters. With respect to the biosphere, comprehensive theories of ecological systems have not been developed, only of greatly simplified aspects, and many doubt that a theory of ecology exists at all. For the anthroposphere, sociologists and historians are generally skeptical that quantitative theories of history are possible. Despite various prominent suggestions there is a complete lack of agreement even on the basic principles that might govern the evolution of societies, and on whether such principles exist. A comprehensive theory of society-environment co-evolution is therefore not within reach.

In this situation, humankind is currently using mainly two tools to support its reflective recognition of the changing state of the planet with respect to its own perspectives for societal development. The first is computer modeling. Models that cover first-order dynamics but average out many of the details, so-called models of intermediate complexity, seem to provide the best chance of understanding the top-level dynamics of the anthropocenic Earth system. The second type of tool are macroscopes, a global system of networks of instruments that allow the construction of global images permitting humankind to perceive the whole of the planet as a system. Satellite images, worldwide networks of buoys and observation stations, global socioeconomic reporting and statistical accounting have all fed the construction of such images of the world and provide the empirical foundation required for theoretical analysis and modeling. Without such a foundation, humankind would be largely blind regarding the world as a whole.

Macroscopes provide empirical feedback of the past and present while earth system model provide projections into the future of the topology of the co-evolutionary phase space defined by societal and environmental dimensions (Lucht and Jaeger 2001). The topology of this space is at present still not very well known. By providing a navigation chart of the future knowing, it would allow directed attempts at steering spaceship Earth through the dangerous straights of global change and avoiding cliffs and shallows, where possible. It is not currently known just how quickly the curvature of a developmental trajectory in this phase space can be changed. For example, rapidly stopping the environmental impacts of societal development would imply that rapid decoupling of societal material reproduction from the environment is possible. Given a certain position of the present in Earth system phase space, certainly important regions of that space exist that are not accessible to trajectories of co-evolution. It is also not known in what regions of that phase space societal conditions are incompatible with environmental states. In turn, understandings of the environmental basis for societies remain partial, despite much research.

What is certainly known is that currently the phase space angle of development between the dimensions of society and environment has a considerable magnitude that reflects the intimate coupling of the two domains in the Anthropocene. If the trajectory continues, a limit of environmental change will be reached that implies both environmental collapse and societal collapse. It is currently not know which of the two systems is less resilient to change and will collapse first. Nor is it how known how much and how quickly the topology of the phase space transversed will be changed by new technologies, though optimism in this respect is wide-spread. Current discourse seems to assume that the environment will reach limits before societies do, and the debate concerns mainly the development of societies in the face of such impending limitations. But there is no particular reason to not believe that social structures are actually as much or more vulnerable to rapid change as the environment, implying that sociocultural organization may degrade before the environment does. Examples of this might be seen in the collapses of a number of historical civilization where environmental change may have played an important role.

Humankind has a very strong instrument available in facing this sustainability challenge. Its knowledge of the dynamics of the world is incredibly much larger than at any previous time, making it possible that the human mind will be able to not just blindly suffer the impacts of global change on its societies, but to predict them and act accordingly to prevent those that by agreement should be prevented. This has become possible as a consequence of the breakthroughs in geoscientific understanding of our planet that have been achieved only in the past century and a half. It is only in a few recent generations that humankind has gained an understanding of the true age of the earth, of the meaning of fossils, of the evolution of species, of the coming and going of ice ages, of human origins in Africa, and of the phenomenon

Air: A Planetary Hybrid

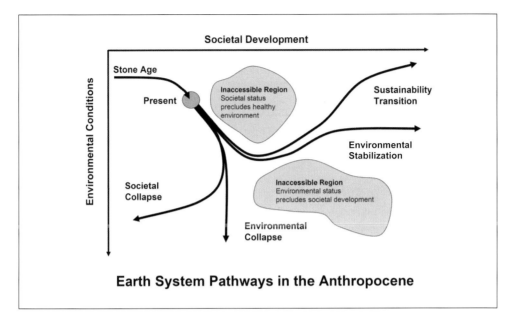

Fig. 3 The co-evolution of the geosphere-biosphere environment with societies reflects the degree of coupling between these two dimensions and their internal dynamics under forcing. While both environmental and societal collapses are clearly part of the space of possible pathways, finding a transition to sustainability is the challenge jointly before the world's societies. Currently, possible pathways and regions not accessible are only known only crudely known as comprehensive theories of the environment and societal evolution, and of their co-evolution, are only very partially available.

that is our current interglacial, the Holocene; and with this knowledge, and understanding of the workings of the Earth system as a whole. The works, to name just a few, of VERNADSKY, of TEILHARD DE CHARDIN, of LOVELOCK and of the scientific global change community of the past 30 years, expressed in the political milestone of the Rio Conference of 1992 and the scientific milestone of the Amsterdam Conference of 2001, have developed the foundations upon which next actions concerning the Earth system can be based.

Earth system analysis is a science emerging from this process that humankind can employ in dealing with global change, but it requires more serious attention to its second dimension, the dynamics of human societies, their information and material exchanges and networks (SCHELLNHUBER et al. 2005). Humankind has the opportunity to use its capacity for science and technology to reach a state of reflective analysis that will allow, for the first time, to take earth system dynamics into account when shaping the intrinsic dynamics of the anthroposphere. Earth system analysis is fundamental as a tool to be used. But it is certainly not enough: societal action is not directly a function of scientific insight but requires recourse to the driving forces of the social system, such as lie in issues of identity, the geographies of place and landscape, and the narratives that shape cultures and their relation to the environment (LUCHT and PACHAURI 2004).

This present essay is a small part in this attempt to construct narratives that provide a scientifically founded perspective on achieving a sustainability transition in the next several decades. In the end, it is about finding a place for humans in the wider world that ultimately will have to reflect the relationship of human to that world (Cosgrove 1984). Traditionally, stories of origin and meaning have been mythological, have been foundation myths and epics of constitution and change. It is in that tradition that air as an element of culture and of science is of such importance. Full of cultural meaning as one of the classical elements, air is also at the center of the accelerating debates about common global futures by being central to the springboard challenge of change in the atmosphere and the climate system. Air can serve as a physical, spiritual and perhaps, then, even divine guide to sustainable paths into the future.

7. Geophysiology

In a play with images taken from ancient cultural connotations one might consider planet Earth to be composed of the classical four elements. The lands are earth, the oceans water, air the atmosphere and fire the mysterious power of life. As one might see the human body as composed of the elements, and its intricate balance of substance, water, breath and spirit among the secrets of its existence, so one could consider the planetary system a body of systemic balances of the elements.

A body is an identifiable whole: as such it cannot be understood as a mere addition of organs, as the result of measured levels of hormones or blood content, or lists of chemical substances. Accordingly, it cannot be reconstructed from a reductionist catalogue of findings. Is the same true for the planet in relation to science? Is it true for the nature of the relationship of humans with their planet? The human body has in it a being that is as difficult to explain as the existence of life on the surface of a planet. As a living form, a body comes together as a whole that has its own place in the world: vulnerable, finite, but also resilient, self-regulated and full of possibilities, within the framework of chemical, physical and genetic dimensions. If the same is true for the planetary body with its life, then the field of knowledge required to study Earth comprehensively and as a whole is geophysiology.

When its elements change, the planetary body changes. As humans collectively transform Earth into a new state, it will be important that none of Earth's vital organs fail (Lenton et al. 2008). Humans have been called an illness to the planetary body, as when global warming is described as a planetary fever, occurring while Earth is already in an interglacial period of uncharacteristic warmth which humans now additionally heat. But if the organ failure that is a consequence can be avoided, humans could probingly perhaps also be seen as agents of a major transition in evolution that leads to a new planetary state with currently unknown properties (Smith and Szathmáry 1995). In either case, it is a transformation of the elements of Earth that is at the centre of scientific and cultural reflection. And with life being one of these elements, the planetary body is certainly one capable of metamorphoses that rival those seen in the evolution of life.

The challenge humankind currently faces is to devise directions of development that ensure the integrity of the planetary body under change, an integrity upon which, more than anything, human cultures depend. From among the elements constituting the body of Earth, air is the unifying element at the forefront of the debate: its geophysiological status has changed greatly both in the environment and in the human mind as air pollution and greenhouse gas emissions are entering collective consciousness as world-changing processes that require reflective attention.

As we have seen, in geophysiological earth system analysis, to be able to engage politically, socially and culturally with the elements and their change, it is part of the unfolding narrative to have a contemporary understanding of the elements. Here, to realize that the planet's air is a co-product of life, more than a phenomenon of a dead planet defined merely by the laws of chemistry and physics, but a hybrid co-construct of the planet and its life, of the geosphere and the biosphere. A hybrid that has now, through the increase in the flow of anthropogenic materials into the air, become a co-construct of the geobiosphere and the recently evolved anthroposphere.

As a consequence, air as an element is now as closely linked as any element to the comprehensive completeness of global processes that together determine the state of the bounded planet. Understanding this joint system and where it will lead both the environment and human societies requires geophysiological views of the elements that do not disentangle, but rather entangle the scientific and the cultural dimensions of air. Scientifically re-shaped cultural views of air now arise with a sense of urgency as part of a wider human enterprise of necessity to find exegetical explanations of the elements, explanations that reflect scientific understanding while resting firmly on the basis of the continued cultural presence of the elements in the life of global societies. Without it, the consequences of global change are likely to mount to a colossally detrimental magnitude due to a lack of foresight and appropriate adjustments in sociocultural practices.

In creation, air gives the spirit of breath to life. Nobody has ever observed the origins of life out of dead matter, but once conscious, it realizes that the world as a process is embedded in material realities that are chemically and physically constrained and intricately linked to its own metabolisms. It is now up to the power of that spirit to see the whole of the planetary body, and to adjust its own position in the Earth system in a manner that allows a joint development within the corridors of habitability for human culture.

References

AJTAY, G. L., KETNER, P., and DUVIGNEAUD, P.: Terrestrial primary production and phytomass. In: BOLIN, N., DEGENS, E. T., KEMPE, S., and KETNER, P. (Eds.): The Global Carbon Cycle; pp. 129–182. Chichester, New York, Brisbane, Toronto: John Wiley & Sons 1979
ALLEY, R. B., MAROTZKE, J., NORDHAUS, W. D., OVERPECK, J. T., PETEET, D. M., PIELKE, R. A. jr., PIERREHUMBERT, R. T., RHINES, P. B., STOCKER, T. F., TALLEY, L. D., and WALLACE, J. M.: Abrupt climate change. Science *299*, 2005–2010 (2003)
AMBROSE, S. H.: Paleolithic technology and human evolution. Science *291*, 1748–1753 (2001)
BERGER, A., and LOUTRE, F. M.: An exceptionally long interglacial ahead? Science *297*, 1287–1288 (2002)
BRAUER, A., HAUG, G. H., DULSKI, P., SIGMAN, D. M., and NEGENDANK, J. F. W.: An abrupt wind shift in Western Europe at the onset of the Younger Dryas cold period. Nature Geosci. *8*, 520–523 (2008)
CANFIELD, D. E.: The early history of atmospheric oxygen. Annu. Rev. Earth Planet. Sci. *33*, 1–36 (2005)
CONWAY MORRIS, S.: Life's Solution: Inevitable Humans in a Lonely Universe. Cambridge: Cambridge University Press 2005
COSGROVE, D.: Social Formation and Symbolic Landscape. Madison: University of Wisconsin Press 1984
COSTANZA, R., GRASUMILCH, L., STEFFEN, W., CRUMLEY, C., DEARING, J., HIBBARD, K., LEEMANS, R., REDMAN, C., and SCHIMEL, D.: Sustainability or collapse: What can we learn from integrating the history of humans and the rest of nature? Ambio *36*, 522–527 (2007)
CRUTZEN, P. J., and STOERMER, E. F.: The anthropocene. IGBP Newsletter *41*, 17 (2000)
DIETRICH, W. E., and PERRON, J. T.: The search for a topographic signature of life. Nature *439*, 411–418 (2006)
FALKOWSKI, P. G.: Tracing oxygen's imprint on Earth's metabolic evolution. Science *311*, 1724–1725 (2006)

Goldblatt, C., Lenton, T. M., and Watson, A. J.: Bistability of atmospheric oxygen and the great oxidation. Nature *443*, 683–686 (2006)

Graczyk, A.: Das literarische Tableau zwischen Kunst und Wissenschaft. München: Fink 2004

Haberl, H., Erb, K.-H., Gaube, V., Bondeau, A., Plutzar, C., Gingrich, S., Lucht, W., and Fischer-Kowalski, M.: Quantifying and mapping the human appropriation of net primary production in the Earth's terrestrial ecosystems. Proc. Natl. Acad. Sci. USA, doi_10.1073_pnas. 0704243104 (2007)

Haberl, H., and Fischer-Kowalski, M. (Eds.): Socioecological Transitions and Global Change. Cheltenham, Camberley, Northampton: Edward Elgar Publishing 2007

Hansen, J., Sato, M., Kharecha, P., Russell, G., Lea, D. W., and Sidall, M.: Climate change and trace gases. Phil. Trans. R. Soc. A *365*, 1925–1954 (2007)

Humboldt, A. von.: Ansichten der Natur. Tübingen: J. G. Cotta 1807

IPCC WG1: Climate Change 2007 – The Physical Science Basis. Cambridge: Cambridge University Press 2007

IPCC WG2: Climate Change 2007 – Impacts, Adaptation and Vulnerability. Cambridge: Cambridge University Press 2007

Kates, R. W.: Sustainability science. Science *292*, 641–642 (2001)

Lenton, T. M.: Gaia and natural selection. Nature *394*, 439–447 (1998)

Lenton, T. M., Caldeira, K. G., Franck, S. A., Horneck, G., Jolly, A., Rabbow, E., Schellnhuber, H.-J., Szathmáry, E., Westall, F., Zavarzin, G., and Zimmermann-Timm, H.: Group report: Long-term geosphere-biosphere coevolution and astrobiology. In: Schellnhuber, H.-J., Crutzen, P. J., Clark, W. C., Claussen, M., and Held, H. (Eds.): Earth System Analysis for Sustainability; pp. 111–140. Cambridge: MIT Press 2004a

Lenton, T. M., Held, H., Kriegler, E., Hall, J.W., Lucht, W., Rahmstorf, S., and Schellnhuber, H.-J.: Tipping elements in the Earth's climate system. Proc. Natl. Acad. Sci. USA *105*, 1786–1793 (2008)

Lenton, T. M., Schellnhuber, H.-J., and Szathmáry, E.: Climbing the co-evolutionary ladder. Nature *431*, 913 (2004b)

Lovelock, J.: Gaia – A New Look at Life on Earth. Oxford: Oxford University Press 1979

Lovelock, J. E., and Griffin, C. E.: Planetary atmospheres: Compositional and other changes associated with the presence of life. Adv. Astron. Sci. *25*, 179–193 (1969)

Lucht, W., and Jaeger, C. C.: The sustainability geoscope: a proposal for a global observation instrument for the anthropocene. In: *German National Committee on Global Change Research*: Contributions to Global Change Research; pp. 138–144. Bonn: NKGCF 2001

Lucht, W., and Pachauri, R. K.: The mental component of the Earth system. In: Schellnhuber, H. J., Crutzen, P. J., Clark, W. C., Claussen, M., and Held, H. (Eds.): Earth System Analysis for Sustainability; pp. 341–365. Cambridge: The MIT Press 2004

Mellars, P.: Archeology and the dispersal of modern humans in Europe: Deconstructing the "Aurignacian". Evol. Anthropol. *15*, 167–182 (2006)

Mithen, S.: After the Ice: A Global Human History 20,000 – 5000 BC. London: Weidenfeld & Nicolson 2003

Petschel-Held, G., Block, A., Cassel-Gintz, M., Kropp, J., Lüdeke, M. K. B., Moldenhauer, O., Plöchl, M., and Schellnhuber, H.-J.: Syndromes of global change – A qualitative modelling approach to assist global environmental management. Environ. Mod. Assessm. *4*, 295–314 (1999)

Raskin, P., Banuri, T., Gallopin, G., Gutman, P., Hammond, A., Kates, R., and Swart, R.: Great Transitions. The Promise and Lure of Times Ahead. Boston: Stockholm Environment Institute 2002

Sachs, W.: Planet Dialectics: Explorations in Environment and Development. New York: Zed Books 1999

Saugier, B., Roy, J., and Mooney, H. A.: Estimations of global terrestrial productivity: Converging toward a single number? In: Roy, J., Saugier, B., and Mooney, H. A. (Eds.): Terrestrial Global Productivity; pp. 543–557. San Diego: Academic Press 2001

Schellnhuber, H.-J.: Earth system analysis and the second Copernican revolution. Nature *402* Suppl., C19–C23 (1999)

Schellnhuber, H.-J., Crutzen, P. J., Clark, W. C., and Hunt, J.: Earth system analysis for sustainability. Environment *47*, 10–25 (2005)

Smith, J. M., and Szathmáry, E.: The Major Transitions in Evolution. Oxford: Oxford Univ. Press 1995

Takács-Santa, A.: The major transitions in the history of human transformation of the biosphere. Human Ecol. Rev. *11*, 51–66 (2004)

United Nations: United Nations Millennium Declaration. Resolution of the General Assembly *55*/2. New York: United Nations 2000

Vernadsky, V. I.:. Biosfera. Leningrad: Nauchoe Khimikoteknicheskoe Izdatelstvo 1926 (The Biosphere. New York: Springer 1991)

Weart, S. R.: The Discovery of Global Warming. Cambridge: Harvard University Press 2003
Zachos, J., Pagani, M., Sloan, L., Thomas, E., and Billups, K.: Trends, rhythms, and aberrations in global climate 65 Ma to present. Science *292*, 686–693 (2001)

        Prof. Dr. Wolfgang Lucht
        Potsdam Institute for Climate Impact Research (PIK)
        PO Box 601203
        14412 Potsdam
        Germany
        Phone: +49 331 2882533
        Fax: +49 331 2882600
        E-Mail: wolfgang.lucht@pik-potsdam.de

Festakt zur Ernennung der Deutschen Akademie der Naturforscher Leopoldina zur Nationalen Akademie der Wissenschaften

Ceremony to Mark the Nomination of the German Academy of Sciences Leopoldina to the National Academy of Sciences

Nova Acta Leopoldina N. F., Bd. *98*, Nr. 362
Herausgegeben vom Präsidium der Deutschen Akademie der Naturforscher Leopoldina
(2009, 76 Seiten, 50 Abbildungen, 21,95 Euro, ISBN: 978-3-8047-2551-5)

Die Deutsche Akademie der Naturforscher Leopoldina wurde am 14. Juli 2008 im Rahmen eines Festaktes in Halle zur Nationalen Akademie der Wissenschaften ernannt. Damit erhielt Deutschland – wie andere europäische Länder oder die USA – eine Institution, die Politik und Gesellschaft wissenschaftsbasiert berät und die deutsche Wissenschaft in internationalen Gremien repräsentiert. Der Band dokumentiert den Festakt mit der Übergabe der Ernennungsurkunde durch die Vorsitzende der Gemeinsamen Wissenschaftskonferenz und Bundesministerin für Bildung und Forschung Annette Schavan. Er enthält die Reden von Bundespräsident Horst Köhler, Sachsens-Anhalts Ministerpräsident Wolfgang Böhmer und Leopoldina-Präsident Volker ter Meulen sowie den Festvortrag „Rolle und Verantwortung nationaler Akademien der Wissenschaften" von Jules A. Hoffmann, Präsident der Académie des sciences, Paris. Der Aufbau einer Nationalen Akademie ist ein richtungsweisender Schritt für die deutsche Forschungslandschaft, da für den kontinuierlichen Dialog von Wissenschaft und Politik eine solche Einrichtung erforderlich wurde. Der Publikation ist eine DVD mit dem Mitschnitt der Festveranstaltung beigefügt.

Wissenschaftliche Verlagsgesellschaft mbH Stuttgart

Soil and Human Impact

Hans-Rudolf Bork ML[1], Christine Dahlke[2], Stefan Dreibrodt[1], and Annegret Kranz[1]

With 13 Figures

Abstract

Organisms, climate, characteristics of topography and rocks at the surface of the earth as well as time determine the formation of soils. Soils are the base of life in all terrestrial ecosystems. Soil formation was intensive during Holocene in ecosystems with a humid and warm climate and thus under dense vegetation and without a significant influence of humans. Since the rise of agriculture, humans determine the processes of soil formation and soil devastation. In urban areas, former natural soils are nearly totally lost and replaced by anthropogenic deposits. In predominantly agriculturally used areas soils changed, too. Since land use was rarely sustainable in the past, soils were often eroded in arable land. Eroded soil material is mainly deposited at the bottom of slopes and in valleys. These deposits are named colluvial and alluvial sediments. Extreme weather events are triggering most of the soil loss in agriculturally used land. In western central Europe, the runoff of the 1000-year precipitation event cut deep gullies and eroded fertile topsoils in July 1342. In general, the fertility and the extension of fertile soils has been dramatically as a result of soil erosion since the rise of agriculture. Soils and sediments have not only been used only used for agriculture and forestry since several thousand years. Brick production and pottery making are examples.

Zusammenfassung

Organismen, das Klima, die Oberflächenformen und das Gestein an der Erdoberfläche sowie die Zeit sind Faktoren der Bodenbildung. Böden sind die Basis aller terrestrischen Ökosysteme. Im Holozän war die Bodenbildung in humiden und warmen Klimaten und unter dichter Vegetation zunächst intensiv. Seit dem Beginn der Landwirtschaft bestimmen Menschen die Prozesse der Bodenbildung und -zerstörung. In urbanen Ökosystemen wurden die zunächst ohne Einfluss des Menschen entstandenen Böden nahezu vollständig durch anthropogene Ablagerungen ersetzt. In landwirtschaftlich geprägten Regionen veränderten sich die Böden ebenfalls. Da die Landnutzung in der Vergangenheit selten nachhaltig betrieben wurde, sind fruchtbare Böden oft erodiert. Das abgetragene Bodenmaterial hat sich überwiegend auf den Unterhängen als Kolluvium und in den Talauen als Auensediment abgelagert. Extreme Witterungsereignisse sind die Auslöser von starker Bodenerosion auf landwirtschaftlich genutzten Böden. Im Westen Mitteleuropas verursachte das 1000-jährige Niederschlagsereignis im Juli 1342 das Einreißen von Schluchten; fruchtbare Oberböden wurden flächenhaft erodiert. Bodenerosion hat die Fruchtbarkeit der Böden seit Beginn der Landwirtschaft deutlich verringert. Menschen nutzen Böden und Sedimente nicht nur für die Land- und Forstwirtschaft. Die Herstellung von Ziegeln und die Produktion von Töpferwaren sind ein Beispiel dafür.

---

1 Department of Ecotechnology and Ecosystem Development, Ecology Centre of the Christian-Albrechts-Universität at Kiel, Olshausenstr. 75, 24118, Kiel, Germany.
2 Johann-Friedrich-Blumenbach-Institute of Zoology and Anthropology, Georg-August University at Göttingen, Bürgerstr. 50, 37073 Göttingen, Germany.

Hans-Rudolf Bork, Christine Dahlke, Stefan Dreibrodt, and Annegret Kranz

1. Introduction

Between atmosphere and rock soils are developing at the surface of the earth. Specific characteristics of topography, stones, climate and organisms over time lead to the formation of soils which are classified in soil types. Soil forming processes structure a soil mostly parallel to the surface in predominantly homogenous layers: soil horizons. They are the expression of the material changes that passed *in situ*. New material is formed, the existent material is transformed (e.g. clay minerals). During periods of intense rainfall, material is dislocated in humid regions with the leakage water from top soil horizons to lower soil horizons, in arid and semi-arid regions with the evaporating soil water from lower soil horizons to top soil horizons. In this way, material is either enriched or depleted.

Draining water that is temporarily flowing in the soil on a less permeable horizon parallel to the surface of a slope enriches material namely at the lower end of the slope and in the neighboring floodplain. If for example soils are covered by sediments on concave-shaped lower slopes, they will be situated in the deeper ground and under the impact of other soil forming processes. Soil erosion on slopes can change the topography of the area, uncovering lower soils and transporting top soil material down the slope. The removal and the import of material and mechanical pressures through land use led to constant changes of soils, too. Modifications of the vegetation e. g. by the implementation of new land use systems modify soil forming processes.

For the first time in billions of years one species steers the development of soils: man. Man has been directly and indirectly determined the processes in and on soils since the beginning of agriculture: first in a few regions only, today on the global scale. In some landscapes agriculture began more than 7,000 years ago, e.g. on the Loess Plateau of northern China (near the middle reaches of the Huang He), in the lowlands of eastern China, in Mesopotamia, in southeast, south and southern central Europe. In some American landscapes agriculture has dominated only for a few decades or centuries.

Soils that developed during early or middle Holocene before the first clearings do not reflect – as an effect of later cultivation activities – the current processes anymore. In fact, the soil forming processes adapt quickly to changed land use conditions. If soils are covered by sediments, some characteristics from the time of their formation will remain permanently preserved.

2. Processes of Soil Formation and Major Soil Types

Which processes form soils? In forests, branches, leaves or needles fall on the ground surface. During autumn, in the non-tropical forests a litter layer accumulates that is partly decomposed during the summer half-year. Favorable nutrient conditions, high temperatures, a high humidity and a long vegetation period enable a decomposition of the organic matter in a relatively short time. Therefore in humid tropics significant litter layers with plant remains cannot be found. Humus on the ground surfaces is the result of decomposition processes. Animals can mix the humus several decimeters deep into the soil. This contributes to the formation of fertile black soils, the so-called Chernozems. Especially in continental, semi-arid steppes with a high production of biomass and a short vegetation period, i.e. a temporary decomposition of organic material only, Chernozems developed. Plants and animals that live on and in the soil enrich the soil with different substances. Earth particles which have traveled through the digestive organs of earthworms are enriched with fibrils and transform single soil particles into crumbly soil aggregates. Also physical and chemical forces merge particles and substances to larger soil aggregates. Root growth and the digging activities of animals compress, loosen or mix parts of the soil.

Some soil particles such as three-layered clay minerals can temporarily absorb water and swell. If they release their water, they will shrink. Between the particles that have been merged to aggregates, shrinkage cracks opened which frequently create polygonal patterns. Soils that contain high concentrations of three-layered clay minerals can develop deep crack grids in the warm, periodically humid regions of the Earth. Silt and sand can, during the dry season, be blown into the cracks, animals and humans that walk over the surface carry material into the narrow, deep cracks. At the beginning of the wet season, the space of the dry cracks decreased to a great extent. The water absorbing three-layered clay minerals push the soil aside and stir it. Mounds well up at the surface between small depressions. This microtopography is called Gilgai. A Vertisol develops.

In humid regions acids that are dissolved in water penetrating through the soil may cause the leaching and translocation of substances and soil particles from horizons of the top soil into lower horizons, of easily dissolvable substances into the deeper underground or into the ground water, into the next flood plain, into rivers, lakes and eventually into the sea.

Fig. 1 Ridge and furrow field system near the deserted village of Nieps (Altmark, Sachsen-Anhalt, Germany). Photo: H.-R. BORK. The Anthrosol on the top of the ridges is the result of frequent ploughing on small and long fields during high Medieval Ages. Below colluvial layers of high Medieval Ages are exposed. At the bottom a fossil Luvisol was found which developed during middle Holocene in sand which was deposited by melting water during the Saalian.

Hans-Rudolf Bork, Christine Dahlke, Stefan Dreibrodt, and Annegret Kranz

Fig. 2 A gully cuts into a partly eroded Ferralsol which has developed over a long time in deeply weathered volcanic rocks on Babeldaob (Palau). Photo: H.-R. BORK

Weathering processes dissolve substances and discharge them, crush earth particles, form clay minerals and oxidize ions. Cambisols are developing. If, in a slightly acid setting, clay minerals are transported from the top soil into the lower soil, a Luvisol develops (Fig. 1). If the acidification is intensive, easily dissolvable organic substances and metal ions (namely iron and manganese ions) can be washed out of the top soil and transported into the lower soil by percolating water where they are oxidized. A Podsol soil develops with a mostly ash grey, sallow top soil, and a lower soil horizon that is either black grey because of the accumulation of organic substances, or deep red by sesquioxides.

If leaching and acidification endure in the constantly and periodically wet tropics for long time periods, SiO_2 is dissolved and discharged. The remaining non-soluble minerals and substances – primarily iron and aluminium oxides – are enriched. A Ferralsol (ferr: ferrum, al: aluminium, sol: soil) that develops can grow to a thickness of several meters, if it has been (partly) eroded or covered by sediments in the Quaternary during phases with a low vegetation cover density (Fig. 2). During regular annual desiccation periods, the topsoil is hardening. Extremely hard crusts (Plinthit) of several decimeters to some meters thickness remain.

At the surface of ground water bodies different ions can oxidize. Iron and manganese oxides color the soil horizon, in which the ground water table fluctuates, red to black ("oxidization colors"). Below, in the permanently reduced setting of a ground water soil, which is called Gleysol, light to greenish colors dominate ("reduction colors"). If, in temperate zones during spring time, soil water fluxes are influenced by a clay-enriched lower horizon of a Luvisol or by an initially present (i.e. not triggered by soil forming processes) less permeable layer, water fills soil pores. Reduction processes are dominating in the top soil and oxidation and reduction processes in the lower horizon. A Pseudogley develops.

At sites in dry regions that are influenced by ground water, soil water ascends through very fine soil pores until the water reaches the surface. There, the water evaporates and substances precipitate, that were carried along. Salt is enriched in the top soil.
On the slopes of arid regions physical, rock-crushing weathering processes prevail. Initial soils dominate. At locations with intensive sand deposition Arenosols develop.

Volcanism is forming new bedrock or granular deposits (Tephra). Especially in tephra deposits fertile soils develop, called Andosols.

3. Sediments

Sediments also document environmental dynamics. Terrestrial sediments which were deposited during the Holocene are for example colluvial layers, flood plain deposits and lake sediments. The processes of soil erosion comprise the erosion, transport and deposition of soil particles. Soil erosion is enabled by humans. Thus soil erosion began with the clearing of the pristine vegetation.

Strong wind and surface water, which flows down on the ground surface during short intense rainfalls or with the rapid melting of water-rich snow covers, erode soil particles. On slopes with no or a very low vegetation cover density, heavy rainfall can cause serious damage by soil erosion. The clearing of pristine vegetation enabled a direct influence of raindrops and runoff.

Runoff that flows over the ground surface erodes small rills under the conditions of moderate intensive precipitation. Management practices on fields flatten the ground surface, the erosion rills vanish and generally the soil surface will be slightly lower. Over decades and centuries the soil surface can decline in relevant dimensions. During extreme heavy rainfall events surface water can flow in natural dents, concentrating at the borders of fields or at waysides in great quantities. Then deep gullies can cut into soils and the under laying rocks that are sensitive to erosion. If the gullies are not filled up by natural deposition processes or by man within a certain time, arable land will be lost permanently.

When soil is eroded by water, the transport of soil particles takes place at the soil surface, and, occasionally, also in larger pore systems in the soil or the subjacent rock (subsurface erosion). Wind carries small soil particles up to heights of several kilometers in the atmosphere and occasionally from one continent to another.

Soil particles which were eroded can be deposited in different parts of the landscape. The correlate depositions of soil erosion are called:

– fan sediments at the lower end of gullies,
– colluvial layers on lower slopes,
– flood plain or alluvial sediments in flood plains,
– detrital-allochthonous lake sediments in lakes and
– marine sediments in the sea.

Hans-Rudolf Bork, Christine Dahlke, Stefan Dreibrodt, and Annegret Kranz

3.1 Fan Sediments and Colluvial Layers

As soon as the velocity of runoff slows down due to decreasing precipitation intensity or recessive slopes, coarser soil particles are deposited first, and later the finer ones on the soil surface. At the end of a rainfall, soil particles are deposited even on upper slopes or in gullies. The recessive slope and the different flow distances condition that sediments of fans and colluvial layers contain, on average, larger particles, and flood plain sediments, lake sediments and marine sediments smaller particles (Fig. 3). Because transport via surface flow is determined by gravitation, the sink area of the soil particles can often be precisely attributed to a fan sediment or a colluvial layer.

Fig. 3 Colluvial layers which were deposited during Medieval and Modern Times on alluvial sediments near Prašice (Slovak Republic). In the terminology of soil science an Anthrosol. Photo: H.-R. BORK

However, agricultural management practices not only homogenize the rills of an erosion area, but also the fan sediments and colluvial layers. If these sediments are less deep than the management depth, they are mixed with the soil horizons below. The initially high, event-related resolution of fan sediments and colluvial layers is altered into a mixed material (BORK 2006).

The characteristics of the sediments, that are deposited on lower slopes and in flood plains and of the soils that were covered there by these sediments, permit far-reaching deductions regarding landscape development, human impact and effects on humans.

Artifacts that were relocated and embedded in the sediments or *in situ* buried archaeological findings contain significant information about the anthropogenic chapter of landscape history.

3.2 Lake Sediments

Lake sediments contain information about the water body of a lake, about its catchment area (e.g. matter inputs), and about the surrounding area (e.g. pollen). They have very different characteristics, predominantly depending on the distance to the lake shore (Håkanson and Jansson 1983). Whereas in the area of the lake shore, incomplete sequences are deposited due to water movement, bio-turbidity and changing water levels, the lake sediments of the deep lake area (the profundal zone) are mostly characterized by a continuous sedimentation.

Different processes determine the formation of sediments in lakes. Lake shore erosion or soil particles which were imported with runoff from the catchment contribute to lake sedimentation only to a small degree. Most soil particles which were eroded on slopes were deposited in fans and in colluvial layers. Biological (microfossils, organic substances) and chemical sedimentation products (lime), which were formed by processes within the lake, are dominating in the lowland lakes of the temperate zone.

The continuity of lake-internal deposits allows the processing of precise chronologies with a high temporal resolution of the environmental history of a lake and its surroundings. Short-lived organic residues such as deciduous leaves that entered from the surroundings and were embedded in lake sediments can, with the use of physical methods for age determination, be dated more precisely than wood or charcoal.

Under special conditions, annual layers (varves) are preserved in the deepest area of a lake (Brauer 2004). These facilitate a very precise age determination by enumeration. Their fine layers represent the seasonally changing conditions of a lake.

Lake sediments often preserve detailed information about life in and around the water body. Reactions of life in the lake upon environmental changes can be reconstructed by the examination of embedded microfossils, and the structure and composition of the sediments. They allow deductions on the governing environmental factors. Frequently examined organism groups are: diatoms, copepods, larvae of non-biting midges and bivalves. On some fossils, the contents of stable isotopes are measured for the reconstruction of the palaeo-environment (Berglund 1986). Pollen analyses of lake sediments provide evidence of the vegetation composition and the land use in the surrounding of a lake (Beug 2004, Dreibrodt et al. 2003).

Soil particles that were deposited on the bottom of a lake often provide further important information on landscape development. Some of these deposits can be dated precisely, because they stem from rare, known events to which they can be unmistakably attributed.

3.3 Flood Plain Sediments

The runoff of heavy rainfall events causes flooding in the valley bottoms. Coarser grains are deposited next to river beds in river bank walls. Beyond river beds and bank walls, the flood plain vegetation slows down the flowing velocity of the flooding water. There, silt grains sink unto the surface. Immediately after flooding, water rich in suspended material fills the small sinks of flood plains. The water seeps away and the fine suspended material, clay minerals with organic substances are deposited on the surface.

Flood plain sediments integrate data of larger water catchment areas. Hardly delineable sediment source areas and homogenization by management practices complicate the identification of the origin and the age of flood plain sediments. Therefore, reconstructions with a high temporal and spatial resolution are rare.

4. Man as a Factor of Soil Formation

Material which was deposited by humans (e. g. deposit of removed soil or rock, of material enriched with organic matter) is named Anthrosol in soil science (Fig. 2, 4, 5). In geomorphology and geology, their names depend on the process of deposition. Probably Anthrosols have been the soil type with the largest extension on earth for a few decades. Anthrosols are widespread today in all landscapes which are used by humans. In many cities the natural soils which had developed during Holocene were nearly totally replaced by Anthrosols. In predominantly agriculturally used landscapes, Anthrosols dominate on concave segments of slopes (i. e. colluvial layers and fan sediments) and in valley bottoms (i. e. alluvial sediments). On convex slope segments partly eroded soils of all types mentioned above are common.

Fig. 4 A small hollow road near the deserted village of Nieps (Altmark, Sachsen-Anhalt, Germany) which is the result of soil erosion on a field road during Medieval Ages. Remnants of wheel tracks were observed. Photo: H.-R. BORK

4.1 Soil Development and Soil Erosion in central Europe

During the first millennia of the Holocene, soils formed in the woodlands of central Europe (Fig. 6, 7). Intensive land use began during Neolithic, Bronze or Iron Ages or in some hilly landscapes as recently as Medieval Times. At most slopes slight hillslope erosion dominated in agriculturally used hilly areas during Neolithic Age. Gully erosion was an exception. Extented and intensified agriculture enabled strong hillslope erosion and at some sites gullying during Bronze Age, Iron Age and Roman Times. In total, the surfaces of arable land were lowered by soil erosion in regions with highly erodible soils several decimeters, in others only a few centimeters during prehistoric times. After the decline of the Roman Empire woodland spread and soil development was intensive. No soil erosion and thus no sedimentation in the flood plains occurred in catchments that were totally wooded. Since the 7th century the wood-

Fig. 5 Extraction and exchange of soils and rocks by humans: huge pits are opened first, coal is extracted then, and mixed rock and soil particles are filled in the pit. Finally fertile material is deposited on the top. The procedure is called "recultivation", the soil type Anthrosol. (Vicinity of Jülich, Rhineland, Germany). Photo: H.-R. BORK

land was cleared and agriculture spread first in basins with fertile soils and then during high Middle Ages in the central German mountains and in the sandy areas of northeast Germany. At a few sites clearings in early or high Medieval Times and subsequent agriculture enabled gullying. In most cleared areas slight hillslope erosion dominated. Population pressure was high at the end of the 13th century, when the period of high medieval warm weather with high yields turned into lower summer temperatures and a growing number of weather extremes, years with disastrous yields, famines, high groundwater levels, surface runoff, floods and dramatic soil erosion. The 1000-year precipitation, runoff, and erosion event hit western central Europe from July 19th until July 25th, 1342: Warm and humid air was flowing from the eastern Mediterranean into western central Europe, causing there an extraordinary high amount of precipitation, runoff and erosion (Fig. 8, 9). More than 30% of the total soil erosion of the past 1500 years occurred during this event.

4.2 Agricultural Terraces

Some prehistoric, medieval or modern land use systems or landscape-forming measures left traces in the soils which have been preserved until the 20th century (JÄGER 1994, BORK 2006). Examples are: agricultural terraces, irrigation and drainage ditches, scratch traces of soil processing tools, cultivation pits, ploughing horizons, hollow roads and ridge-and-furrow field systems (Fig. 2, 4, 10). Changes in the landscape, however, have deleted many traces, especially during the second half of the 20th century namely due to the reallocation of land.

Agricultural terraces provide a segment of the long history of soil use and human action (SANDOR 2006). Until today, agricultural terraces are only systematically analyzed in a few cases, even if they are located in almost all agricultural regions. During archeological surveys

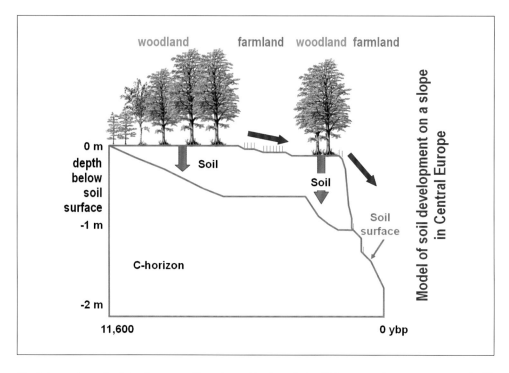

Fig. 6 Interactions of soil development, soil erosion and land use during Holocene on slope segment covered with loess in central Europe (BORK et al. 1998, modified). In periods with a dense vegetation cover on the slope (woodland) soil formation dominated (e.g. formation of a Cambisol or a Luvisol), in periods with no or only a low vegetation cover density due to agricultural land use soil particles were eroded and transported to the foot of the slope or into the floodplain.

in Greece, old agricultural terraces probably from Greek-Roman times were found (PRICE and NIXON 2005). The authors are discussing problems in dating these terraces. They used additional written sources for their analysis. HARD and RONEY (1998) are describing 3000 year old terraces in Mexico. VILLALON (1995) assumes that some rice terraces of the Philippine Cordillera are 200 years old. WILKINSON (1997) found agricultural terraces about 4000 years old in Yemen. BRAY (1984) assumed the introduction of agricultural terraces on the Chinese Loess Plateau more than 1100 years ago. ROBERTS et al. (2001) identified a 2500 to 2070 year old agricultural terrace in Duowa on the western margin of the Chinese Loess Plateau in the Qinghai province. BORK and DAHLKE investigated a terrace near Yan'an on the Chinese Loess Plateau which started to grow more than 4500 years before today.

The development of terraces modifies soil formation processes and changes the landscape structure. The genesis of agricultural terraces shows a large variety of processes depending on several factors such as topographical parameters, soil conditions, land-use systems (BORK et al. 2006). BEACH and DUNNING (1994) are describing principles of terrace construction in Guatemala. TREACY and DENEVAN (1994) show examples from Peru. Functions of constructed terraces are also drivers, including stable bases for farming on steep slopes, a reduction of soil loess or water management with irrigation, ponding (VILLALON 1995), drainage and microclimatic effects.

Soil and Human Impact

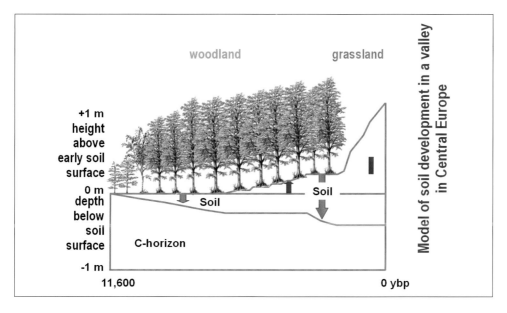

Fig. 7 Interactions of soil development, soil erosion and land use during Holocene at the border of the floodplain of a small river in central Europe (BORK et al. 1999, modified). In periods with a dense vegetation cover in the catchment of the valley soil formation dominated (e.g. development of a shallow Cambisol or a Gleysol), in periods with no or only a low vegetation cover density in parts of the catchment due to agricultural land use soil particles were eroded on the slopes and transported into the flood plain (deposition of alluvial sediments).

Fig. 8 Sediments of the fan of the gully Kirschgraben which were deposited during Medieval Times on alluvial sediments of the river Elsava (excavation of castle Mole, Spessart, Germany). *In the foreground*: Large stones which were deposited very probably during the 1000-year rainfall, -runoff and -erosion event in July 1342. Photo: H.-R. BORK

Fig. 9 Stones which were extracted from a debris layer in the fan of the gully Kirschgraben (excavation of castle Mole, Spessart, Germany). Very probably the stones were eroded and deposited during the 1000-year rainfall, -runoff and -erosion event in July 1342. The largest one has a weight of approx. 650 kg. Photo: H.-R. BORK

Fig. 10 During the last years a path has been eroded into Pleistocene Sediments on a steep slope of Changunarayan Hill (Kathmandu Valley, Nepal). The soil which has formed during early and middle Holocene under woodland was eroded already centuries ago. Photo: H.-R. BORK

5. The Use of Soil by Humans

Humans are using soil not only for agriculture and forestry but for several other purposes, too. One of the oldest techniques of soil use are construction techniques such as brick and pottery production.

5.1 Bricks and Adobe

Buildings represent a long history of soil and man. In regions where wood and stones were rare, for example in landscapes with loess deposits, people found other solutions for constructing houses (HILLENBRAND 1981, CAMPELL and PRYCE 2003). One of the oldest materials used by man is soil and sediment with a silty or loamy texture, which includes different mineral components (cf. MÄMPEL 2002, POTTER et al. 2005). The Chinese Loess Plateau, the largest loess area on earth with an extension of about 500,000 km², is characterized by a thick loess cover of up to several hundred meters in thickness. The silty, calcareous material builds steep and stable walls in the landscape. In the Yan'an region (Shaanxi Province) people are using the loess walls directly for building cave-like houses (Fig. 11). These houses are well adapted to the local environmental conditions according to the temperature regulation. In the semiarid region temperature varies between + 40 °C in summer and –30 °C during winter.

Fig. 11 Different stages of house construction on the Chinese Loess Plateau (Yangjuangou, Shaanxi Province; China). Photos: C. DAHLKE

In other regions of the earth, people are using soil for the production of adobe. Sun-dried soil or sediment is the oldest and cheapest building material on earth (SCHUNK 2000, CAMPELL and PRYCE 2003). Relicts of adobe walls can be found for example in Jericho since about 8000 BC (MÄMPEL 2002) or in Anatolia in Catal Hüyük since about 7000 BC (SCHUNK 2000). In some regions adobe walls are still constructed until now as in San Pedro de Atacama in northern Chile. First the wet loamy soil or sediment is homogenized by trampling or other techniques. Then the wet material will be filled in forms and dried in the sun. Sometimes straw or other plants are mixed together with the wet material in order to produce more stable adobe. In former time adobe were also used for inside house walls in Germany (Fig. 12).

Fig. 12 Adobe of a wall inside a farm-worker house (Müncheberg, Brandenburg Province, Germany). Photo: C. Dalchow

Fig. 13 The City Hall in Pietermaritzburg (KwaZulu-Natal, South Africa) is the largest red-brick building on the Southern Hemisphere. In South Africa this house construction style is an exception and is linked to the colonial history of the country. Photo: C. Dahlke

In semi-humid and humid landscapes where stone and wood is rare, bricks (burned silt or loam) were often used for house building (Fig. 13). Around 5000 years ago in Mesopotamia, loam was first burned with a low temperature of 550–600 °C (Schunk 2000). During the centuries people improved the techniques by introducing new components such as sand and calcareous material or new burning methods. Bricks were used for roofs as well as for walls. In Roman Times brick production spread over the Roman Empire (Schunk 2000). The development of bricks gave new opportunities for permanent settlement.

5.2 Pottery and Ceramics

Most clay minerals in soils and sediments are weathering products of feldspar rocks (Mämpel 2003). Clay deposits can be found *in situ* where the material weathered (e. g. in Cornwall, UK, or in Westerwald, Germany) or in secondary deposits, where clay was transported with runoff from the place of formation for example to valleys (Potter et al. 2005). Clay and other soil particles are used for making pottery. Early figures made of soil more than 20,000 years ago were found. The famous ceramic statuette Venus of Dolni Vestonice was found in the south of the Czech Republic. Some of the oldest known pots date back to 11,000 BC. They were found in the Fukui-caves in Japan (Mämpel 2003). The analysis of pottery provides important information of man-soil interactions, too.

References

Beach, T., and Dunning, N. P.: Ancient Maya terracing and modern conservation in the Peten rain forest of Guatemala. Journal of Soil and Water Conservation *50*/2, 138–145 (1994)

Berglund, B. E. (Ed.): Handbook of Holocene Palaeoecology and Palaeohydrology. Chicester: John Wiley & Sons Inc 1986

Beug, H.-J.: Leitfaden der Pollenbestimmung für Mitteleuropa und angrenzende Gebiete. München: Pfeil 2004

Bork, H.-R.: Landschaften der Erde unter dem Einfluss des Menschen. Darmstadt: Primus Verlag 2006

Bork, H.-R., Bork, H., Dalchow, C., Faust, B., Piorr, H.-P., und Schatz, T.: Landschaftsentwicklung in Mitteleuropa. Gotha: Klett-Perthes 1998

Bork, H.-R., Dahlke, C., und Yong, L.: Entdeckungen in einer 4750 Jahre alten Gartenterrasse im nordchinesischen Lößplateau. In: Bork, H.-R. (Ed.): Landschaften der Erde unter dem Einfluss des Menschen. S. 25–29. Darmstadt: Primus Verlag 2006

Brauer, A.: Annually laminated lake sediments and their palaeoclimatic relevance. In: Fischer, H., Kumke, T., Lohmann, G., Flöser, G., Miller, H., Storch, H. von, and Negendank, J. F .W. (Eds.): The Climate in Historic Times. Towards a Synthesis of Holocene Proxy Data and Climate Models; pp. 109–127. Berlin, Heidelberg, New York: Springer 2004

Bray, F.: Agriculture. In: Needham, J. (Ed.): Science and Civilisation in China. Vol. 6 Biology and Biological Technology. Part II. Cambridge: Cambridge University Press 1984

Campell, J. W. P., und Pryce, W.: Backstein: Eine Architekturgeschichte – von den Anfängen bis zur Gegenwart. München: Knesebeck 2003

Dreibrodt, S., Bork, H.-R., Brauer, A., und Negendank, J. F. W.: Der Einfluss der Landnutzung im Einzugsgebiet auf die Sedimentation jahresgeschichteter Oberflächensedimente in zwei Becken des Woseriner Sees (Mecklenburg-Vorpommern). In: Bork, H.-R., Schmidtchen, G., und Dotterweich, M. (Eds.): Bodenbildung, Bodenerosion und Reliefentwicklung im Mittel- und Jungholozän Deutschlands. Forschungen zur Deutschen Landeskunde. S. 229–249. Flensburg: Dt. Akad. für Landeskunde 2003

Hard, R., and Roney, J. R.: A massive terraces village complex in Chihuahua Mexico: 3000 years before present. Science *279*, 1661–1664 (1998)

Håkanson, L., and Jansson, M.: Principles of Lake Sedimentology. Berlin: Springer 1983

Hillenbrand, K: Volkskunst der Ziegelbrenner. Stempel, Symbole und Heilszeichen in Ton. München: Callwey 1981

Hans-Rudolf Bork, Christine Dahlke, Stefan Dreibrodt, and Annegret Kranz

JÄGER, H.: Einführung in die Umweltgeschichte. Darmstadt: Wissenschaftliche Buchgesellschaft 1994
MÄMPEL, U.: Keramik. Kultur- und Technikgeschichte eines gebrannten Werkstoffs. Beiträge zur Wirtschafts-, Sozial- und Technikgeschichte der Porzellanindustrie, Bd. 6 (Schriften und Kataloge des deutschen Porzellanmuseums, Bd. *80*). Hohenberg: Zweckverband Dt. Porzellanmuseum 2002
POTTER, P. E., MAYNARD, B. J., and DEPETRIS, P. J.: Mud & Mudstones. Introduction and Overview. Heidelberg, Berlin: Springer 2005
PRICE, S., and NIXON, L.: Ancient Greek agricultural terraces: Evidence from texts and archeological survey. American Journal of Archaeology *109*/4, 665–694 (2005)
ROBERTS, H. M., WINTLE, A. G., MAHER, B. A., and HU, M.: Holocene sediment-accumulation rates in the western Loess Plateau, China, and a 2500-year record of agricultural activity revealed by OSL dating. The Holocene *11*/4, 477–483 (2001)
SANDOR, J. A.: Ancient agricultural terraces and soils. In: WARKENTIN, B. P. (Ed.): Footprints in the Soil. People and Ideas in Soil History; pp. 505–534. Amsterdam, Boston, Heidelberg: Elsevier 2006
SCHUNK, A.: Vom Lehmziegel zum Meißner Porzellan. In: SCHUNK, A., und THEWALT, U.: Historische anorganische Werkstoffe. Lebendige Geschichte der Naturwissenschaften. Bd. *4*, S. 37–64. Ulm: Universitätsverlag Ulm 2000
TREACY, J. M., and DENEVAN, W. M.: Creation of cultivalbe land through terracing. In: MILLER, N. F., and GLEASON, K. L. (Eds.): The Archeology of Garden and Field. Philadelphia: University of Pennsylvania Press 1994
VILLALON, A. F.: The cultural landscape of rice terraces of the Philippine Cordilleras. In: DROSTE, B. VON, PLACHTER, H., and RÖSSELER, M. (Eds.): Cultural Landscapes of Universal Value-Comments of a global Strategy; pp. 198–114. Jena, Stuttgart, New York: Fischer 1995
WILKINSON, J. T.: Holocene environments of the High Plateau, Yemen. Recent geoarchaeological investigations. geoarcheology *12*/8, 833–864 (1997)

        Prof. Dr. Hans-Rudolf BORK
        Christian-Albrechts-Universität zu Kiel
        Ökologie-Zentrum
        Fachabteilung Ökotechnik und Ökosystementwicklung
        Olshausenstraße 75
        24118 Kiel
        Germany
        Phone: +49 431 8803953
        Fax: +49 431 8804083
        E-Mail: hrbork@ecology.uni-kiel.de

The Fifth Element:
On the Emergence and Proliferation of Life on Earth

Max M. von Tilzer (Konstanz)

With 5 Figures

Abstract

Life-formation, 3.5–3.8 billion years ago, is seen as a succession of steps of self-organization that were only possible in the absence of oxygen in the atmosphere and in the presence of liquid water. It is a matter of debate whether primitive replicator molecules or simple metabolic pathways were the first steps leading to life. The *last universal common ancestor* (LUCA) of all life must have had all properties that are shared by all extant organisms. Photosynthetic microbes evolved from chemolithotrophic ancestors. Perhaps the greatest breakthrough during the evolution of life was the emergence of oxygenic photosynthesis at least 2.7 billion years ago, whereby molecular oxygen was released to the environment as a waste product of water oxidation. Oxygen first appeared in the atmosphere ca. 2.4 billion years ago and subsequently increased in leaps. Molecular oxygen as universal oxidant in respiration made the evolution of higher forms of life possible. The *Eukaryotes* first emerged 1.4–1.6 billion years ago. They are characterized by a highly complex cell architecture that was generated by the fusion of microbial cells. The next major evolutionary breakthrough was the evolution of *multi-cellular* plants and animals, about 600 million years ago. About 200 million years later the *continents* were colonized which led to an unprecedented diversification of habitats and organisms. The *evolution of cognition* and self-awareness represented evolutionary steps equal in rank with the previous ones.

After the end of the last glaciation ca. 10,000 years ago, biological evolution of humanity came to a halt and was replaced by *cultural evolution*. The latter is characterized by an acceleration of progress because successive generations can build on the achievements of the preceding ones. The past 200 years have been considered a new geologic era that is characterized by noticeable human impacts on a global scale (the *Anthropocene*). Human civilization causes the extinction of biological species that is equal in magnitude to previous mass extinctions. Indigenous species in part are replaced by species with high reproductive rates and little ecological specialization.

Zusammenfassung

Die Entstehung des Lebens auf der Erde vor 3,5–3,8 Milliarden Jahren kann als eine Abfolge von Schritten der Selbstorganisation angesehen werden. Wichtigste Voraussetzungen dafür waren das Vorhandensein von flüssigem Wasser und die Abwesenheit von molekularem Sauerstoff in der Atmosphäre. Die Frage, ob die Speicherung von Erbinformation durch spezifische Moleküle oder die Herausbildung eines autonomen Stoffwechsels den entscheidenden Schritt der Lebensentstehung dargestellt hat, ist umstritten. Der letzte gemeinsame Vorfahre allen Lebens müsste sämtliche heute lebenden Organismen gemeinsamen Eigenschaften auf sich vereint haben. Die ersten autotrophen Organismen waren chemolithotrophe Mikroorganismen. Photosynthetische Bakterien entwickelten sich aus diesen. Der wahrscheinlich größte Durchbruch in der Evolution des Lebens stellte die Herausbildung der oxygenen Photosynthese vor mindestens 2,7 Milliarden Jahren dar. Molekularer Sauerstoff entsteht dabei als Abfallprodukt bei der Spaltung von Wasser. Erste Spuren von Sauerstoff traten vor ca. 2,4 Milliarden Jahren auf. Als Folge stand der Biosphäre nicht nur ausreichend organische Substanz, sondern auch genügend Energie zur Verfügung. Erst dies erlaubte die Herausbildung von höheren Formen des Lebens. Die folgenden Schritte sind dabei entscheidend gewesen: Die Entwicklung der Eukaryoten, vor 1,4–1,6 Milliarden Jahren, die Evolution mehrzelliger Pflanzen und Tiere (letztere vor ca. 600 Millionen Jahren) und die Eroberung des Festlands vor etwa 430 Millionen Jahren. Die Entstehung des zu

Max M. von Tilzer

diskursivem Denken befähigten menschlichen Gehirns steht gleichrangig neben den vorgenannten Errungenschaften der biologischen Evolution.

Nach dem Ende der letzten Vereisung vor etwa 10 000 Jahren wurde die biologische Evolution des Menschen durch die zivilisatorische und kulturelle Entwicklung abgelöst. Da jede Generation auf den Errungenschaften der vorhergehenden aufbauen konnte, kam es seit dem Beginn der Neuzeit zu einer starken Beschleunigung der naturwissenschaftlich-technischen Entwicklung. Diese, sowie das seit 1800 einsetzende beschleunigte Wachstum der Weltbevölkerung haben zu einer Beeinflussung der Umwelt auf globaler Ebene geführt. In Würdigung dieser Tatsache wurde vorgeschlagen, eine neue geologische Epoche, das Anthropozän, zu definieren. Als Folge der menschlichen Zivilisation kommt es zur Auslöschung von biologischen Arten in einem Ausmaß, das mit den größten Massensterben im Verlaufe der Erdgeschichte vergleichbar ist. Die Verluste an heimischen Arten werden zum Teil durch die weltweite Ausbreitung von ökologisch anspruchslosen Arten mit hohen Vermehrungsraten ersetzt.

1. Introduction

Many pre-requisites had to be met for life to emerge and thrive on Earth. For the evolution of higher forms of life including intelligent ones, extended periods of time were required, and additional conditions had to be fulfilled. From this we conclude that complex life in the Universe is an unlikely occurrence in terms of statistics (Ward and Brownlee 2004).

It is the aim of this article to provide an overview of the steps leading to life on Earth and on the evolution of the diversity among the organisms we encounter today. Moreover, we will give some background concerning the historic development of the theory of evolution. The article ends with a brief description of the influence of our own species on the biosphere.

2. The Formation of the Earth System

The evolution of the Universe, from its very beginnings, as well as the formation of the solar system and of planet Earth, provided fundamental prerequisites for the emergence and the subsequent evolution of life on Earth.

2.1 The Creation of the Universe

In 1927, the Belgian astronomer and priest Georges Lemaître (1894–1966) has proposed that the Universe was created by the explosion of a "primeval atom". In 1929, Edwin P. Hubble, based on the observation of receding motions of distant galaxies, supported Lemaître's hypothesis: By extrapolating cosmic expansion back, the conclusion was drawn that the entire cosmos in its beginning was concentrated at one single point, termed the *Big Bang Singularity* (Hawking 1988). The current best estimate of the timing of the Big Bang, and hence the age of the Universe, is 13.7 billion years.

2.2 Pre-Requisites for Life Based on the History of the Universe

Roughly 380,000 years after the Big Bang, the average temperature of the Universe had cooled to 3,000 K. This allowed the formation of atoms. We can imagine the Universe at this time as a vast and almost empty space, filled only with highly dilute hydrogen and helium gas. The afterglow of the process, by which the Universe became transparent, is the Microwave

Background which today, owing to the continued expansion of the Universe, has a temperature of ca. 2.7 K.[1] A remarkable feature of the *Microwave Background* is its highly homogeneous spatial distribution which is also reflected in the large-scale distribution of galaxy clusters (WEINBERG 1994).

2.2.1 Heterogeneous Distribution of Matter in the Early Universe

However, slight heterogeneities in the distribution of matter in the early Universe, as evident from temperature variations by 10^{-5} K in the microwave background do exist (SMOOT 1991). These were sufficient for the accretion of matter under the influence of gravity.

2.2.2 The Synthesis of Chemical Elements Heavier than Helium

As material coalesced, the hydrostatic pressure increased at the centers of the primordial aggregates of matter, leading to rising temperatures that finally led to nuclear fusion of hydrogen to helium. To the present day, this process is the dominant source of energy in the cores of *main-sequence stars*. About one billion years after the Big Bang, large galaxies had formed (LOEB 2006).[2]

Following the formation of first-generation stars, chemical elements heavier than helium were synthesized.[3] Elements up to iron are synthesized by energy-yielding nuclear fusion. Elements heavier than iron are formed by neutron capture in massive stars and during explosions of supernovae. At least part of this material subsequently was ejected into interstellar space by ageing stars[4] and during supernova explosions.

2.2.3 The Origin of the Solar System and of the Earth

Interstellar material thus formed provided the raw material for the accretion of *second-generation stars* such as the Sun and its planets. Dust particles within a disk surrounding the Sun attracted more material by gravitational pull, leading to the formation of planets. The overall accretion process of the Earth took no longer than 33±2 million years (KLEINE et al. 2002).[5]

2.2.4 The Establishment of Clement Conditions on Earth

The advent of life on Earth required the cessation of life-threatening impacts, the formation of a solid Earth surface, a temperature range suitable for life, and the formation of an atmosphere

---

1 The decrease in the temperature of the radiation background can also be related to the Doppler red shift due to the receding motion of the outer boundaries of the observable Universe (HAWKINS 1988).
2 The most powerful telescopes at present allow the observation of objects as they were created about one billion years after the Big Bang.
3 In cosmochemistry all chemical elements heavier than helium are called „metals". They comprise only 2% of the total mass of the visible universe. (Up to 90% of the total mass of the Universe is believed to consist of "dark matter".)
4 When stars have consumed the hydrogen fuel in their cores, they expand to form *Red Giants*, and eject ca. 30% of their mass into space.
5 By definition, the formation of 99% of a planet's metal core is considered to represent the termination of the accretion process.

and a hydrosphere. Life could possibly have formed more than once during extended quiescent periods between massive impacts during the preceding period (BADA 2004). Around 3.8 billion years ago, however, impact frequency significantly leveled off, and life-supporting conditions persisted (SLEEP et al. 1989).

3. What is Life?

Two fundamental positions exist concerning the specific ontological definition of life: The *vitalists* claim that in addition to the physical and chemical characteristics, organisms are controlled by principles that cannot be explained in terms of natural sciences (*entelechy*).[6] Vitalist positions have dominated from Antiquity (ARISTOTLE) to the eighteenth century. However, the development of experimental natural sciences during the Renaissance and the subsequent *Age of the Enlightenment* have prompted the *mechanist* position, which states that, at least in principle, all characteristics of life can be explained in terms of natural sciences. This is the generally accepted view in natural sciences today.

3.1 Hierarchical Complexity

ARISTOTLE has distinguished three essential features of life. They represent a hierarchical order of increasing complexity:

– *Anima vegetativa*: All life is characterized by growth. This applies to all organisms, but is the only definition applicable to plants.
– *Anima sensitiva*: Some organisms are capable of sensing external stimuli and respond to them. This characterization applies to both animals and humans.[7]
– *Anima rationalis*: Only we humans are aware of our own selves and are capable of cognition.

3.2 Definition of Life by its Most Salient Features

Erwin SCHRÖDINGER, in his influential booklet *What is Life* (1945), has focused on two particularly astounding features of organisms: (*i*) Their enormous complexity, which, in thermodynamic terms, can be characterized by extremely low entropy (or, conversely, high content of information, "*neg-entropy*"). (*ii*) SCHRÖDINGER moreover emphasized the importance of genetic information encoded in the chromosomes. His views have paved the way for the development of molecular biology and the deciphering of the genetic code by Francis CRICK and James WATSON in 1953 (MOORE 1994).

Autonomous metabolism and *replication* represent the most salient characteristics of life. However, we can speak of life only if all of the following criteria are met:

---

[6] The term "entelechy" was first used by Greek philosophers including ARISTOTLE. In modern history it was reintroduced by W. LEIBNIZ (1646–1716). The German biologist and philosopher H. DRIESCH (1867–1941) used the term to define a fundamental feature of life according to a neo-vitalist position.
[7] It has since then become abundantly clear that plants are capable of responding to a great many of external influences, in particular to light and gravity.

- *Confined to boundaries*: Organisms are enclosed within body walls that separate the interiors of an organism from its environment.
- *Metabolism and internal regulation*: Within its boundaries, an organized system of cellular components allows complex metabolic pathways.
- *Exchange with the environment*: Energy and nutrients are taken up from the environment, and waste products are released to the environment.
- *Sensibility to external stimuli*: Receptors enable organisms to receive information about their respective physical environments that can trigger responses by these organisms.
- *Reproduction*: Organisms generate offspring, thereby transferring genetic information to subsequent generations.
- *Growth and differentiation*: Comparatively simple germs (spores or fertilized gametes), during the juvenile development of organisms, differentiate into body architectures of considerably greater complexity.
- *Evolution*: The interplay of mutations of the genetic makeup and selection by the environment (both non-living and living), leads to the emergence of new forms of life.

4. The Origin of Life

The *Genesis* of the Bible explains the origin of life as an act of God, who, at one point, has been breathing life into clay: By contrast, natural scientists from ARISTOTLE to Isaac NEWTON had assumed that life is continuously emerging spontaneously from non-living matter. This hypothesis was falsified by experiments during the 17th and 18th centuries. Based on this evidence, Louis PASTEUR in 1864 formulated the principle *omne vivum ex vivo* (all life comes from life). Nevertheless, the view that living entities did emerge from non-living precursors in principle is still held today. However, we now think that life was formed only once, and that this process took place in the distant past. Two principle scenarios concerning the origin of life have been proposed.

4.1 Spontaneous Generation of Life on Earth

In a lecture series of 1867/68, Ernst HAECKEL postulated the existence of a "primordial slime" (*Urschleim* in German)[8] as an intermediary between non-living and living matter. The Russian A. I. OPARIN in 1924 and the British biochemist J. B. S. HALDANE in 1929 independently (and probably also not aware of Ernst HAECKEL's work), put forward the hypothesis that life was formed at an early stage of Earth history by self-organization from a "pre-biotic soup" consisting of non-biologically synthesized organic molecules, and that this process required an atmosphere free of molecular oxygen.

4.2 The Panspermia Hypothesis

This hypothesis states that life has originated *somewhere else* in the Universe, and that Earth was seeded with complex forms of life, once it had become habitable (WARMFLASH and WEISS

---

8 In fact, the term "Urschleim" was first used by another eminent German biologist Lorenz OKEN (1779–1851) who considered vesicles filled with primordial slime as the material organisms are made of, thus anticipating the living cell.

2005). The term *Panspermia* originally is based on the teachings of the Greek philosopher ANAXAGORAS (ca. 500–428 BCE) who suggested that all things, including life, originated form the combination of tiny seeds pervading in the cosmos. Panspermia was revived by Lord KELVIN (1824–1907) and Hermann VON HELMHOLTZ (1821–1894), and was made into a scientific hypothesis by Svante ARRHENIUS (1908). The main argument in favor of the Panspermia Hypothesis is that the time-span between the establishment of clement conditions on Earth and the first occurrence of life would have been insufficient for the evolution of even the most primitive organisms (LINE 2002).[9] Advocates of Panspermia included eminent scientists such as Francis CRICK and Leslie ORGEL (CRICK and ORGEL 1973), and Fred HOYLE (1983).

However, the dominant view today is that life did emerge on Earth. The following arguments can be put forward in favor of this conclusion:

– Panspermia does not explain the origin of life as such, but just puts this process to a far-away place and possibly a more distant past that would not have been possible on Earth.
– Space transit of "*seeds of life*" is difficult to imagine, given the environmental conditions in interstellar space that are hostile to any living structures.[10]
– Due to the vastness of the Universe, it statistically is highly improbable that seeds of life could have reached Earth, unless they were exceedingly abundant or transmitted to Earth on purpose by an intelligent civilization ("directed panspermia") as proposed by HOYLE.

4.3 Steps Leading to Living Structures

Several successive processes are believed to have occurred towards the emergence of living structures on Earth. They can be subdivided into a pre-biotic evolution, and the emergence of life itself.

4.3.1 Pre-biotic Evolution

Pre-biotic evolution provided the building-blocks of living matter.

– *Synthesis of pre-biotic macromolecules.* The generation of pre-biotic molecules on Earth from simple precursor molecules, as suggested by OPARIN and HALDANE, was first simulated in the famous Miller-Urey-*Experiment* (MILLER 1953).[11] Meteorites and comets could have acted as additional vehicles for the introduction to Earth of organic molecules that previously had formed in interstellar space (CHYBA and SAGAN 1992). From small organic precursor molecules, macromolecules on Earth could either have been synthesized on clay minerals in hydrothermal systems (HAZEN 2001), or in aerosols in which high concentrations of an array of small organic molecules are found. Aerosol particles found in today's atmosphere contain a high variety of organic molecules at concentrations sufficiently high to undergo chemical reactions (DOBSON et al. 2000).
– *Chiral selection. Handedness* (chirality) of asymmetric biological molecules is one of the most distinctive features of all living structures (BADA 1997). In organisms, only one of

---

9 For example, Lynn MARGULIS (1996) has succinctly stated that (as quoted after LINE 2002): "... to go from bacteria to people is less of a step than to go from a mixture of amino acids to that bacterium".
10 Particularly life-threatening would have been ultraviolet and ionizing radiation in interstellar space.
11 The Miller-Urey Experiment possibly is flawed on the grounds that the assumed chemical composition of the atmosphere at the time of life-formation, especially its high ammonia content, thus far has not been confirmed.

the two possible optically active enantiomers of organic macromolecules such as proteins, sugars, and nucleic acids is found. Thus far, no 100-% effective selection process of one enantiomer over the other could be simulated experimentally (BADA 1995). The origin of handed biological molecules in biological material is among the major as yet unresolved enigmas in connection with the emergence of life.
– *Formation of biological membranes*. Experiments with the aim of simulating the generation of pre-biotic structures from a mixture of precursor molecules under the presumed environmental conditions at the time such as the Miller-Urey-Experiment yielded small droplets that sometimes evolved into membrane-enclosed *micro-spheres*. More recently, the formation aerosol droplets surrounded by bi-layer membranes has been suggested as possible pre-biological reactors (DOBSON et al. 2000).

4.3.2 Autonomous Pre-biotic Structures

As result of pre-biotic evolution, structures must have formed initially that did not fulfill all criteria of organisms. They were called *Progenotes* by WOESE (1998). Progenotes must have had an autonomous metabolism and some form of replication. It is a matter of scientific debate which step came first.

– *The Replication-first scenario*. By far the most popular candidate for a primordial replicator is RNA. The strongest argument in favor of this scenario is that RNA is more readily synthesized than DNA, and that DNA can easily evolve from RNA (ORGEL 1994). In fact, some RNA molecules ("*ribozymes*", CECH 1986) turned out to be self-replicating. Later the role of RNA as the first replicator was challenged because of the difficulties in synthesizing molecules of this level of complexity by non-biological reactions, and alternative scenarios have been proposed. For example, CAIRNS-SMITH (1982, 1985) has suggested that the first hereditary information was stored by mineral crystals such as clays. By *genetic takeover*, autonomous replicators could have subsequently been formed by heterogeneous catalysis by the adsorption of organic molecules onto the clay surfaces.
– *The Metabolism-first scenario*. Recently SHAPIRO (2007) has suggested that simple cyclic reaction chains had formed within primordial vesicles. The first transfer of information from one generation to the next could have been based not on molecules that encoded this information, but by the molecules themselves, which represented a "*compositional genome*".

Although replication-first scenarios are more widely accepted, the metabolism-first scenario appears more attractive, based on probability considerations.

4.3.3 The Emergence of Life: The Annealing Hypothesis[12]

According to WOESE (1998), pre-biotic structures (*progenotes*) only contained simple genomes on small chromosomes, capable of transcription, however inaccurate. As evolution progressed, new and more effective proteins and novel metabolic pathways emerged. Internal organization became progressively more rigid. Replication and translation became more

---

[12] The term „annealing" is derived from technology and means hardening of steel by which a more organized metal lattice is formed.

accurate, giving rise to offspring that closely resembled the parent generation. Changes in the genetic makeup (mutations) allowed the evolution through vertical gene transfer. *At this stage, all criteria of life were met.*

4.4 The Last Universal Ancestor

All forms of life share a large number of common features.[13] Charles DARWIN was the first to design a *Tree of Life* which described the evolution of life descending from a single universal ancestor. This concept, still held today, implies that life, as we know it today, emerged only once during Earth history. Nevertheless, the modern concept of the Last Universal Common Ancestor (LUCA) in essence is a construct. It represents no more than the point at which the lines suggesting the hierarchy of similarities and differences between organisms meet (Fig. 1).

If LUCA indeed had all features that are common to all organisms of today, it would represent the *largest common denominator* of all life. However, only about 60 proteins are shared by all extant organisms (KOONIN 2003). These would be by far too few to allow all cellular structures and functions. KOONIN arrived at an estimate of the gene-pool size of LUCA of ca. 600. Many of the genes of LUCA therefore since then must have been lost. Conversely, it cannot be excluded that some of the genes shared by all organisms of today might have been acquired only later by horizontal gene transfer. WOESE 1998 asserts that LUCA was not yet a true organism, but was a tight consortium of progenotes.

4.5 The Tree Domains of Life

Based on the analysis of ribosomal RNA sequences, WOESE and FOX (1977) have distinguished between *Three Domains of Life* and designed a *Universal Tree of Life* (Fig. 1; WOESE et al. 1990). Although the Three-Domain-Concept at present is widely accepted, eminent evolutionary biologists think that the distinctions between *Prokaryotes* (*Archaea* plus *Eubacteria*) on the one, and *Eukaryotes* on the other, more appropriately reflect the relationships between the main groups of organisms. This is because the difference in the level of complexity of the Prokaryotes *vis á vis* the Eukaryotes is to be considered to represent a more significant distinctive criterion than the differences between the Archaea and the Eubacteria (MAYR 1998).

5. Evolution of Our Views on the Evolution of Life

Since ARISTOTLE the dominant view of science was that organisms, once they had come into existence, remained unchanged. The idea that life has undergone an evolution once it was created, was slow in coming.

---

13 Features common to all extant organisms include: 20 left-handed amino acids, DNA-based replication using the same genetic code, energy conservation by ATP synthesis, pathways of intermediate metabolism, and biosynthesis of major cellular components.

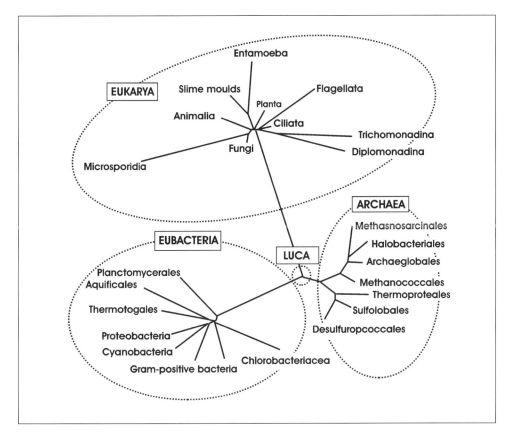

Fig. 1 The Universal Tree of Life, based on base sequence analysis of ribosomal RNA. The Last Universal Common Ancestor (LUCA) is positioned where the lines leading to the *Tree Domains of Life* meet. Although molecular similarities also imply phylogenetic relationships, this graph does not claim to represent a reconstruction of the macro-evolution of life. To make this clear, no time-axis is shown. Modified from DAWKINS 2004.

5.1 The Natural System and the Tree of Life

The Swedish natural scientist Carl VON LINNÉ (LINNAEUS; 1707–1778) was the first to systematically categorize the plant and animal kingdoms in 1735. Whereas his first system was rather artificial, he later attempted to base his system on similarities and differences in the morphological features of the organisms. Although he believed that organisms were constant entities like minerals, his natural system later became the basis for the first phylogenetic *Tree of Life*.

5.2 Catastrophism

The French naturalist Georges CUVIER (1789–1832) observed that some fossil remains of vertebrates belonged to species that no longer exist. From this he rightly concluded that these forms were eliminated by catastrophic events during Earth history. His *Catastrophism* was strongly opposed by the British geologist Charles LYELL (1797–1875) who's *Principle of Actualism* stated that mechanisms and processes acting today also have been in effect during the geologic past, implying continuity during Earth history. The role that mass extinctions might have played in the evolution life subsequently was neglected.

5.3 Lamarckism

Jean-Baptiste DE LAMARCK (1744–1829) was the first to propose that organisms have descended from ancestral forms by an evolutionary process. As mechanism leading to the emergence of new forms of life he postulated the heritability of acquired characteristics in response to environmental influence. In modern terms this implies that the environment directly acts on the gene pool of organisms to make them more suitable for survival. To the present day, the enormous scientific merits of LAMARCK have been significantly underrated.[14]

5.4 Darwinism

The Theory of Evolution, founded by the British geologist Charles DARWIN (1809–1882) is considered to be one of the most important breakthroughs of modern natural sciences. The essence of DARWIN's theory is that biological variability is generated by chance processes, and that the direction of evolution is due to the survival of the fittest. DARWIN's theory later was corroborated by both genetics and ecology: (*i*) Mutations (changes in the genome) in principle are random and non-directional. (*ii*) Selective pressure is exerted by the environment, including competing organisms (Fig. 2).

Darwinist principles are criticized, mainly based on two arguments: (*i*) Darwinism postulates a gradual change in the properties of organisms. In the fossil record, intermediate forms are rarely found ("missing links"). (*ii*) It is difficult to imagine the development of highly complex forms from simpler ones simply by chance processes. Albert EINSTEIN has succinctly expressed this sentiment by stating that "*God does not play dice*".[15]

To overcome these shortcomings, the French priest and palaeontologist Pierre TEILHARD DE CHARDIN (1881–1955) has attempted to reconcile evolutionist theory with Christian faith by proposing that evolution is directional and is following a preconceived divine blueprint with the ultimate goal to generate humanity (teleology).

5.5 Teleonomic Interpretation of Evolution

The British/American biologist Colin S. PITTENDRIGH (1958), by contrast, has introduced the term *teleonomy*, which can be defined by "*purposefulness toward survival*": It is reasonable to describe biological structures and functions by their respective purposes without implying that they have evolved with this goal in mind. The concept of teleonomy is strictly based on modern Darwinism.

---

14 In fact, *Lamarckism* almost is a bad word. This is mainly due to the Russian agronomist Trofim D. LYSSENKO (1898–1976) who, by adopting Lamarckism and forging experimental results, during the period of Stalinism, for several decades prevented any progress of the biological sciences in the Soviet Union.
15 Quoted from DAWKINS 2006, p. 41

The Fifth Element: On the Emergence and Proliferation of Life on Earth

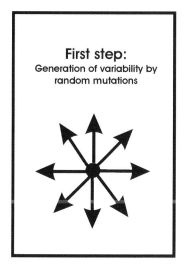
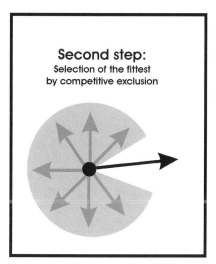
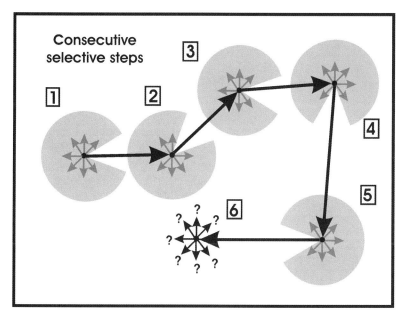

Fig. 2 The principle of Darwinian evolution. *Upper panels*: Generation of variability by random mutations (*Step 1*), and selection of the fittest due to pressures from the environment and by competitive exclusion between individuals having slightly different genomes (*Step 2*). Most individuals are eliminated (grey arrows, grey shading), and only few survive (black arrows). *Lower panel*: Successive non-directional evolutionary steps as a consequence of random generation of genetic variability and subsequent selective elimination. Competition is effective between individuals and *not* between genotypes, and is greatly influenced by the physical environment. Environmental change, therefore, will alter the direction of evolutionary pressure. Evolutionary innovations can only occur in small populations and will spread to larger populations if the reproductive success of the mutant greatly exceeds that of its competitors. Original.

5.6 The Evolution of Behavior

The Austrian behavioral biologist Konrad LORENZ (1903–1989) has demonstrated that animal and human behavior result from biological evolution, as do any other properties of organisms. This hypothesis since then has been strongly supported by neurobiology. LORENZ in addition has proposed the term *fulguration* (from Latin *fulgur*, lightening) to describe major evolutionary leaps (LORENZ 1973).

5.7 Punctuated Equilibrium

The American paleontologist Stephen Jay GOULD (1941–2002), together with Niels ELDREDGE, has proposed a hypothesis which is based on evidence derived from the fossil record: Certain forms of life can be found over extensive periods of time and at one point are replaced by novel forms without any intermediates ("missing links"). It is suggested that periods with only minor evolutionary change (*equilibrium*) are *punctuated* by rapid transitions to new forms.[16] Evolutionary change occurs in small and isolated sub-populations. If successful, novel traits might spread to larger populations (GOULD and ELDREDGE 1993). The Punctuated Equilibrium hypothesis at least in part overcomes the first of the above-mentioned major points of criticism put forward against Darwinian Evolution Theory.

5.8 Catastrophism Revisited

Recently the view that mass extinctions played a major role in the evolution of life has gained new ground (RAUP 1991). We have to distinguish between two types of extinctions: (*i*) *Background Extinctions* are caused by degenerative processes within existing species or by environmental changes, some species cannot cope with. In general, most biological species have life-spans of between five and ten million years. Lost species are continuously replaced by new ones; hence total species numbers remain about the same. (*ii*) *Mass extinctions* are caused by rapid deteriorations in the environmental conditions. As a consequence, the numbers of species decrease significantly. The ecological niches[17] thus vacated subsequently are occupied by newly emerging species by a process called *adaptive radiation*, whereby species numbers rise again. The recovery of the species inventory from mass extinction takes between five and ten million years. This is shown by the fossil record (Fig. 3).

5.9 Evolution within One Generation: Endosymbiosis and Horizontal Gene Transfer

The tacit assumption of the classical Theory of Evolution is that features essentially are transferred from one generation to the next. Lynn MARGULIS (1970) has proposed that the eukaryotic cell is the result of the inclusion of symbiotic microbial cells into a host cell in which they subsequently formed cell organelles (*endosymbiosis*). Microbes moreover are capable of

---

16 Because of small population size and the swiftness of evolutionary change, the preservation of intermediary forms ("links") is unlikely. However, some intermediate forms have been found, such as the primitive amphibian *Ichthyostega* which apparently is derived from lung-fishes, and *Archaeopteryx* which has properties of both dinosaurs and birds.
17 By an *ecological niche* we understand the specific combination of environmental conditions under which a given organism lives, including their temporal and spatial variations. The term "niche" not necessarily implies a defined spatial entity.

The Fifth Element: On the Emergence and Proliferation of Life on Earth

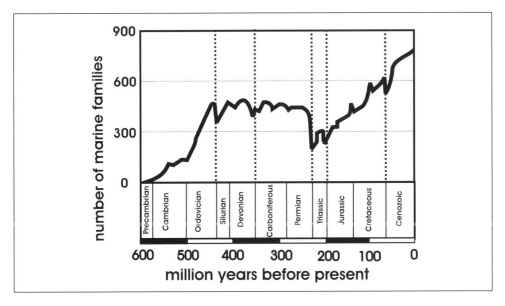

Fig. 3 Changes in the numbers of marine families during the past 600 million years. Sudden drops in the numbers of families are caused by mass extinctions (vertical dotted lines), followed by increases in the numbers of families due to the occupation of the vacant ecological niches by newly emerging species (*adaptive radiation*). The graph suggests an increasing trend in biological diversity which, however, might only be suggested by more complete preservation of fossils in more recent sediments. Modified from COURTILLOT 1999, after RAUP and SEKOSKI.

transferring genes, not only to subsequent generations, but also to cells that are not their own offspring in a process now called *horizontal (or lateral) gene transfer* (SYVANEN 1984). As a consequence, it is no longer possible to design a generalized tree of life such as DARWIN's. Ancestral lineages consequently can only be designed for specific cellular components or features (Fig. 4). To our present knowledge, horizontal gene transfer essentially is restricted to Prokaryotes, whereas gene transfer from one generation to the next (*vertical gene transfer*) continues to be viewed as the only mechanism of heredity in higher organisms. This means that the principles of Darwinist Evolution remain applicable.

5.10 Conclusions

The Belgian biochemist Christian DE DUVE (1995) has proposed that the generation of biological variability is far from random because the degrees of freedom in any process of self-organization, both in the non-living[18] and the living worlds,[19] is highly constrained and far from random in the statistical sense.

---

18 The *Periodic Table of Elements* represents a good non-biological example how self-organization works: The complex arrangement of the chemical elements in groups and periods is the result of simple principles based on the physical properties of protons, neutrons, and electrons, as well as on their interactions. Based on both, the physical properties of elements, as well as their interactions in chemical compounds, can easily be explained.
19 Evolution creates similar shapes in response to selective pressure in unrelated groups of organisms that cannot be explained by common phylogenetic origin. Two examples: *Succulent growth* occurs in cactuses and euphorbia in response to dry conditions. *Bullet shape* is found in both fishes and birds, by which drag by the medium as a consequence of rapid movement is minimized.

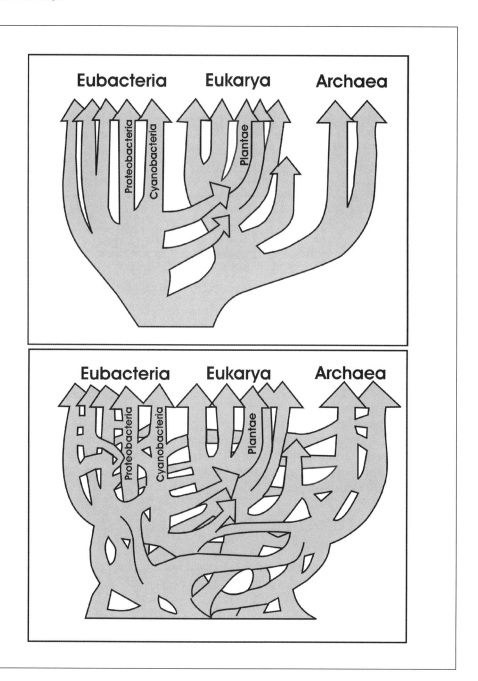

Fig. 4 Two alternative models of the evolution of the *Three Domains of Life*: The vertical axis represents time, but is not drawn to scale. Most groups of organisms within the Domains of Life are not indicated for the sake of simplicity. *Upper panel*: Tree of life, based on the Serial Endosymbiosis Hypothesis. *Lower panel*: Genealogical relationships (no longer to be called a "tree") showing the complexity of the genetic linkages between the Three Domains of Life as a consequence of horizontal gene transfer (HGT). The role of HGT between Eukaryotes is unknown, but assumed to be much smaller to negligible. Simplified after DOOLITTLE 1999.

To avoid frequent misunderstandings with modern Darwinian theory, moreover the following in addition has to be kept in mind:

- *Darwinian selection is effective on the individual organism*, not on the genome. The primary outcome of any selective event, therefore, is the survival of an individual and not of a biological species.
- *The emergence of new features by mutations is independent of function.* This implies that many changes don't have any selective significance. New functions are generated only as a secondary step. As a result of this, *the variability of structures within a group of organisms is significantly greater than the variability of functions*. Many changes will occur, but only few will be superior, and will provide selective advantages (Fig. 2).
- *Biological evolution as a whole is non-directional*. It therefore does not necessarily lead to more complex forms of life from simpler ones. However, the emergence of complex life forms is a function of the time that altogether has been available for evolution. Whereas primitive forms of life might already have emerged 100–200 million years after Earth provided conditions conductive to life, the evolution of our own species required 3.5–3.8 billion years.
- Genetic information is not only transferred from one generation to the next but, at least in micro organisms, also between organisms within one generation.[20]

6. Turning Points during the Evolution of Life

Analysis of the history of life on Earth has revealed that a strictly gradualist view of evolution as reflected by the Darwinist theory in its original form has to be revised. Evolutionary leaps ("fulgurations") basically were triggered by changes in environmental conditions on Earth that have significantly altered the direction of selective pressure. Drastic environmental changes were caused either by external forcing or indirectly by the biosphere itself. They have led to *changes of paradigm* during the course of evolution.

6.1 The Rise of Atmospheric Oxygen and its Consequences for Early Evolution

Almost half of the history of life has been the history of life without oxygen. During this time, the supply with energy was the most severe limiting factor for the proliferation of life on Earth, mainly because, unlike today, no universal oxidant was present to power energy-yielding chemical reactions. As a consequence, biological activity on early Earth was small. It is likely that the very first organisms were heterotrophic, utilizing organic molecules produced by non-biological reactions. Autotrophy, which must have evolved soon thereafter, required the capability to reduce carbon dioxide to organic matter. The energy demands of autotrophic carbon assimilation initially were met by oxidation reactions of inorganic compounds (*chemolithotrophy*). This, however, required redox disequilibria in the environment which mainly were generated by the escape of hydrogen to space by which the atmosphere became slightly more oxidizing as compared to the highly reduced gases exhaled by volcanoes. *Anoxygenic photosynthesis* evolved from chemosynthesis 3.2–3.4 billion years ago. Inorganic and organic

---

[20] Viral infections have some similarities with horizontal gene transfer.

matter were used as primary sources of reductant. However, even after solar radiation could be utilized to power the reduction of inorganic carbon, the problem of small energy supply for driving cellular metabolism persisted. *Oxygenic photosynthesis* evolved from anoxygenic photosynthesis at least 2.7 billion years ago, but might be considerably more ancient (Des Marais 2000). It is characterized by a significantly higher level of complexity than anoxygenic photosynthesis (Xiong and Bauer 2003, Blankenship et al. 2007).

6.2 Evolutionary Breakthroughs during the Period of Rising Atmospheric Oxygen

After oxygenic photosynthesis had evolved, there was an inexhaustible supply of primary reductant (the water molecule) for building organic molecules from carbon dioxide and inorganic nutrient salts. Moreover once oxygen was present in the environment, sufficient oxidant (molecular oxygen) became available for the combustion of organic molecules as energy source. As a result, overall biological production increased by two to three orders of magnitude (Des Marais 2000). However, free atmospheric molecular oxygen only appeared at least ca. 300 million years after the evolution of oxygenic photosynthesis. The reasons for this were that at first large quantities of reduced volcanic gases such as hydrogen and hydrogen sulphide (Holland 2006) were oxidized, as well as massive deposits of reduced iron, which was leading to *banded iron formations*.[21] A high level of uncertainty still exists concerning the atmospheric oxygen content during early Earth history but it is well established that oxygen rise occurred in leaps. It is estimated that between 2.4 and 2.05 billion years ago, atmospheric oxygen increased by ca. three orders of magnitude (*The Great Oxidation Event*, Goldblatt et al. in press). Another surge in atmospheric oxygen occurred between 900 and 400 million years ago (Lenton 2003; Fig. 5). The overall controlling mechanisms of the early oxygen rise as yet are poorly understood.

In the oxygenated environment of today, 99.9 % of all oxygen that evolves during oxygenic photosynthesis is consumed for the breakdown (decomposition) of organic matter. Liberation of molecular oxygen to the environment is only possible if an equivalent amount of organic carbon is buried and thereby spared from decomposition.

6.2.1 The Evolution of the Eukaryotes

The first eukaryotes appeared 1.4–1.6 billion years ago. Bacteria cells were incorporated into an Archaean cell as symbionts and later formed cell organelles by a process called endosymbiosis (Margulis 1970; Fig. 4). However, not all structures of the eukaryotic cell such as the flagella have been explained by the endosymbiosis hypothesis.

The prokaryotes which developed long before the advent of molecular oxygen in the environment are characterized by a large diversity of metabolic pathways, both for the synthesis of their body substance, as well as for gaining energy, but are morphologically simple. Eukaryotes, by contrast, essentially are aerobic. Their metabolism is comparatively uniform but they are characterized by high morphological complexity, especially after the development of multi-cellular organization which was restricted to eukaryotes. In addition, eukaryotes are characterized by predominantly bisexual reproduction (Lane 2005).

---

21 There is evidence that some oxidation of ferrous iron already occurred by the action of anoxygenic bacteria (Widdel et al. 1993), prior to the release of molecular oxygen to the environment by oxygenic photosynthesis. Banded iron formations, therefore not necessarily are indicative of the presence of molecular oxygen in the atmosphere.

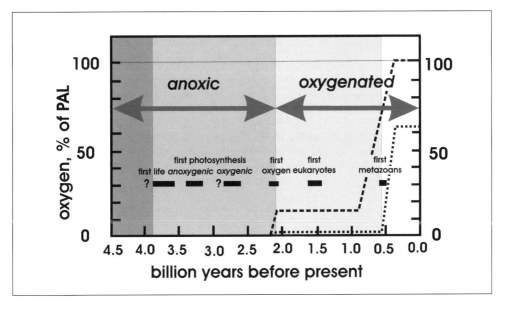

Fig. 5 The rise of atmospheric oxygen (as percent of the present atmospheric level PAL of 0.21 atmospheres at sea level) and major evolutionary breakthroughs during Earth history. Estimates are based on geochemical and biological evidence. The broken line gives the upper limit; the dotted line gives the lower limit of oxygen estimates. The dark shading on the left of the graph oxygen represents the period of heavy meteorite bombardment which at best would have made life possible deep inside the Earth crust during the last phase of this period (the *Hadean*). Intermediate grey shading denotes the anoxic period during the early history of life (the *Archaean*). The *Phanerozoic*, beginning ca. 570 million years ago, is shown in white. Major evolutionary breakthroughs during this time are also indicated: Anoxygenic photosynthesis is estimated to have appeared 3.2–3.4 billion years, oxygenic photosynthesis at least 2.7 years ago from anoxygenic precursors. The first traces of molecular oxygen showed up between 2.4 and 2.05 billion years ago. During most of the Archaean, estimates of atmospheric oxygen partial pressures are uncertain by one order of magnitude, but it is well established that atmospheric oxygen concentrations increased in leaps. Original, after LENTON 2003 and other sources.

6.2.2 The Evolution of Multi-cellular Animals

The first metazoans (the *Ediacara fauna*) appeared 550–600 million years ago. Large bodies for geometric reasons have diminished surface-to-volume ratios. Respiration through the body surface requires high ambient oxygen concentrations because molecular diffusion in water is almost negligible.[22] In multi-cellular organisms, cells differentiate into different types, forming organs that fulfill the different vital tasks. In addition, in both animals and plants, systems for the transport of nutrients and gases in aqueous solution have developed. It is not clear whether the representatives of the *Ediacara fauna* belonged to animal phyla that subsequently became extinct. Much less is known about the evolution of multi-cellular plants, especially of the vascular plants.

---

22 All representatives of the *Ediacara fauna* were soft-bodied. It moreover is likely that their bodies were flat (similar to plant leaves) which allowed considerably higher surface-to-volume ratios, despite large body size: This can be seen as an adaptation to the presumably low prevailing atmospheric oxygen partial pressures at the time, as compared to present levels.

Max M. von Tilzer

6.3 Evolutionary Breakthroughs during the Phanerozoic

Uncertainty exists concerning the oxygen content of the atmosphere at the beginning of the Cambrian, but a partial pressure of ca. 15 atmospheres at sea level appears realistic (Fig. 5).

6.3.1 The Cambrian Explosion

Provided that the *Ediacara* fauna consisted of animal phyla that soon thereafter became extinct, it is remarkable that all 20 animal phyla existing today probably evolved within a comparatively brief time-span between 540 and 500 million years ago.[23] Animal phyla represent different types of body architecture that are arranged in several more closely related groups. Over the course of the Palaeozoic, ecosystems of increasing complexity evolved. The early metazoan fauna consisted of soft-bodied animals.[24] However, soon thereafter also organisms with hard body shells developed, which is seen as a protection against predation. No new animal phyla emerged since then (WARD and BROWNLEE 2004).

6.3.2 The Colonization of the Continents

Life was restricted to the oceans during almost 90% of its history. During the Silurian, the continents finally were colonized, first by the plants ca. 430 million years ago, and subsequently by the animals, ca. 400 million years ago.[25] It would be tempting to speculate that the conquest of land was made possible by the establishment of a stratospheric ozone shield as a protection against solar UV-radiation. However, a fully effective ozone shield was already established at atmospheric oxygen contents of no more than 1% of the present level (KASTING and CATLING 2003).

Specific adaptations to terrestrial life: The most significant differences between aquatic and terrestrial habitats are due to the different physical properties of water and air, especially the following:

– *The density of water* is 830 times higher than that of air, therefore providing buoyancy. Aquatic organisms have densities only a few percent higher than the ambient water. This allows communities in the free water (the *pelagial*), that either are passively floating (the *plankton*), or actively swimming (the *nekton*) in the medium, in addition to organisms living on the ground (the *benthos*). On land, owing to the lack of buoyancy, only benthic communities are possible.[26] Moreover, supporting tissues (such as skeletons) are required especially in larger terrestrial organisms. Terrestrial animals frequently move by the help of legs.[27]

---

[23] The only exceptions are the bryozoans and the vertebrates which are not represented in the fossil record of the early Cambrian. They might have already existed at the time, but were not preserved.
[24] Although originating from the middle Cambrian, the *Burgess-Shale* of British Columbia allows a glimpse at early soft-bodied metazoans that are exceptionally well preserved here (GOULD 1989).
[25] Primitive forms might have lived on land much earlier such as cyanobacteria and lichens (which in fact represent tight symbioses between fungi and unicellular autotrophs).
[26] On land, organisms such as insects, flying reptiles, birds, and bats, only spend limited periods of time in the air because of the great energy expenditures of flying.
[27] Legs are only found in two animal phyla, the arthropods (both in water and on land) and in terrestrial vertebrates (tetrapods).

- The *viscosity of water* is 750 times greater than that of air, generating a considerably stronger drag to which rapidly moving animals had to adapt. Overall, velocities of locomotion are considerably greater in air than in water.
- *Osmotic stress* in water is replaced by desiccation stress in air. Drought-resistant terrestrial organisms posses cuticles or an exoskeleton to avoid water loss. Among the animals, only some arthropods and the vertebrates were able to make a complete transition to dry terrestrial habitats. Representatives other groups either are restricted to moist environments, or move from water to land during their lives. The best-known example is the amphibians. Some terrestrial organisms have returned to the aquatic environment as a secondary adaptation.
- *Gas diffusivity in air* is ca. four orders of magnitude larger in air as compared to water. This has vast consequences for both photosynthesis and respiration, the latter of which in water requires either movement relative to the surrounding medium or turbulent diffusion. Respiration in air is moreover favored by the ca. 30 times higher oxygen content of air as compared to water, even if the latter is saturated with molecular oxygen.
- *Transparency of air* is significantly greater than that of even the clearest water. This makes considerably more solar radiation available to photosynthesis. Unlike in water also the red absorption peak of chlorophyll can be utilized which is not possible in water, except at the very surface because red light is rapidly absorbed by water.

Advantages of terrestrial life: Although only few groups of both animals and plants were able to make the transition from water to land, species numbers on the continents by far exceed those in the sea. Some 250,000 marine species and 1.5 million terrestrial species are known to science.[28] This is because land offers a much broader spectrum of different habitats than the sea, to which organisms could adapt. Respiration in air can support considerably higher overall metabolic rates. Terrestrial organisms with cuticles that prevent gas exchange across the body surface have evolved either tracheal systems (arthropods) or lungs (vertebrates) to overcome this shortcoming. Both essentially enlarge the surface through which gas exchange is possible. This is particularly important in large animals.[29] Today, 71 % of the Earth surface is covered by ocean which contributes only 40 % of the global primary productivity (FALKOWSKI and RAVEN 1997). In other words, on a per-unit-area basis, land is 3.7 times more productive than are oceans.

Atmospheric oxygen reached its all-time high during the Carboniferous when primary productivity on land reached exceptionally high levels, and oxygen losses were minimized by high rates of carbon burial.[30] Whereas BERNER and CANFIELD (1989) estimated maximum levels of atmospheric oxygen of 35 %, LENTON (2001), by assuming additional oxygen sinks, estimated maximum levels of 25 %.[31] The carboniferous oxygen maximum is coincident with

---

28 Source: *U. N. Atlas of the Oceans* 2007.
29 Gas exchange including water vapor in terrestrial plants is regulated by stomata, specialized epidermis cells that open and close because of the cuticle covering the epidermis which in most plants prevents direct gas exchange.
30 Organic matter deposited over the course of Earth history has been preserved in fossil fuel deposits and as organic matter in shale.
31 Estimates concerning the upper limit of atmospheric oxygen, controlled by the spontaneous ignition of organic material such as plant stands, range from 25 to 35 % due to contradictory experimental evidence. At atmospheric oxygen contents fewer than 13 % fire is impossible. These boundary conditions define the "fire window" (LENTON 2001).

the emergence with exceptionally large winged insects which has been interpreted as a consequence of increased oxygen supply to meet their energy demands during flight. Since about 250,000 years, the atmospheric oxygen remained rather stable at little over 20%, mainly due to negative feedback mechanisms (LENTON 2001, FALKOWSKI 2002; Fig. 5).

7. Biological versus Cultural Evolution

The emergence of consciousness, self-awareness, and rational thinking are to be considered to represent evolutionary breakthroughs equal in rank with the other major evolutionary leaps (fulgurations) discussed in the preceding section. It is likely that these innovations evolved over an extended time span. This is made plausible by the fact that in the animal kingdom, brain structures as well as behavioral patterns exist over a broad range of complexity.

7.1 The Palaeolithic: A long Era of Hardship

Our species emerged ca. 170,000 years ago[32] during the past interglacial (the *Eehmian*) that was warmer than it is today. The last glaciation began about 100,000 years ago. It is generally accepted that, in contrast to most other animals, hominids are characterized by a lack of specialization and consequently a high level of flexibility. These capabilities could have been a prerequisite for the survival under the extremely harsh conditions during the past glacial during which humanity nevertheless might almost have become extinct ca. 73,000 years ago in the aftermath of a major volcanic eruption (*Toba, Indonesia*; AMBROSE 1998).[33]

7.2 Life during the Holocene

The development of human civilization and culture on a broad scale could only set in after the end of the last glaciation.[34] It is estimated that about five million people lived on Earth at the end of the Ice Age. In terms of the development of humanity as a biological species, and of its impact on the global environment, two transitions stand out in particular: The *Neolithic* and the *Industrial Revolutions*.

7.2.1 The Neolithic Revolution

About 10,000 years ago, as a consequence of the improvement of the living conditions after the end of the last ice age, it is likely that population growth accelerated significantly, which in turn must have caused an increased demand for food. In order to cope with this, the economy shifted from hunting and gathering to agriculture. During the late Antiquity, the world population was between 250 and 300 million and remained unchanged until ca. 1000 CE.

---

32 The first proven fossils of *Homo sapiens* are 100,000 years old.
33 It is estimated that only 3,000–10,000 humans survived this crisis.
34 This is not to neglect earlier great artistic achievements, such as the 35,000 years old *Venus of Willendorf* (Austria) and the *cave paintings of Lascaux* (France) which date back 16,000 years.

7.2.2 The Industrial Revolution

The *Industrial Revolution* after 1800 CE was triggered by the invention of the steam engine by James WATT in 1765 which allowed the utilization of external energy sources. By this time the world population had reached one billion (Wikipedia). Mainly as a consequence of the availability of abundant energy, the rate of technological innovation has accelerated considerably during the past 200 years. This suggests that the principles governing the development of natural sciences and technology differ fundamentally from those controlling biological evolution. The main reason for this is that scientific findings and technological innovations are directional and build on previous ones. In other words, the development of natural sciences and consequently technology is *progressive*.

7.2.3 Development of Humanities and Arts

The time-course of the history of our spiritual culture, as manifested in philosophy, humanities, but above all, in the fine arts, is in stark contrast to that of science and technology. Here the development is characterized by brief episodes of highest achievement. For example, the life-spans of the greatest Greek philosophers (SOCRATES, PLATO, and ARISTOTLE) cover no more than 148 years. The life spans of the greatest painters of the Renaissance (Leonardo DA VINCI, MICHELANGELO, and Raffaello SANTI) fall within 112 years, and the greatest composers of all times (J. S. BACH, MOZART, and BEETHOVEN) lived within 142 years. A conspicuous feature of this development is the fact that, with the exception of the climax of the Greek culture during Antiquity, maximum achievements in the different art forms are out of phase: For example, painting and sculpture peaked centuries earlier than European music.

7.2.4 Did the biological Evolution of Humanity come to an End?

Let us briefly ask whether the highest levels of intellectual and artistic achievement are the result of a continued biological evolution of humanity. Most authors agree that this is not the case. Three reasons can be given for this:

– *Theory of evolution*: The pace of biological evolution in general is slower by several orders of magnitude than that of cultural evolution. It is extremely unlikely that any evolutionary steps are possible within millennia or even centuries.
– *Population genetics*: The overall size of the human population (close to seven billion at present), in conjunction with the high degree of genetic exchange due to the high global mobility of humanity, make it impossible for any genetic change to spread to the collective gene pool of humanity as a whole.
– *Cultural history and Sociology*: The current ethical standards, which are based on religion and human rights that at least in principle are universally accepted, forbid the application of the Darwinian principles of selection to human society. We all hope that barbarian ideologies such as that of the Nazis will never be repeated. Nevertheless, frequently observed behavioral patterns leading to "ethnic cleansing" or even genocide can be interpreted as strategies towards applying a primitive form of Darwinist selection.

How, then, is the emergence of individuals whose accomplishments greatly exceed those of the population at large possible? The explanation for this paradox is the fact that the intellectual capabilities of human individuals vary over an enormously wide range.

8. The Anthropocene: Human Impacts on the Earth System and the Biosphere on a Global Scale

As a consequence of the improvements in the living conditions of the population at large in the aftermath of the Industrial Revolution, an unprecedented acceleration of human population growth and technological development set in. Since 1800, the human population has grown almost sevenfold. At the same time, the world-wide average of *per capita* resource consumption has increased by a factor of ca. 20 (VALLENTYNE 2006).[35] The overall rate of resource consumption depends on the product population size multiplied by the *per capita* resource consumption. In other words, the results of human population growth and of the increase in human wealth are multiplicative. The combination of both has drastically altered the fluxes of matter on Earth. The consequences of this development have prompted the suggestion that by 1800, a new geological era has begun that is characterized by human interference with the environment on a global scale. In particular, global biogeochemical cycles, the global climate, and the survival of biological species are affected. This era has been called *Anthropocene* by CRUTZEN and STOERMER (2000).[36]

In natural ecosystems, fluxes of matter as a rule are predominantly cyclic. As a consequence, rates of permanent deposition of waste products generated by the cumulative metabolism of the biosphere are only a minute proportion of the overall flux rates. For example, as mentioned, only 0.1 % of the organic carbon produced by net primary production is permanently deposited as "organic waste". As a consequence of the predominantly cyclic fluxes of matter, the system as a whole is rather stable and undergoes directional change only over extended periods of time (Fig. 5). This is in strong contrast to fluxes of matter induced by the human civilization. They as a rule are unidirectional: Resources are withdrawn from the environment, and waste products are released to the environment. As a result, resources are depleted. This is true even if resources in principle are renewable, if the rates of resource withdrawal exceed the rates of resource regeneration. On the other hand, waste products are deposited at a rate that usually exceeds rates of natural deposition by several orders of magnitude. As a consequence, end products of the human community metabolism accumulate in the environment. Some of the results of this development can be summarized as follows (CRUTZEN and STOERMER 2000):

- The current world population of cattle is 1.4 billion (one cow per family).
- The SO_2 emission of 160 Tg[37] per year by coal and oil burning is two times the magnitude of all natural emissions combined.
- Between 30 and 50 % of all land area on Earth has been transformed by human action.
- More nitrogen is fixed synthetically and is applied as fertilizers for food production than is fixed naturally in all terrestrial ecosystems.

---

[35] As a first approximation energy consumption is a good measure of overall resource consumption. Prior to the Industrial Revolution, the main energy source of the human civilization was muscle power, mainly of man himself (mean energy production: 100 Watts). The world-wide average capita energy consumption is 2,000 Watts. Maximum values close to 10,000 Watts are reached in North America. Energy demands are met by using external energy sources, mainly fossil fuels (VALLENTYNE 2006).

[36] One might argue that the *Anthropocene* already set in about 10,000 years ago with the *Neolithic Revolution* which has led to the conversion of land for human use that by now affects almost all land area, with the exception of polar ice caps and deserts.

[37] Terragram: 10^{12} g, or 10^6 metric tons.

- The release of nitrous oxides from fossil fuels and biomass combustion to the atmosphere is larger than natural inputs.
- The atmospheric levels of CO_2 have risen by over 35 % since 1800.
- The atmospheric level of CH_4 has doubled during this time-span.

There can be little doubt that these changes have also drastically affected the biosphere. Let us briefly elaborate on the influence of our own species on the biosphere, and what this influence means both for the biosphere as a whole and for humanity itself.

8.1 What is Biodiversity?

Consensus exists that the loss of biological species will most probably represent the longest lasting consequence of human interference with the natural environment. Whereas, at least in principle, climate change due to the emission of greenhouse gases could be reversed within decades, biological species that have become extinct cannot be replaced (LEAKEY and LEVINS 1995). Unfortunately, the term "biodiversity" frequently is used without a clear idea about its meaning. According to GASTON and SPICER (1998), there exist more than 12 published definitions.

Here we use the definition given by Article 2 of the *Convention on Biological Diversity*[38] (CBC) that was established during the Rio Conference (UNCED) of 1992: "Biological diversity means the variability among living organisms from all sources including inter alia terrestrial, marine and other aquatic ecosystems and the ecological complexes of which they are par; this includes diversity within species, between species, and of ecosystems" (quoted from GASTON and SPICER 1998). In other words, the term "biodiversity" is not only related to the number of biological species within a given environment or ecosystem.

8.1.1 The Elements of Biodiversity

According to HEYWOOD and BASTE (1995) the following elements of biodiversity can be distinguished: (*i*) *Ecological diversity* including biomes, bioregions, landscapes, ecosystems, habitats, ecological niches, and populations, (*ii*) *Genetic diversity*, related to populations, individuals, chromosomes, genes, and nucleotides, and (*iii*) *Organismic diversity* which includes the hierarchy of taxonomic categories ranging from kingdoms, phyla, families, genera, species, subspecies, populations, and individuals. Whereas this set of definitions might be helpful for the taxonomist (who admittedly is most concerned with biodiversity) and takes into consideration genetic differences within species and populations, it might not always be helpful to the ecologist who is concerned with the consequences of species loss for natural ecosystems. It therefore makes sense to add *Functional diversity* as a category, meaning the functions that one or a several species have within an ecosystem, mainly on the flow of matter and the structure of the community.

8.1.2 How Many Species of Organisms Live on Earth?

Species diversity allows comparisons of different geological époques as shown in Figure 3, or within different groups of organisms as well as environments. As mentioned, it is believed

---

[38] The terms „biodiversity" and „biological diversity" here are used interchangeably.

that the number of genera never was as high as it is today. At present, 1.75 million biological species have been described by taxonomists. This is only a small fraction of all species that are expected to actually live today. Estimates of the total species number range from 3.6 to 112 million. A species number of 13 million species is considered a reasonable "working figure" (HAWKSWORTH and KALIN-ARROYO 1995). The reliability of species number estimates varies widely between groups of organisms, depending mainly on knowledge of their respective taxonomies. Species number estimates can be expected to be reasonably accurate in vertebrates and, to a lesser extent, in vascular plants, but for example, are extremely unreliable in bacteria and unicellular animals and plants. Species inventories are better known in terrestrial than in marine environments.

8.2 Susceptibility of Biological Species to Extinction

The main mechanism by which biological diversity on the level of biota (genus or species) is affected is by *extinction*. By this we understand the death of all representatives of the taxonomic category in question. The geologic record shows that even during periods characterized by massive losses of species or genera, some groups have readily survived. This was true in the past and is true in the present situation. The main factors influencing the chances of survival can be summarized as follows:

- *Abundance*: Species with high numbers of individuals are less susceptible to extinction than are rare species. Because small species in general are more abundant than large ones, large organisms are to a greater extent threatened by extinction than small ones.
- *Reproductive strategy*: Organisms under continuous pressure from other organisms (for instance, by predation) as a rule produce more offspring in compensation (*r-selection*), than organisms that have no or few natural enemies (*K-selection*). Under life-threatening conditions r-selected species are more likely to survive than K-selected ones.
- *Geographic distribution*: The geographic distribution of organisms mainly depends on their capability for dispersal and their ecological requirements. Species with wide geographic distribution are more likely to survive than organisms which, by definition, are found only in highly restricted areas. The most extreme case of restricted geographic distribution of species is *endemism* which means that some species are found only in one particular location. Endemic species are particularly common in ancient and/or isolated ecosystems. Well-known examples are the Galápagos Islands, Lake Baikal and the East African Rift Lakes.
- *Degree of specialization*: Generalists are organisms that can tolerate a wide range of environmental conditions (for example, temperature or humidity in terrestrial biota), or can utilize a wide range of essential resources, such as food. Generalists in general have significantly greater chances of survival under adverse conditions than specialists. Well-known examples of extreme food specialization are the Panda bear and the Koala bear.

8.3 The Sixth Extinction

LEAKEY and LEWIN (1995) liken the current loss of biological species with the previous five most severe mass extinctions (*"The Big Five"*) during Earth history (Fig. 3). Consequently, they introduced the term "The Sixth Extinction". Given the high degree of uncertainty concerning the total number of biological species on Earth, it is not surprising that estimates concerning the extent of the current overall species loss are characterized by an extremely high degree of uncertainty, although the fact as such is undisputed (Leakey and Lewin 1995, *WBGU* 2000, Wilson 2004). In order to give an idea about the presumed extent of the human-induced extinction let us give two figures: Some estimates suggest that the normal "background" extinction rate of 10 biological species per year has risen to 130 species per day, suggesting an increase in extinction rates by a factor close to 5,000. Other estimates go so far as to state that the current mass extinction rate caused by humanity might be about four orders of magnitude greater than to be expected without human interference (Gaston and Spicer 1998).

8.3.1 Human-induced Causes for the Loss of Biological Diversity

In particular the following factors are responsible for the human-induced loss of biological species (*WBGU* 2000, Wilson 2004):

– *Loss of natural habitats by conversion*: The transformation of natural ecosystems into land that is used to meet human needs. Particularly dramatic is the loss of forests whose total area has decreased by over 50% since the Neolithic revolution. Because primary forests are characterized by high species diversity, forest loss is among the dominant causes of species loss.
– *Loss of natural habitats by ecosystem degradation*: Exceeding the uptake capacity of natural ecosystems for waste products leads to a deterioration of the living conditions of organisms. Ecosystem degradation is particularly effective in coastal marine ecosystems, lakes and streams where already fertilization leads to environmental degradation, mainly due to the depletion of oxygen in deeper water layers (*eutrophication*). Another severe cause of ecosystem degradation is the deposition of slowly or not degradable noxious substances that frequently accumulate in organisms. Endemic species in affected habitats are threatened by extinction. Also global climate change can lead to loss of species if the living conditions in the remaining habitats are altered beyond the tolerance limits of some biota.
– *Over-exploitation of living resources*: Species are threatened whenever their rate of exploitation exceeds their rate of reproduction. In general, mainly species without natural enemies are threatened by extinction due to elimination, as a consequence of their small rates of reproduction (*K-selection*). Particularly well-known examples include the large whales, as well as large terrestrial animals, both predatory and herbivorous. Less well known is the fact that also sharks are threatened because they are characterized by low reproductive rates. Most bone-fishes, by contrast, have high reproduction rates and, although their stocks are notoriously overused, in general are not in danger of extinction.[39] However, both pelagic and benthic ocean fisheries also lead to the capture of unwanted species ("by-catch") such as dolphins, whose reproduction rates are small.
– *Introduction of alien species*: The introduction of alien species, either voluntarily or by accident, frequently leads to their excessive reproduction in the new environment which by competition affects indigenous species to the point of their elimination. Global mobil-

---

[39] One species of bonefish, the bluefin thuna (*Thunnus thynnus*), recently had to be added to the list of threatened species. Main cause is the popularity of their meat as *Sashmi* in Japan and elsewhere in the world (source: R. Ellis, Scientific American March 2008, pp. 59–65)

ity has increased the long-range transport of species and their introduction by accident. Particularly noteworthy is ballast water in ocean-going vessels. In addition, species that can be utilized are introduced at will such as freshwater fishes in streams. A particularly dramatic recent example is the introduction of the Nile perch (*Lates niloticus*) into Lake Victoria which has led to the extinction of hundreds of endemic fish species. As a consequence, the entire ecosystem structure of Lake Victoria was irreversibly altered (Goldschmidt 1997).

– *Fractionation of habitats*: As a rule, the number of species in any given ecosystem is a function of the size of the available habitat. This is particularly true for islands and lakes that can also be considered insular habitats surrounded by insurmountable boundaries. Species number in any given environment is the result of equilibrium between immigration of species and extinction (MacArthur and Wilson 1967). Extinction rates in small populations are higher, mainly because genetic defects are more effective on the overall population than in large populations. Immigration rates, by contrast, are independent of island size (Smith and Smith 2006). Large "islands" moreover allow a larger diversity of ecological niches that can accommodate higher numbers of species (Krebs 2001). Only large undisturbed terrestrial habitats can accommodate organisms requiring large territories for foraging and mating. Human activities, especially ecosystem conversion for land-use and road construction, have drastically transformed these ecosystems by creating isolated small pockets. As a consequence, species were lost, even in cases where living conditions otherwise remained favorable.

8.3.2 Consequences of Species Loss for Ecosystem Structures and Functions

It is frequently argued that the stability of an ecosystem is a function of the total magnitude of its species inventory. In other words, it is believed that ecosystems with many species are more stable than ecosystems with few species. This is a gross oversimplification for two reasons: (*i*) It is necessary to distinguish between structural and functional stability. *Structural stability* implies that the species inventory of an ecosystem remains unchanged. *Functional stability* means that major ecosystem processes such as overall productivity and food-chain dynamics remain virtually unaffected, in spite of changes in species composition. Therefore, it is necessary to specify which type of "stability" is meant in any particular case. (*ii*) The consequences of species loss for ecosystems depend on the role of the species in question for the functioning of an ecosystem. The following functional types of organisms can be distinguished (*WBGU* 2000):

– *Dominant species* are characterized by high abundance and/or high biomass. As a consequence, such species have a strong influence on the fluxes of matter within the ecosystem as a whole. As a rule, dominant species are not threatened by elimination due to their high abundance.
– *Keynote species* may not be particularly abundant, but play a dominant role in the interactions between different components within the ecosystem. For example, they may act as top predators. Their impact as a rule is not restricted to the components of the ecosystem that are immediately affected, but also cascades down to components that depend on directly affected species. As a consequence, loss of keynote species influences both community structure and function.
– *Redundant species* are species whose functions within the ecosystem as a whole overlap to a great extent.

Ecosystem stability can be defined by the threshold beyond which significant alterations in ecosystem structure and functions occur. As a rule, impacts on ecosystem structure and functioning are strongest if keynote species are eliminated, and weakest if redundant species are eliminated.

- In *resistant ecosystems* the structure (species inventory) does not show significant change in response to external impacts. We can expect this to happen if the dominant players within the ecosystem are resistant to elimination.
- *Resilient ecosystems* are those whose functioning is restored, after some transitory period, by redundant species that take over the role(s) of eliminated species.
- *Fragile ecosystems* are those whose structure is characterized by a subtle balance between species whose susceptibility to elimination is high. Tropical rain forests and coral reefs have the highest species diversity of all ecosystems on Earth. It could well be that despite of this, both are highly fragile. An explanation for this could be that they have evolved and matured under comparatively stable (climatic) conditions that did not select organisms for flexibility in their ecological requirements.

8.3.3 The Future of a Biosphere Dominated by Humanity

It is tempting to expect that in regions that have been utilized heavily by man over extended periods of time, total species numbers will drop. This is not necessarily the case, as an analysis of the plant species inventory of Germany since the onset of the Neolithic reveals (SCHERER-LORENZEN et al. 1999). However, there has been a shift from indigenous species to introduced species that has accelerated since the Industrial Revolution and especially since about 1950, with the mechanization of agriculture. From this the conclusion can be drawn that species richness alone is insufficient to characterize biodiversity.

Given the increasing demand for food and hence the continued expansion of land area used to meet his demands, it can be expected that habitats of indigenous species will continue to decrease, leading to an acceleration of the human-induced mass extinction of biological species. The ecological niches created thereby will be occupied by generalists with high reproductive rates and wide geographic distribution. Some of these generalists will be high-yield crops, but many newcomers will be weeds and pests that cannot be utilized by humanity and will out-compete both usable crops and indigenous species. Since habitat diversity will continue to drop, overall species richness will substantially decrease on a world-wide scale in the foreseeable future.

8.3.4 Ethical Considerations Concerning the Protection of Biodiversity

Ecological ethics is an attempt to characterize and evaluate our positions towards the living world around us in view of the demands of a growing human population (OTT 1994). Basically, we can distinguish between anthropocentric and biocentric positions. The first set of positions takes into consideration human needs, not only for the sake of meeting material demands, but also for recreational and cultural reasons. Different value categories have been defined (*WBGU* 1999):[40]

---

[40] These categories are derived from those proposed by OTT (1994), but to a greater extent are based on economic considerations.

– *User value*: Value of biota to directly meet human demands such as for nutrition, raw materials (for example, fabrics), health (for example, medications), and recreation.
– *Symbolic value*: Religious and spiritual value of organisms such as sacred trees and animals.
– *Functional value*: Values based on function of organisms that might serve human needs such as flood prevention, or within ecosystems, such as keynote species.
– *Optional value*: Value, based on the prospect that an organism might eventually be of use for humans, such as for producing medications.
– *Intrinsic value*: All of the above categories, with the exception perhaps of some of the functional values are entirely anthropocentric. Only the intrinsic value category is biocentric in that it grants an organism the right to exist for its own sake.

Humanity is facing a dilemma that, in its briefest form, can be summarized as follows: We are morally obliged not only to look after our own generation, but also after those who will succeed us. This, among other things, means that we have to continue to increasingly exploit living resources. On the other hand, we should be obliged to preserve the biosphere for its own sake. As the only beings on Earth that are capable of reflecting the results of their own actions, we not only have the power of control, but also heavy burden of responsibility for all organisms which whom we share this planet.

References

AMBROSE, S. H.: Late Pleistocene human population bottlenecks, volcanic winter, and differentiation of modern humans. Journal of Human Evolution *34*, 623–651 (1998)
ARRHENIUS, S.: Worlds in the Making: The Evolution of the Universe. New York: Harper & Row 1908
AVIEZER, N.: In the Beginning. Biblical Creation and Science. Hoboken, NJ: KTAV Publishing House 1990
BADA, J. L.: Extraterrestrial handedness? Science *275*, 942–943 (1997)
BADA, J. L.: Origins of homochirality. Nature *374*, 594–595 (1995)
BADA, J. L.: How life began on Earth: a status report. Earth and Planetary Science Letters *226*, 1–15 (2004)
BERNER, R. A., and CANFIELD, D. E.: A new model for atmospheric oxygen over Phanerozoic time. American Journal of Science *289*, 333–361(1989)
BLANKENSHIP, R. E., SADEKAR, S., and RAYMOND, J.: The evolutionary transition from anoxygenic to oxygenic photosynthesis. In: FALKOWSKI, P. G., and KNOLL, A. H. (Eds.): Evolution of Primary Production in the Sea. Chapter 3, 21–35. London 2007
CAIRNS-SMITH, A. G.: Genetic Takeover – and the Mineral Origin of Life. Cambridge, UK: Cambridge University Press 1982
CAIRNS-SMITH, A. G.: Seven Clues to the Origin of Life. Cambridge, UK.: Cambridge University Press 1985
CECH, T. R.: A model for the RNA-catalyzed replication of RNA. Proceedings of the National Academy of Sciences USA *83*, 4360–4363 (1986)
CHYBA, C., and SAGAN, C.: Endogenous production, exogenous delivery and impact-shock synthesis of organic molecules: An inventory for the origins of life. Nature *355*, 125–132 (1992)
COURTILLOT, V.: Evolutionary Catastrophes. The Science of Mass Extinctions. Cambridge, UK: Cambridge University Press 1999
CRICK, F., and ORGEL, L.: Directed panspermia. Icarus *19*, 341 (1973)
CRUTZEN, P. J., and STOERMER, E. F.: The anthropocene. Global Change Newsletter *41*, 17–18 (2000)
DAWKINS, R.: The Ancestor's Tale. A Pilgrimage to the Dawn of Life. London: Weidenfeld & Nicolson 2004
DAWKINS, R.: The God Delusion. London: Trans World Publishers 2007
DES MARAIS, D. J.: Evolution: When did photosynthesis on Earth? Science *289*, 1703–1705 (2000)
DOBSON, C. M, ELLISON, G. B., TUCK, A. F., and VAIDA, V.: Atmospheric aerosols as prebiotic chemical reactors. Proceedings of the National Academy of Sciences USA *97*, 11864–11868 (2000)
DOOLITTLE, W. F.: Phylogenetic classification and the universal tree. Science *284*, 2124–2128 (1999)

Duve, C. de: Vital Dust. Life as a Cosmic Imperative. New York: HarperCollins Publishers 1995
Falkowski, P. G.: On the evolution of the carbon cycle. In: Williams, L. B., Thomas, D. T., and Reynolds, C. S. (Eds.): Phytoplankton Productivity. Carbon Assimilation in Marine and Freshwater Systems; pp. 318–449, Oxford: Blackwell Science 2002
Falkowski, P. G., and Raven, J. A.: Aquatic Photosynthesis. Malden, Mass.: Blackwell 1997
Gaston, K. E., and Spicer, J. I.: Biodiversity. An Introduction. Oxford: Blackwell Science 1998
Goldblatt, C., Watson, A. J., and Lenton, T. M.: Bistability of atmospheric oxygen and the Great Oxidation: implications for life detection. ASP Conference Series. (2009, in press)
Goldschmidt, T.: Darwins Traumsee. Nachrichten von meiner Reise nach Afrika. München: Beck 1994
Gould, S. J.: Wonderful Life. New York: Norton 1989
Gould, S. J., and Eldredge, N.: Punctuated equilibrium comes of age. Nature *366*, 223–227 (1993)
Hawking, S.: A Brief History of Time. From the Big Bang to Black Holes. New York: Bantam Books 1990
Hawksworth, D. L., and Kalin-Arroyo, M. T.: Magnitude and distribution of biodiversity. In: Heywood, V. H. (Ed.): Global Biodiversity Assessment; pp. 107–191. Cambridge: University Press 1995
Hazen, R. M.: Life's rocky start. Scientific American *284*/4, 76–85 (2001)
Heywood, V. H., and Baste, I.: Introduction. In: Heywood, V. H. (Ed.): Global Biodiversity Assessment; pp. 1–19. Cambridge: Cambridge University Press 1995
Holland, H. D.: The oxygenation of the atmosphere and the oceans. Philosophical Transactions of the Royal Society B *361*, 903–915. doi: 10.1098/rstb.2006.1838 (2006)
Hoyle, F.: The Intelligent Universe. London: Michael Joseph Limited 1983
Kasting, J. F., and Catling, D.: Evolution of a habitable planet. Annual Review of Astronomy and Astrophysics *41*, 429–463 (2003)
Kleine, T., Münker, C., Metzger, K., and Palme, H.: Rapid accretion and early core formation on asteroids and the terrestrial planets from Hf-W chronometry. Nature *418*, 952–955 (2002)
Krebs, C. J.: Ecology. The Experimental Analysis of Distribution and Abundance. Sixth Edition. San Francisco, CA: Benjamin Cummins 2001
Koonin, E. V.: Comparative genomics, minimal gene-sets and the last universial common ancestor. Nature Reviews/Microbiology *1*, 127–136 (2003)
Lane, N.: Power, Sex, Suicide. Mitochondria and the Meaning of Life. Oxford: Oxford University Press 2005
Lazcano, A., and Miller, S. L.: How long did it take for life to begin and evolve to cyanobacteria? Journal of Molecular Evolution *39*, 546–554 (1994)
Leakey, R., and Lewin, R.: The Sixth Extinction. Patterns of Life. New York: Doubleday 1995
Lenton, T. M.: The role of land plants, phosphorus weathering and fire in the rise and regulation of atmospheric oxygen. Global Change Biology *7*, 613–629 (2001)
Lenton, T. M.: The coupled evolution of the atmospheric oxygen. In: Rothschild, L. J., and Lister A. M. (Eds.): Evolution on Planet Earth. The Impact of the Physical Environment; pp. 35–53. Amsterdam: Academic Press 2003
Line, M. A.: The enigma of the origin of life and its timing. Microbiology *148*, 21–27 (2002)
Loeb, A.: The dark ages of the universe. Scientific American *295*, 23–29 (2006)
Lorenz, K.: Die Rückseite des Spiegels. Versuch einer Naturgeschichte der menschlichen Erkenntnis. München: Piper 1973
MacArthur, R. H., and Wilson, E. O.: The Theory of Island Biogeography. Princeton, N. J.: Princeton University Press 1967
Margulis, L.: Origin of the Eukaryotic Cell. New Haven: Yale University Press 1970
Mayr, E.: Animal Species and Evolution. Cambridge: Harvard University Press 1963
Mayr, E.: Two empires or three? Proceedings of the National Academy of Sciences USA *95*, 9720–9723 (1998)
Miller, S. L.: Production of amino acids under possible primitive earth conditions. Science *117*, 528–529 (1953)
Moore, W. A: Life of Erwin Schrödinger. New York: Cambridge University Press 1994
Orgel, L. E.: The origin of life on Earth. Scientific American *271*, 53–61 (1994)
Ott, K.: Ökologie und Ethik. Ein Versuch praktischer Philosophie. Tübingen: Attempto 1994
Pittendrigh, C. S.: Adaptation, natural selection, and behaviour. In: Rose, A., and Simpson, G. G. (Eds.): Behaviour and Evolution; pp. 390–416, New Haven: Yale University Press 1958
Raup, D. M.: Extinction. Bad Genes or Bad Luck? New York: Norton 1991
Scherer-Lozenzen, M., Elend, A., Nöllert, S., and Schulze, E. D.: Plant invasions in Germany. In: Mooney, H. A., and Hobbs, R. J. (Eds.): The Impact of Global Change on Invasive Species. Corvallis: Corvallis Island Press 1999
Schrödinger, E.: What is Life. New York: The Macmillan Company 1945

SHAPIRO, R.: A simpler origin of life. Scientific American 296, 25–30 (2007)
SLEEP, N. H., ZAHNLE, K. J., KASTING, J. F., and MORWITZ, H. J.: Annihilation of ecosystems by large asteroid impacts on the early Earth. Nature 342, 139–142 (1989)
SMITH, T. M., and SMITH, R. L.: Elements of Ecology. Sixth Edition. San Francisco, CA: Pearson Education Inc. 2006
SMOOT, G. F.: New sky maps of the early universe. Astrophysics and Space Science Library 169, 281 (1991)
SYVANEN, M.: Cross-species gene transfer: Implications for a new theory of evolution. Journal of Theoretical Biology 112, 333–343 (1985)
U. N. Atlas of the Oceans 2007
VALLENTYNE, J. R.: Tragedy in Mouse Utopia. An Ecological Commentary on Human Utopia. Victoria: Trafford Publishing 2006
WARD, P. D., and BROWNLEE, D.: Rare Earth. Why Complex Life is Uncommon in the Universe. New York: Copernicus 2004
WARMFLASH, D., and WEISS, D.: Did life come from another world? Scientific American 293, 40–47 (2005)
WEINBERG, S.: Life in the universe. Scientific American 271, 22–27 (1994)
WIDDEL, F., SCHNELL, S., HEISING, S., EHRENREICH, A., ASSMUS, B., and SCHINK, B.: Ferrous iron oxidation by anoxygenic phototrophic bacteria. Nature 362, 834–835 (1993)
WILSON, E. O.: The Future of Life. New York: Knopf 2004
WBGU (Wissenschaftlicher Beirat Globale Umweltveränderungen): Welt im Wandel: Umwelt und Ethik. Marburg: Metropolis 1999
WBGU (Wissenschaftlicher Beirat Globale Umweltveränderungen): Welt im Wandel: Erhaltung und nachhaltige Nutzung der Biosphäre. Jahresgutachten 1999. Heidelberg: Springer 2000
WOESE, C. R.: The universal ancestor. Proceedings of the National Academy of Sciences USA 95, 6854–6859 (1998)
WOESE, C. R., and FOX, G. E.: Phylogenetic structure of the prokaryotic domain: the primary kingdoms. Proceedings of the National Academy of Sciences USA 74, 5088–5090 (1977)
WOESE, C. R., KANDLER, O., and WHEELIS, M. L.: Towards a natural system of organisms: proposals for the domains Archaea, Bacteria and Eukarya. Proceedings of the National Academy of Sciences USA 87, 4576–4579 (1990)
XIONG, J., and BAUER, C. E.: Complex evolution of photosynthesis. Annual Review of Plant Biology 53, 503–521 (2002)

Prof. Dr. Max M. VON TILZER
Universität Konstanz
Fachbereich Biologie
78457 Konstanz
Germany
Phone: +49 7533 97663
E-Mail: max.tilzer@uni-konstanz.de

Continents

Europa: Umwelthistorische Determinanten

Rolf Peter Sieferle (St.Gallen)

Mit 1 Abbildung und 4 Tabellen

Zusammenfassung

Der Beitrag betrachtet die spezifischen natürlichen Umweltbedingungen Europas, vor allem im Hinblick auf Fragen der Bevölkerung, Voraussetzungen einer ertragreichen Landwirtschaft und das Auftreten von Naturkatastrophen (z. B. Erdbeben, Vulkanausbrüche, Extremwetterereignisse), unter verschiedenen historischen Kontexten. Europa erweist sich dabei als besonders begünstigt, so dass von hier die Industrialisierung ausgehen konnte.

Abstract

This contribution explores the specific natural environmental conditions of Europe, above all with regard to population issues, the conditions for a productive agriculture and the incidence of natural disasters (e.g. volcanic eruptions, extreme weather events), in different historical contexts. Europe appears to be particularly favourable in this respect, so that it was from here that industrialisation could develop.

In einer älteren, auf die hippokratische Säftelehre zurückgehenden Erklärungstradition für historische Prozesse und kulturelle Differenzen spielten die natürlichen Umweltbedingungen eine große Rolle (Glacken 1967). Hier handelte es sich um den klassischen Fall von limitierenden Bedingungen, und ihr wesentliches Merkmal war, dass sie als unveränderlich angesehen wurden. Ein seit dem 19. Jahrhundert populär gewordenes Standardargument gegen diese Interpretation lautet daher, dass historischer Wandel nicht aus invarianten Faktoren erklärt werden kann. Damit war dieser Erklärungsansatz allerdings noch nicht erledigt. Schließlich können auch invariante natürliche Faktoren historisch wirksam werden, wenn sie in eine neue spezifische Konstellation geraten. Hinzu kommt aber vor allem, dass nicht alle Naturbedingungen als unveränderlich anzusehen sind. In den letzten Jahren hat sich eine dynamischere Sicht von Naturprozessen durchgesetzt, und es wurde deutlich, dass weit mehr Elemente der Natur in Bewegung sind, als der ältere Naturdeterminismus gemeint hat. Letzteres gilt vor allem für das Klima (Pfister 1999), für Krankheitserreger (McNeill 1976, Ewald 1994) und für die Gleichgewichtszustände von Ökosystemen (Botkin 1990, Worster 1990).

Die Umweltsituation in Europa ist vor allem in Hinblick auf die von Europa ausgehende Industrialisierung von Interesse. Bevor hierauf näher eingegangen werden kann, muss geklärt werden, was genau unter „Europa" zu verstehen ist. Einer auf die Antike zurückgehenden geographischen Begriffsbildung zufolge wird mit Europa der westlich des Urals liegende Teil des eurasischen Kontinents verstanden. Dieser Definition entspricht aber keine ökonomische, kulturelle oder politische Realität in der Geschichte, weshalb wir den Begriff Europa in Anlehnung an MARQUARDT (2005) auf den Raum der *christianitas*, also der westlichen oder römischen Christenheit beschränken wollen. Europa in diesem eher kulturell-gesellschaftlich-politischen Sinne umfasst daher das Gebiet, das nach Osten von Skandinavien, Polen-Litauen, Böhmen, Ungarn-Kroatien begrenzt wird. Im Unterschied zur antiken-hellenistischen Welt liegt das Zentrum dieses mittelalterlichen und neuzeitlichen Europa nicht mehr im Mittelmeer, sondern nordwestlich der Alpen. Von diesen Räumen ging die Industrialisierung aus, d. h., als „Europa" könnte auch das Gebiet bezeichnet werden, in dem sich die Industrialisierung mit den geringsten Widerständen und Verzögerungen durchsetzte.

In der wirtschaftsgeschichtlichen Literatur hat vor allem Eric JONES (1987) die spezifischen natürlichen Umweltbedingungen in Europa (im Gegensatz zu Asien, vor allem zu China) als Faktoren betont, die den Weg in die Industrialisierung begünstigten. Als invariante Bedingungen können die Strukturen der Geomorphologie gelten, also die Küstenlinien, die Verteilung von Gewässern und Landmassen, die Flussläufe, die Berge, Täler, Ebenen. Diese Faktoren spielen eine dominante Rolle in der Tradition der historischen Geographie, vor allem in dem bedeutenden Werk von Fernand BRAUDEL (1949, 1986). Als Erklärungsmuster sind sie in universalgeschichtlich bzw. wirtschaftsgeschichtlich orientierten Arbeiten rezipiert worden, die nach naturalen Voraussetzungen der industriellen Transformation fragen, etwa dem populären Werk von Jared DIAMOND (1997).

In geographischer Hinsicht ist bemerkenswert, dass Europa eine kleinräumig gegliederte Halbinsel am westlichen Rand des eurasischen Kontinents bildet. Diese Randlage bedeutet, dass westlich von Europa nur noch der Atlantik (und schließlich Amerika) liegt. Dies hatte etwa militärische Folgen. Europa war vom Westen nicht bedroht; Invasionen konnten nur vom Osten kommen, aus Zentralasien, aus Anatolien oder vom südlichen Mittelmeer. Das Gleiche galt spiegelbildlich für die Zivilisationen Ostasiens (China, Korea, Japan), die mit Europa die Randlage gemein haben.

Die Lage Europas am äußersten westlichen Rand des eurasischen Kontinents hatte zur Folge, dass nur von hier aus eine Entdeckung Amerikas, d. h. der Anschluss des amerikanischen Kontinents an das eurasische System möglich war. Von keinem anderen Ort Eurasiens aus ist der Weg nach Amerika kürzer als von Westeuropa. Von den anderen eurasischen Zivilisationen fallen als Kandidaten einer möglichen Entdeckung Amerikas die mittleren Mächte von vornherein aus: Indien und die muslimischen Reiche blieben auf den Indischen Ozean verwiesen, den Osmanen wurde das westliche Mittelmeer versperrt. Russland war eine reine Kontinentalmacht und expandierte auf dem Landweg kräftig nach Osten.

China oder Japan, die die kontinentale Randlage teilen, hätten dagegen theoretisch über den Pazifik nach Osten bis nach Amerika expandieren können. Man sollte dabei jedoch nicht übersehen, dass eine Segelfahrt von Asien nach Amerika über den Pazifik navigatorische Anforderungen stellte, die erst im 18. Jahrhundert von den europäischen Seemächten wirklich und dauerhaft gelöst wurden. Zuvor blieb dies ein Abenteuer mit ungewissem Ausgang. Die Spanier schickten zwar seit dem 16. Jahrhundert jährlich eine Flotte von Peru nach Manila, doch nutzten sie dafür ein sehr kleines Zeitfenster. Die Reise von Europa nach Amerika dauerte im 17. Jahrhundert vier bis sechs Wochen. Für eine Überquerung des Pazifiks wurde ein halbes Jahr benötigt. Die polynesische Expansion von Südostasien bis Südamerika hangelte sich von Insel zu Insel und hatte große Probleme bei der Ost-West-Reise (KIRCH 1984). Dies bedeutet, dass schon aus geographischen Gründen eine Einbeziehung der Neuen Welt in das eurasische System nur von Europa aus möglich war, mit wichtigen Konsequenzen für das Weltwirtschaftssystem.

Die Entdeckung und Annexion Amerikas durch europäische Mächte hatte weitreichende historische Konsequenzen. Aufgrund dieser Expansion ist die den Europäern zur Verfügung stehende Fläche seit dem 16. Jahrhundert massiv gestiegen. Der Gewinn Amerikas führte zu einer drastischen Verbesserung des Flächenbestands, was in einem Vergleich zwischen den europäischen Kolonialmächten und den übrigen agrarischen Zivilisationen Eurasiens besonders deutlich wird.

Tab. 1 Bevölkerungsdichte pro km², 1500 und 1800. Daten nach WEBB 1952, JONES 1987, MOLS 1974, LIVI-BACCI 1997

Jahr	Indien	China[1]	China + Zentralasien[2]	Anatolien	Europa[3]	Amerika	Europa + Amerika
1500	23	23	–	8	14	2	3
1800	42	70	27	12	30	0,6	3,6

In den von Europa kontrollierten Räumen ist also während der Neuzeit trotz eines beträchtlichen Bevölkerungswachstums die Bevölkerungsdichte drastisch zurückgegangen, wenn man die kolonialen Territorien hinzurechnet. Ein wichtiger Grund hierfür ist das Massensterben der indigenen Bevölkerungen in den Kolonien. Allerdings sollte man im Auge behalten, dass die hinzugewonnenen Flächen zunächst weitgehend ungenutzt blieben und ihr Besitz kaum Einfluss auf die Lebenssituation für die Masse der Bevölkerung in Europa hatte. Daher ist die These WALLERSTEINS (1974), die Ernährung der europäischen Bevölkerung sei schon in der frühen Neuzeit auf Importe aus den Kolonien angewiesen gewesen, nicht plausibel. Bis zum 18. Jahrhundert wurden aus Übersee fast ausschließlich Luxusgüter und Genussmittel (Zucker, Tabak, Rum, Gewürze) importiert, die für die Ernährungsbilanz der breiten Masse ohne Bedeutung waren. Lebensmittelimporte aus Amerika, Südafrika und Australien gewannen dagegen erst im 19. Jahrhundert an Gewicht. Die amerikanischen Räume waren aber insofern von Bedeutung, als sie eine gewaltige Ressourcenreserve verfügbar machten, die dann seit dem 19. Jahrhundert nutzbar gemacht werden konnte, was die Industrialisierung physisch unterfütterte.

Von Bedeutung für die politische und wirtschaftliche Entwicklung kann auch die kleinräumige Gliederung Europas sein. Es gibt keine großen homogenen Flächen, sondern eine Abwechslung von Mittelgebirgen und kleineren Ebenen, von Flüssen und Tälern. Europa ist als Halbinsel fast überall vom Ozean umgeben, d. h., der Weg zum nächsten Seehafen ist nirgendwo sehr weit. Dies hatte zur Folge, dass Transporte zwischen den einzelnen europäischen Regionen immer relativ einfach und kostengünstig waren, da sie zum überwiegenden Teil auf dem Wasserweg stattfinden konnten, sei es über Flüsse und Seen, sei es über Kanäle, die nur kurze Zwischenstrecken zu überwinden hatten, sei es durch Küstenschiffahrt (vgl. die Beiträge in SIEFERLE 2008).

Auch mit mineralischen Bodenschätzen ist Europa gut versorgt. Sämtliche Metalle, die vom Menschen genutzt werden können, finden sich in Europa. Vor allem Eisenerze sind allgegenwärtig, so dass keine Gebietsmonopole entstehen konnten. Auch Edelmetalle, vor allem

---

1 „China" umfasst die 18 Provinzen des chinesischen Kaiserreichs.
2 „Zentralasien" umfasst die Mongolei, Mandschurei, Tibet, und Singkiang.
3 „Europa" ist das Gebiet westlich von Russland, Weißrussland und der Ukraine.

Silber, finden sich in verschiedenen Gebieten. Wichtig für die Industrialisierung war dann das Vorkommen von Steinkohle in guten Qualitäten und in leicht (häufig zunächst im Tagebau) zugänglichen Lagerstätten, jedenfalls in Nordwesteuropa. Im Mittelmeerraum gibt es keine Kohle, was einer der Gründe dafür sein kann, dass es in der Antike zu keiner industriellen Transformation gekommen ist.

Europa hat sehr gute, geologisch junge und fruchtbare Böden. Landwirtschaftlich gehört es global zu den privilegierten Zonen. Die Böden sind außerordentlich robust, denn ihr geringes Alter macht immer wieder Mineralien durch Verwitterung verfügbar. Dies hatte zur Folge, dass in Europa über 7000 Jahre hinweg Landwirtschaft betrieben werden konnte, ohne dass es zu größeren Bodenverlusten durch Erosion oder Versalzung kam (mit gewissen Ausnahmen im Mittelmeerraum).

Das Relief mit seiner Abwechslung von (Mittel-) Gebirgen und Ebenen begünstigte eine Kombination von extensiver und intensiver Landwirtschaft. Es gibt Areale, die für den Anbau von Feldfrüchten wenig geeignet sind und daher als Weideland oder Wald genutzt werden können. Dies hatte zur Folge, dass kein Druck zur völligen Rodung der Wälder und zur Umwandlung von Weideland in Ackerland bestanden hat (wie es in Südchina der Fall war). Aus diesen Gründen existierten immer Landreserven, vor allem konnte man über Holz und über Arbeitstiere verfügen, die niemals (wiederum wie in China) vom Menschen verdrängt werden konnten.

Das europäische, besonders das mittel- und westeuropäische Klima des Holozäns war für menschliche Lebensprozesse, vor allem aber für die Landwirtschaft geradezu ideal. Die Winter sind nicht sehr kalt, die Sommer sind nicht zu heiß, vor allem aber gibt es regelmäßige Niederschläge. Klimaextreme sind äußerst selten. Abgesehen von der iberischen Halbinsel spielten Trockenheit und daher Maßnahmen zur künstlichen Bewässerung keine Rolle. Dies vereinfachte die Landwirtschaft, stabilisierte vor allem die Böden, die nicht der Versalzung ausgesetzt waren.

1. Landwirtschaft

Die nordwesteuropäische Landwirtschaft war flächen- und ressourcenextensiv sowie energieintensiv. Während der Römerzeit dominierte im südlichen Europa der Anbau von Weizen und Gerste, kombiniert mit Sonderkulturen wie Olivenöl oder Wein. Im nördlichen Europa überwog noch die Viehhaltung, kombiniert mit sehr extensiven Formen des Getreideanbaus (Brandrodung). Im frühen Mittelalter kam es dann zu einer „Agrarrevolution" (MITTERAUER 2001), in deren Zentrum ein Prozess der „Vergetreidung" stand: der Anbau von Roggen als Brotgetreide und von Hafer als Nahrung für die Pferde. Organisatorische Basis dessen wurde ab dem 6. Jahrhundert die Dreifelderwirtschaft. In diesem spezifisch nordwesteuropäischen System der Landwirtschaft wurde das gesamte Land in die folgenden Elemente eingeteilt: Neben dem Dorf mit den Hofstätten und den Gärten gab es die Feldmark, die in drei rotierende Abschnitte geteilt war (Sommerfeld, Winterfeld, Brache) sowie die Allmende, die nicht rotierte. Jeder Bauer besaß Anteile auf jeder Flur. Diese Anteile befanden sich in Gemengelage, waren also nicht von den anderen Losen durch Grenzen, Zäune etc. getrennt, so dass Flurzwang bestand, d. h. die Felder zur gleichen Zeit auf gleiche Weise angebaut werden mussten.

Als Hauptnachteil der Gemengelage gilt häufig der hohe Transportaufwand, etwa des Pfluggespanns vom Hof zum Acker. Da man aber am Tag nur etwa einen Morgen mit Ochsen pflügen konnte, verringerte sich dieses Problem. Außerdem waren die Entfernungen so kurz, dass der tägliche Mehraufwand für Transport bei etwa 15 min gelegen haben dürfte. Die einzelnen Lose waren von etwa 1 m breiten Rainen umgeben. Dies könnte als Flächenverschwendung gelten, doch warf man auf diese Raine aufgepflügte Steine, wendete dort den Pflug und nutzte das Gras als Grünfutter. Immerhin machten die Raine etwa 10 % der Fläche aus.

Die Brache wurde regelmäßig beweidet, wobei eine Beschränkung des Viehs auf das je eigene Los nicht möglich war (keine Einzäunung). Gewöhnlich wurde die Menge des einzutreibenden Viehs proportional zur Größe der jeweiligen Lose festgelegt und überwacht. Zuweilen definierte die Dorfversammlung auch die Menge des auf der Allmende weidenden Viehs und versteigerte das Recht an die Meistbietenden.

McCloskey (1976) interpretiert dieses System als eine Institution zur Risikominimierung. Es gibt eine Vielzahl von Eigenschaften des Bodens, die mit Risiken verbunden sind, welche kleinräumig auftreten können, wie stauende Nässe, Neigung des Landes, Bodenstruktur, Auswirkungen von Frost, Sonne und Wind. So verträgt etwa ein sandiger, geneigter Boden starke Niederschläge, während lehmiger Boden in einer Niederung Feuchtigkeit gut hält und damit weniger gegen Trockenheit anfällig ist. Auf einem dem Wind ausgesetzten Gelände legt sich das Getreide nieder, wenn es während der Erntezeit regnet und stark windig ist; dagegen besteht hier wenig Gefahr von stauender Nässe oder Fäulnis. Umgekehrt droht windgeschütztem Gelände wenig Gefahr durch Sturmschäden, wohl aber durch Fäulnis bei Feuchtigkeit. Viele Gefahren können kleinräumig eintreten: Überschwemmung, Feuer, Insekten, Vögel, Mehltau, Kaninchen, Maulwürfe, Felddiebe, Hagel, durchziehende Armeen. Die spezifischen Kosten des Dreifeldersystems, zu denen auch der Abstimmungsaufwand in den Gemeinden gehört, können daher als Versicherungsprämie in einer Umwelt mit niedrigen und schwankenden landwirtschaftlichen Erträgen verstanden werden.

Im Unterschied zu den klassischen Zonen der Bewässerungslandwirtschaft handelte es sich in Nordwesteuropa um Regenlandwirtschaft, bei der eher das Problem der Entwässerung bestand. Die feuchten Sommer ermöglichen die Nutzung von Wiesen, d. h. die Verfügung über Flächen, die regelmäßig gemäht werden konnten, so dass einfaches Viehfutter für den Winter bereitstand. In Kombination mit der Weide (Brache, Marginalböden, Stoppelfelder) gestattete diese extensive Landnutzungsform die Haltung eines hohen Anteils von Großvieh in der Landwirtschaft, vor allem Rinder und Pferde, die als Zug- und Reittiere genutzt werden konnten.

Der schwere Räderpflug mit Streichbrett kam bereits im frühen Mittelalter auf und setzte sich seit dem 11. Jahrhundert durch. Er bildet die Voraussetzung für die Bearbeitung schwerer, nasser Böden (Schwemmland). Er erforderte größere Gespanne, das Wenden wurde schwieriger, so dass man die Parzellen verlängerte. Der schwere Pflug mit Gespannen war eine teure Investition, die nur von größeren Gütern oder Genossenschaften aufgebracht werden konnte. Der Pflug wurde mit der Egge kombiniert, die den gewendeten, umgebrochenen Boden glättete und die Saat mit Erde bedeckte.

Die europäische Landwirtschaft beruhte somit auf der Nutzung von Extensivflächen, was eine dauerhafte Haltung von Nutztieren gestattete, aber auch Nährstofftransfer von den Extensiv- zu den Intensivflächen ermöglichte. Daher wurde hier die Tendenz zur Re-Hortikulturalisierung gebrochen, die im Nahen Osten und in China zu beobachten war. Dadurch bestanden Mechanisierungspotentiale. Das Transportwesen etwa konzentriert sich auf Karren und Wagen, auch wenn es bis ins 18. Jahrhundert noch immer einen großen Anteil von Packtieren gab (Esel, Maultiere), die so gut wie keine Infrastrukturanforderungen stellten.

Rolf Peter Sieferle

Die europäische Landwirtschaft war also in starkem Maße mit der Nutzung von Tieren verbunden. Diese erfüllten eine Reihe von Funktionen:

– Lieferung von mechanischer Energie, Zugkraft;
– Nahrung (Fleisch und Milch);
– gewerbliche Rohstoffe (Leder, Knochen, Horn);
– Transfer von Nährstoffen aus Extensivflächen;
– Umwandlung von Biomasse, die der Mensch nicht direkt nutzen kann (Stoppelfelder, organische Abfälle).

Der Gebrauch von Tieren machte es möglich, dass Marginalflächen genutzt wurden, die für den direkten (hortikulturellen) Gebrauch untauglich waren. Tiere traten nicht immer als Nahrungskonkurrenten des Menschen auf, nämlich dann, wenn sie mit Pflanzenresten (Stoppelfelder) oder Abfällen (Schweine) gefüttert wurden. Dies hatte aber zur Folge, dass die europäische Landwirtschaft bezogen auf die gesamte landwirtschaftliche Nutzfläche relativ geringe Hektarerträge hatte.

Tab. 2 Flächenproduktivität in Europa und China (kg Getreide pro Hektar). Quelle: HELBLING 2003

	1750	1850
England + Wales	1570	2250
Österreich	800	1000
China: 4 Provinzen mit der geringsten Produktivität	770	970
China: 4 Provinzen mit der höchsten Produktivität	2800	3600

Die europäische Agrargesellschaft, die im frühen Mittelalter entstanden ist, hatte einen Bestand von über 1000 Jahren, mit recht geringfügigen Modifikationen des Grundmusters. Das „vergetreidete" Europa mit seinen zentralen Merkmalen Dreifelderwirtschaft, hohem Viehanteil, recht extensiver Landwirtschaft, Grundherrschaft, dezentraler Herrschaftsverteilung bildet wahrscheinlich eine historische Voraussetzung für die industrielle Transformation, doch war diese nicht zwangsläufig in ihm angelegt. Das alte agrarische Europa bildete vielmehr ein recht stabiles Muster, das immerhin mehr als 1000 Jahre alt wurde und erst in einer recht „unwahrscheinlichen" Kombination von Einflussgrößen seit dem 18. Jahrhundert verschwunden ist.

Dennoch scheint es plausibel, dass eine Reihe von Merkmalen, die der nordwesteuropäischen agrarischen Zivilisation zugerechnet werden, wichtige, wenn nicht unabdingbare Voraussetzungen für den Weg in die Industrialisierung bildeten. Das agrarische alte Europa war nicht zwangsläufig selbsttranszendierend, d. h. auf seine eigene Auflösung angelegt, sondern scheint ein recht stabiles Muster gebildet zu haben. Dennoch besaß es Merkmale, die als präadaptiv für die Industrialisierung gelten können. Einige dieser Merkmale hängen direkt mit der Organisation der Landwirtschaft unter den spezifischen Umweltbedingungen Nordwesteuropas zusammen.

Der schwere Pflug erforderte starke Pferde (oder Ochsen). Solche Pferde mussten erst gezüchtet werden, doch bildeten sie dann die Basis der gepanzerten Ritterheere. Diese aber waren (vielleicht als einzige Militärmacht des Mittelalters) der Invasion von leichten Reitereien aus Zentralasien gewachsen – ein Grund, weshalb Europa im Gegensatz zu China, Indien, Persien, Mesopotamien, Anatolien etc. in den letzten 1000 Jahren nicht mehr von außen erobert wurde. Zudem begünstigte das schwere Schlachtross eine dezentrale Verteilung des Militärs im Sinne des „Feudalismus", was vielleicht eine Wurzel für die europäische „Fragmentierung", also die Abwesenheit einer despotischen Zentrale war. Hinzu kommt, dass das schwere Pferd die Basis für den Überlandtransport mit Wagen bildet, den es weder in China noch im Nahen Osten gegeben hat. Die Transportrevolution, die sich daraus dann in der Neuzeit entwickelt hat, war fraglos ein Aspekt der industriellen Transformation, und es handelte sich hierbei um recht „zähe" Voraussetzungen, die nicht nach Belieben geschaffen werden konnten (POPPLOW 2008).

Schließlich gab es Folgen für die technische Entwicklung: Die Kombination von Brotbacken und Verfügung über fließendes Wasser mit ausreichendem Gefälle begünstigte die Ausbreitung der Wassermühlen, deren Ubiquität eine Voraussetzung für zahlreiche Innovationen war und aus denen sich die gesamte Räder- und Maschinentechnik seit dem Mittelalter entwickelte. Wir sollten aber nicht übersehen, dass der technische Innovationsprozess, der mit der Mühlenmechanik verbunden war, bereits im späten Mittelalter zuende ging. Im technologischen Sinne war die Zeit zwischen 1500 und 1750 eine Stagnationsperiode, in der es nur minimale technische Verbesserungen gab (abgesehen von der Dampfpumpe). Dies spricht dafür, dass es eine Sonderentwicklung des agrarischen Europa gegeben hat, die im späten Mittelalter bereits ihre Reife erreicht hatte. Die industrielle Transformation seit dem 18. Jahrhundert war dann ein ganz anderer, neuartiger Prozess.

Die europäische Landwirtschaft beruhte primär auf der Nutzung von Getreide, kombiniert mit Viehwirtschaft. Künstliche Bewässerung war eher die Ausnahme. In der Regel verließ man sich auf natürliche Niederschläge. Diesem Typus der Landwirtschaft lag die Mechanisierung, d. h. der Einsatz von Arbeitstieren und mechanischen Geräten nahe. Ein Getreidefeld bildet im Idealfall eine homogene Pflanzengesellschaft, die synchron wächst und geerntet werden kann. Hier bot sich daher schon früh der Einsatz von Geräten wie Pflug oder Egge an, woran sich der Einsatz von Maschinen zur Ernte und zur Weiterverarbeitung (Dreschen) anschließen konnte. Wie leicht sich die Getreidelandwirtschaft mechanisieren lässt, wird im Vergleich zum Garten- oder Weinbau besonders deutlich. Hortikulturelle Anbauformen (mit der Hacke, dem Spaten und viel Handarbeit) widersetzten sich der Mechanisierung, und es ist kein Zufall, dass in diesen Bereichen erst spät im 20. Jahrhundert in beschränktem Umfang Arbeitsmaschinen eingesetzt wurden. Hier mag ein wichtiger Unterschied zum Nassreisanbau liegen, wie er im südlichen China verbreitet war. Ähnliches gilt auch für den Anbau von Knollenfrüchten. Der Weg zur Mechanisierung ist unter diesen Bedingungen wesentlich weiter als beim Getreideanbau.

Ein wichtiges Merkmal der europäischen landwirtschaftlichen Revolution lag schließlich darin, dass mit ihr ein Pfad eingeschlagen wurde, der zur Steigerung der Arbeitsproduktivität führte. Dies ist vor allem in Amerika (allerdings erst seit dem 19. Jahrhundert) zu beobachten, wo große Flächen zur Verfügung standen, während Arbeitskräfte tendenziell knapp waren.

2. Naturkatastrophen

Vor allem Eric JONES (1987) hat auf die prägende Rolle unterschiedlicher objektiver Gefahrenniveaus in verschiedenen geographischen Räumen hingewiesen. JONES geht es um die Erklärung der Tatsache, weshalb die Industrialisierung mit ihrem charakteristischen wirtschaftlichen Wachstum ausgerechnet in Europa einsetzte und nicht etwa in China, das ja die ältere agrari-

Rolf Peter Sieferle

sche Zivilisation war und auf eine lange Geschichte der landwirtschaftlichen und gewerblichen Produktion zurückblicken konnte, mit ausgesprochenen Blütezeiten rascher technischer und ökonomischer Innovation, etwa zur Zeit der Song- oder der frühen Ming-Dynastie.

Diese unterschiedliche historische Entwicklung will er auf differentielle Unsicherheitsinzidenz in Europa und in China zurückführen, die ihrerseits unterschiedliche Strategien der Ruinvermeidung, also unterschiedliche Formen der Gefahrenbewältigung, provoziert hat. Die Formen der Risikoverarbeitung werden daher als Adaptationen an unterschiedliche objektive Unsicherheitslagen interpretiert, was in mehrerer Hinsicht problematisch ist: Einerseits wird die Unsicherheit auf Gefahren reduziert, also die Dimension der Chancen weitgehend ignoriert; andererseits wird der Komplex der aktiven Gefahrenbewältigung im Sinne der Ausprägung unterschiedlicher Risikostrategien unterschätzt und nicht zwischen primären und sekundären Risiken unterschieden, so dass auch das Problem der Bewältigungstrajektorien wie der (unterschiedlichen) Risikospiralen nicht ins Visier gerät (vgl. SIEFERLE 2006).

Trotz dieser Beschränkung öffnet JONES eine wichtige Dimension historischer Interpretation, so dass sich eine nähere Beschäftigung mit seinen Thesen lohnt. Als ersten groben Indikator für Unterschiede der Unsicherheitsinzidenz nimmt er das Auftreten von tödlichen (Natur-) Katastrophen, also von schädlichen Extremereignissen (*natural disasters*), wohl unter anderem auch deshalb, weil diese am ehesten dokumentiert und zu quantifizieren sind. Er ist sich aber dessen bewusst, dass Naturkatastrophen physische und ökonomische Dimensionen zugleich haben, d. h. dass sich ihre Bedeutung mit der Nutzungsform des betroffenen Gebietes ändert.

Für die Analyse einzelner Gesellschaften kann es sinnvoll sein, von der Existenz unterschiedlicher Gefahren von Naturkatastrophen auszugehen. Generell wäre das folgende Muster zu erwarten: Angenommen, das allgemeine Gefahrenniveau bzw. die zu erwartende Schwankungsbreite vitaler Naturprozesse ist in zwei Gesellschaften, die unter unterschiedlichen geographisch-ökologischen Bedingungen leben, unterschiedlich groß. In diesem Fall wäre zu erwarten, dass beide Gesellschaften unterschiedliche Methoden der Ruinvermeidung entwickeln, was bedeuten kann, dass sie sich auf unterschiedliche Entwicklungstrajektorien begeben, da die jeweiligen Formen des Risikoverhaltens unterschiedliche Risikospiralen in Gang setzen. Allerdings könnte daraus nicht der Schluss gezogen werden, dass die jeweilige soziale Strategie von der objektiven Risikoinzidenz determiniert ist, sondern es wäre damit zu rechnen, dass es eine Reihe von funktionsäquivalenten Reaktionsmustern auf vergleichbare Unsicherheitslagen gibt.

Die objektive, naturale Risikoinzidenz kann daher nur ein einzelner Erklärungsfaktor für historische Prozesse und Strukturen sein. Im konkreten Fall handelt es sich hierbei jedoch um ein schwieriges Problem, da Unsicherheitsinzidenz und Strategien zur Ruinvermeidung eng miteinander verwoben sind und unter Bedingungen der Landwirtschaft primäre Gefahren kaum von sekundären Risiken zu unterscheiden sind. Beobachtete Katastrophenereignisse wie etwa Hungersnöte können daher naturale wie kulturelle Dimensionen besitzen, und es dürfte analytisch schwerfallen, beide Elemente voneinander zu isolieren, wenn sich bereits eine Risikospirale aufgebaut hat.

Bereits Henry Thomas BUCKLE (1857) hat die Vermutung geäußert, die natürliche Umwelt in Europa sei weniger katastrophenträchtig als in Asien, was er als einen Grund dafür ansah, dass sich in Europa eine stabile technisch-ökonomisch ausgerichtete Zivilisation ausprägen konnte. In der Tat ergibt eine Auflistung überlieferter großer Naturkatastrophen ein Bild, das diese Meinung bestätigen kann (Tab. 3).

Tab. 3 Todesopfer großer einzelner Naturkatastrophen, 1000–1800[4]

Jahr	Art	Ort	Opferzahl
1556	Erdbeben	China	830 000
1737	Orkan	Indien	300 000
1662	Erdbeben	China	300 000
1642	Überschwemmung	China	300 000
1693	Erdbeben	Italien	153 000
1730	Erdbeben	Japan	137 000
1290	Erdbeben	China	100 000
1099	Überschwemmung	England/Niederlande	100 000
1730	Erdbeben	China	100 000
1667	Erdbeben	Kaukasus	80 000
1727	Erdbeben	Iran	77 000
1268	Erdbeben	Kleinasien	60 000
1755	Erdbeben/Tsunami	Portugal	50 000
1287	Überschwemmung	Niederlande	50 000
1707	Tsunami	Japan	30 000
1780	Orkan	Karibik	20 000
1669	Vulkanausbruch	Italien	20 000
1783	Vulkanausbruch	Island	10 000
1421	Überschwemmung	Niederlande	10 000
1618	Lawine	Schweiz	1500

Hierbei sollte nicht übersehen werden, dass für die Räume außerhalb Europas (wohl mit Ausnahme Chinas) größere Überlieferungslücken bestehen als für Europa selbst. Berücksichtigt man besser dokumentierte Zahlen seit dem 19. Jahrhundert, so werden die Unterschiede zwischen Asien und Europa deutlicher, was natürlich auch auf den wachsenden Bevölkerungsanteil Asiens zurückzuführen ist. Grundsätzlich wäre jedenfalls zu erwarten, dass bei solch unterschiedlicher Gefahreninzidenz unterschiedliche Methoden des Umgangs mit Gefahren entwickelt worden sind, sei es im Sinne der Prävention, sei es im Sinne der Schadensverarbeitung. Wenn Risikostrategien im adaptiven Sinn verstanden werden können, läge eine Prämie darauf, diese Gefahren in dem Sinne in Risiken zu verwandeln, dass ein verstetigender kultureller Umgang mit ihnen entwickelt wird.

Ein natürliches Extremereignis wird erst dann und insofern zur historisch relevanten Katastrophe, wenn das betreffende Gebiet von Menschen genutzt wird. Rein natürliche Katastrophen sind dagegen geschichtlich belanglos. Das größte historisch gesicherte Extremereignis war wohl der Impakt, der am 30. Juni 1908 in Tunguska, Sibirien, einschlug. Es hat sich vermutlich um einen Meteoriten oder Asteroiden gehandelt. Mehrere hundert Quadratkilometer Tundra wurden vernichtet, doch ist nicht bekannt, ob dabei Menschen zu Schaden kamen. In der gewöhnlichen historischen Erinnerung kommt dieses Ereignis nicht vor. Es wäre jedoch denkbar gewesen, dass der Einschlag in dichter bewohnten Gebieten stattgefunden hätte. Dann wäre mit mehreren Millionen Toten zu rechnen gewesen, es hätte sich also um den größten Unglücksfall der menschlichen Geschichte gehandelt.

---

4 Zahlen nach CORNELL 1982 sowie LATTER 1968/1969, Tab. 4.

Generell lassen sich in Anlehnung an JONES die folgenden Typen von Naturkatastrophen im Sinne von schädlichen Extremereignissen nennen, auf die im Folgenden näher eingegangen werden soll:

– Geophysische Katastrophen: Erdbeben, Vulkanausbrüche, Bergrutsche, seismische Wogen (Tsunami), evtl. Einschläge von Himmelskörpern (Impakte).
– Klimatische Katastrophen: Orkane, Sturmfluten, Hagelschlag, Lawinen, Überschwemmungen, Dürren, Waldbrände.
– Biologische Katastrophen: Epidemien, Viehseuchen, Pflanzenkrankheiten, Heuschrecken.

Als vierter Komplex können rein soziogene Katastrophen genannt werden, wie Kriege, innere Unruhen, Feuersbrünste in Siedlungen, gewerbliche Unfälle. Hierauf soll hier aber nicht näher eingegangen werden.

In unserem Kontext stellt sich die Frage, ob und wie weit es in Europa (im Verhältnis zu anderen Räumen agrarischer Zivilisationen) eine spezifische Wahrscheinlichkeit von Naturkatastrophen gab. In einem zweiten Schritt wäre dann zu fragen, ob und wie weit sich in verschiedenen agrarischen Zivilisationen unterschiedliche Trajektorien der Risikobewältigung (im Sinne von Risikospiralen) aufgebaut haben, wobei zu unterscheiden wäre, was daran im adaptiven Sinne, und was im autopoietischen Sinne als kulturspezifische Tradition zu verstehen ist.

Was geophysische Katastrophen betrifft, wird in der Literatur vor allem die Rolle von Erdbeben als historischer Faktor thematisiert, wobei man dazu neigt, ihren Einfluss auf die Geschichte eher zu überschätzen, da sie so spektakulär sind. Schätzungsweise gibt es global jährlich etwa 16 000 potentiell schädliche Erdbeben, vor allem in unbewohnten Gebieten oder unter dem Meeresspiegel. Spürbare Schäden werden nur von 0,16 % der Beben angerichtet. Die meisten Erdbeben finden am Rand des Pazifiks statt, doch konzentrieren sich die schädlichen Beben auf eine Zone, die vom Mittelmeer bis zum Himalaya reicht. Dies hat geologische und soziale Ursachen: Hier liegen die Epizentren nahe an der Erdoberfläche, und dieses Gebiet ist besonders dicht besiedelt. Zwischen 1949 und 1969 fanden nur etwa 1 % der potentiell schädlichen Erdbeben im Iran statt, dagegen gab es dort nicht weniger als 41 % der Erdbebenopfer. Insgesamt schwankt die Zahl der Erdbebenopfer in den Zeiträumen, für die genauere Daten existieren, ganz beträchtlich. 1926 – 1950 gab es 351 914 Tote (= 14 000 im Jahr), 1951 – 1968 waren es 67 507 Tote (= 3750 im Jahr). Auch sind die Unterschiede zwischen dem Jahr 1932 mit der größten (78 000) und 1937 mit der kleinsten Opferzahl (77) erheblich (LATTER 1968/1969). Sicherlich wäre es sinnvoll, sich nicht auf die ganz großen Katastrophen zu konzentrieren, sondern auch Fälle mit geringerem Schadensausmaß zu berücksichtigen. Allerdings ist es schwierig, aussagekräftige historische Daten für kleinere Schadensereignisse zu erhalten.

Es ist zweifelhaft, ob Erdbeben ganze Kulturen vernichtet haben können. Vulkanausbrüchen wird eine größere Wirkung zugeschrieben. Sie ereignen sich weit seltener als Erdbeben, finden in eng begrenzten Gebieten statt und sind im Gegensatz zu Erdbeben aufgrund ihrer Emissionen (Lava, Asche) historisch rekonstruierbar. Erdbeben und Eruptionen, die unter dem Meeresspiegel oder in Meeresnähe stattfinden, können große seismische Wogen (Tsunami) verursachen, die die eigentlichen Schäden bewirken. Der Ausbruch des Krakatau 1883 hat eine 41 m hohe Woge erzeugt, und in der Ägäis sind Spuren einer Woge von etwa 250 m Höhe archäologisch nachgewiesen, die vielleicht um 1400 v. Chr. zum Untergang des minoischen Kreta geführt hat.

Tab. 4 Gesamtzahl der Opfer von Erdbeben im letzten Jahrtausend[5]

Gebiet	Anzahl der Opfer
China	2 030 000
Indien	300 000
Europa	193 000
Naher Osten	217 000
Japan	280 000

Trotz aller Überlieferungsprobleme und trotz der unterschiedlichen Bevölkerungsdichte scheint also die Inzidenz von Erdbeben in Asien größer gewesen zu sein als in Europa. Europa hatte in der Neuzeit etwa 20 % der Bevölkerung Eurasiens. Zu erwarten wären also etwa 600 000 von rund 3 Millionen Toten gewesen.[6] Stattdessen hatte Europa nur etwa ein Drittel der zu erwartenden Opferzahl zu beklagen. Hinzu kommt, dass sich die europäischen Erdbebenschäden auf Südeuropa konzentrieren. In Italien sind für den Zeitraum zwischen 464 v. Chr. und 1980 nicht weniger als 350 Erdbeben quellenmässig nachgewiesen. Nach 1300 gab es zwischen 207 und 225 Erdbeben in Italien, also im Schnitt alle drei Jahre eines (GUIDOBONI 1994, BOSCHI 1995). Auch in Spanien und Griechenland sind zahlreiche Erdbeben dokumentiert, doch ist die genaue Zahl im gesamten Mittelmeerraum nicht bekannt (QUESADA 1999, PAPAZACHOS und PAPAZACHOU 1997).

Im nördlichen Europa dagegen hat es in den letzten Jahrhunderten so gut wie keine nennenswerten Erdbeben gegeben. Basel wurde am 18. 10. 1356 von einem Erdbeben zerstört, mit etwa 300 Toten (BORST 1981, BARTLOME und FLÜCKIGER 1999). Auch die Opferzahlen anderer Erdbeben in Mitteleuropa blieben marginal (HERRMANN 1936, S. 135f.): 1601 in Nidwalden 8 Tote, 1689 in Innsbruck und Mürzzuschlag jeweils 19 Tote, 1756 bei Aachen ein Toter, vergleicht man sie mit Indien (1618 in Bombay 2000 Tote, 1819 in Katsch 3000 Tote, 1828 in Kaschmir 1000 Tote, 1905 im Kangratal 19 000 Tote) oder gar China (1556 etwa 80 000 Tote, 1662 und 1731 etwa 400 000 Tote, 1920 etwa 5000 Tote).

Erdbeben oder Vulkanausbrüche verursachen nicht nur Personenschäden (wie dies etwa bei Epidemien der Fall ist), sondern sie zerstören auch bauliche Anlagen. Auf den ersten Blick könnte dies bei Agrargesellschaften keine besondere Rolle gespielt haben, doch sind hierbei die Unterschiede zwischen verschiedenen Typen der Landwirtschaft von Bedeutung. Die in Europa, besonders in Nordwesteuropa dominante Form der relativ extensiven, dezentralen Regenlandwirtschaft nimmt vergleichsweise geringe Eingriffe in die Agrarlandschaft vor. Die Felder und Weiden sind gegenüber äußeren Schocks sehr elastisch, und nach einer gravierenden Störung kann die Produktion bald wieder aufgenommen werden. Anders ist dies bei Bewässerungslandwirtschaft oder bei Landwirtschaft in Flusstälern, wo fließendes Wasser eingedämmt wird. Wenn durch ein Erdbeben Staudämme, Bewässerungskanäle oder

---

5 Nach LATTER 1968/1969, Tab. 4. LATTER zählt nur die Erdbeben, die mindestens 785 000 Opfer gefordert haben, das ist die Summe der Erdbebenopfer zwischen 1949 und 1969. Allerdings gab es in Europa nur ein einziges Erdbeben dieser Größenordnung (Messina 1908 mit 83 000 Toten). Das berühmte Erdbeben von Lissabon (1755) forderte etwa 50 000 Opfer, viele davon durch einen Tsunami. Für die älteren Beben ist eine solche Beschränkung auf große Zahlen sinnvoll, da ohnehin nur sehr pauschale Daten zur Verfügung stehen.
6 Die Gesamtzahl der Erdbebenopfer der letzten 1000 Jahre liegt laut LATTER (1968/1969, S. 362) bei 5 Millionen. Die Zahl von 3 Millionen betrifft nur die 16 größten Beben.

Aquädukte zerstört werden, dauert es Jahre, bis die Schäden wieder behoben sind. Die chinesische Landwirtschaft musste daher nicht nur öfter mit Erdbeben rechnen, die von ihnen verursachten Schäden an den hydraulischen Anlagen bildeten auch einen länger anhaltenden Rückschlag der Produktion, und ihre Reparatur erforderte größere Aufwendungen. Ähnliche Probleme konnten in Nordwesteuropa bei Sturmfluten und Deichbrüchen auftreten (Jakubowski-Tiessen 1992). Prävention mag bei Erdbeben als plötzlichen, unvorhersehbaren Ereignissen sehr schwierig sein, wenn es auch bauliche Maßnahmen gibt, die zumindest die Schäden verringern. Diese sind aber auf jeden Fall mit höheren Kosten verbunden, die hier als Versicherungsprämien angesehen werden können.

Schon Buckle (1857) hatte bei seinem Vergleich zwischen Europa und Asien nicht so sehr die spektakulären Erdbeben und Vulkanausbrüche im Auge, sondern extreme Klimaereignisse wie Dürren oder niederschlagsbedingte Überschwemmungen, deren Folgen in der Gestalt von Hungersnöten spürbar wurden, und auch Anderson und Jones (1988) beziehen sich vorwiegend auf klimatische Variationen. In Abhebung von älteren umweltdeterministischen Klimatheorien vom Schlage Ellsworth Huntingtons (1915) möchten sie nicht so sehr den kleineren kontinuierlichen Klimawandel betrachten, sondern klimatische Extremereignisse, die als externe Schocks wirken, auf welche die betroffenen Gesellschaften reagieren müssen.

Die Grundvermutung zielt dahin, dass es in Asien größere Klimaschwankungen als in den gemäßigten Zonen gibt, so dass es häufiger zu Klimaextremen wie Dürren, heftigen Niederschlägen mit Überschwemmungen sowie zu Wirbelstürmen kommt. Sollte dies der Fall sein, so wäre es dort schwieriger, sich dauerhaft einem bestimmten Durchschnittswert der landwirtschaftlichen Produktion zu nähern.[7] Klimaschwankungen sollten eine umso größere ökonomische Rolle spielen, je homogener die betroffenen Gebiete ökologisch sind. Wenn wie in Europa zahlreiche differenzierte Kleinklima-Zonen existieren, sind weitaus seltener Schwankungen zu erwarten, die flächendeckend in die gleiche Richtung gehen, als etwa auf dem indischen Subkontinent oder in China, wo Klimaextreme große Gebiete betreffen. Hinzu kommt, dass sich die Ausweichmöglichkeiten durch Transport von Nahrungsmitteln unterscheiden. Der auf klimabedingte Missernten zurückgehende Hunger ist daher sowohl ein Ergebnis der Abhängigkeit von unkontrollierbaren Naturfaktoren wie auch der beschränkten Möglichkeiten zur Portfolio-Bildung durch Ferntransport.

Witterungsbedingte Missernten und darauf folgende Hungersnöte sind der agrarischen Produktionsweise prinzipiell inhärent. Sie bilden klassische „sekundäre Gefahren", die aus der primären Gefahrenbewältigung durch Vorratshaltung und aktive Nahrungsproduktion resultieren. Da die Landwirtschaft das Nahrungsspektrum drastisch einschränkt und zugleich die Bevölkerungsdichte steigert, geraten die Menschen in Abhängigkeit von den Wachstumsbedingungen der wenigen favorisierten Pflanzen, vor allem von Gräsern (Getreide), Mais, Reis und Knollenfrüchten, und können im Notfall kaum mehr auf andere Nahrungsmittel ausweichen. Bauern versuchen zwar, diese Wachstumsbedingungen ihrer Nutzpflanzen zu kontrollieren (Rodung, Bewässerung, Entwässerung, Unkrautbekämpfung, Düngung), doch findet diese Kontrolle an Klimaschwankungen eine natürliche Grenze. Die agrarische Naturbeherrschung durch Kolonisierung schlägt daher in eine sekundäre Abhängigkeit von Naturfaktoren um.

---

7 Hinweise zu Klimaextremen in China finden sich bei Wang Shao-Wu and Zhao Zong-Ci 1981. Die Autoren enthalten sich jedoch weitergehender Schlussfolgerungen.

Hungersnöte in agrarischen Zivilisationen sind historisch vielfach überliefert, doch gibt es wenig genaueres Material über Ursachen und Zahl der Opfer. In China soll es nach Jones (1987) zwischen 108 v. Chr. und 1911 mindestens 1818 Hungersnöte aufgrund von Dürren oder Überschwemmungen gegeben haben, also fast jedes Jahr eine, jedoch in verschiedenen Provinzen. Auch aus Indien sind spektakuläre Fälle überliefert, wie die Hungersnot in Bengalen von 1769/1770 mit 10 Millionen Toten – das ist ein Drittel der Gesamtbevölkerung.

Freilich gab es auch in Europa Hungersnöte von nicht geringem Ausmaß, und auch von klimatischen Extremereignissen blieb der Kontinent nicht verschont (Pfister 1999, Glaser 2001). Es wurde sogar bezweifelt, ob die Fluktuation der Niederschläge in Asien und Europa tatsächlich signifikant verschieden ist. Rezente Daten zeigen jedenfalls keine großen Unterschiede (Pryor 1985). Immerhin waren auch in Europa die Opferzahlen aufgrund von Hungersnöten beträchtlich. In Ostpreußen sollen 1708/11 nicht weniger als 250 000 Personen verhungert sein, das sind 41 % der Bevölkerung. In Frankreich fielen 1692/94 etwa 2 Millionen einer Hungersnot zum Opfer, das sind etwa 10 % der Bevölkerung. In Finnland soll 1696/97 ein Viertel bis ein Drittel der Bevölkerung verhungert sein. Laut Braudel (1974, S. 39) gab es in Frankreich vom 10. bis 18. Jahrhundert regelmäßige Hungersnöte, deren Häufigkeit allerdings zwischen 2 im 12. Jahrhundert und 26 im 11. Jahrhundert stark schwankte. In der Regel konnte man aber alle 7–10 Jahre mit einer gravierenden Hungersnot rechnen, die auf witterungsbedingte Ernteausfälle zurückging. In Europa verschwand der Hunger erst im 19. Jahrhundert durch den Anbau neuer Feldfrüchte (Kartoffeln), höhere Erträge und verbesserte Transportmöglichkeiten mit der Eisenbahn (Abel 1978, Montanari 1993), doch sollte nicht vergessen werden, dass die Hungersnot in Irland 1845/1846 noch einmal Millionen von Todesopfern forderte.

Ein Grundproblem besteht nun allerdings darin, dass das Auftreten von Hungersnöten nicht nur Ausdruck der klimatisch-ökologischen Situation ist, sondern auch soziale und politische Aspekte besitzt. Die Klimaschwankung ist nur ein Element, weitere Faktoren werden aber von ihrer sozialen und ökonomischen Verarbeitung bestimmt, also vor allem von der Organisation von Vorratshaltung, Handel und Verteilung. Dies gilt gerade für agrarische Zivilisationen jenseits der Schwelle der Subsistenzwirtschaft, wo Fernhandel, Vorratsbildung und die Möglichkeiten der räumlichen Portfoliobildung eine große Rolle spielen. Hier ist auf den ersten Blick nicht zu entscheiden, ob eine reale Hungersnot primär auf Missernten oder auf Missmanagement zurückgeht. Dieses methodische Problem erschwert jedenfalls den Umgang mit den Daten erheblich.

Es gab eine Reihe von Möglichkeiten, auf diese neue Abhängigkeit zu reagieren: Die Menschen konnten technische Massnahmen ergreifen, konnten Methoden der Pufferung bzw. der sozialen Verarbeitung von Schäden entwickeln, doch ist auch an weiterreichende Folgen zu denken, wenn nämlich bestimmte (soziale oder ökonomische) Entwicklungspfade begünstigt bzw. Alternativen vermieden wurden. Einige Züge asiatischer Gesellschaften können vielleicht aus dem erhöhten Zwang zur Risikominimierung erklärt werden. Jones gibt dafür einige Beispiele. So wurden in Indien große Flächen mit ertragsarmen, aber trockenheitsresistenten Getreidearten bebaut, was als Vorsorge, falls der Monsun ausfällt, gedeutet werden kann. Die der Natur gezahlten „Versicherungsprämien" waren hier höher als in Europa. Dies hatte aber zur Folge, dass der Durchschnittsertrag niedriger war, so dass geringere Spielräume für Innovation, aber auch zur Surplusabschöpfung existierten. Eine ähnliche Funktion könnte die in Indien verbreitete Praxis gehabt haben, Kühe besser als Ochsen zu versorgen, obwohl diese die Feldarbeit leisteten. Dies könnte als Versuch interpretiert werden, auf jeden

Fall mit den Kühen das fruchtbare Potential der Arbeitstiere zu erhalten, auch wenn dies zu Lasten der Zugkraft ging.

Die größte Naturkatastrophe, die Europa je getroffen hat, war der Schwarze Tod, der in der Mitte des 14. Jahrhunderts zum ersten Mal auftrat und danach in kurzen Abständen immer wiederkehrte. In manchen Gebieten Englands und Italiens ist die Bevölkerung um 70–80 % gefallen, und HERLIHY (1997) schätzt, dass die europäische Gesamtbevölkerung um 1420 etwa auf ein Drittel der Bevölkerung von 1320 zurückgegangen ist.

Es gibt Indizien dafür, dass die Pest auch in China ausgebrochen ist. Für 1331 ist ein großes Seuchenereignis dokumentiert. Schließlich kam es im fraglichen Zeitraum zu einem beträchtlichen Bevölkerungsrückgang, von ca. 123 Mio. um 1200 auf ca. 65 Mio. im Jahre 1393 – also praktisch eine Halbierung. Dennoch ist die Ursache (oder sogar das Faktum) dieses Bevölkerungszusammenbruchs ungeklärt. Einwände gehen dahin, dass es sich um geschönte Zahlen handelt, die gezielt zum Zweck der Steuerhinterziehung fabriziert wurden (vgl. zum Datenproblem KOLB 2003).

Die seuchenbedingte Sterblichkeit war in Europa längerfristig vermutlich nicht geringer als in Asien. JONES (1987) sieht den wirksamen Unterschied zwischen Europa und Asien daher auch weniger in der absoluten Sterblichkeit als in den Kollateralschäden von Naturkatastrophen: Seuchen lassen im Gegensatz zu Erdbeben Sachanlagen intakt, so dass sie das Wirtschaftsleben nicht nachhaltiger stören. Wenn daher Massensterben in Europa eher durch Seuchen als durch Zerstörungen verursacht wurden, könnte dies ein Grund dafür sein, weshalb die Wirtschaftsentwicklung in Europa stabiler war in Asien.

Die Identifikation unterschiedlicher Neigung zu Naturkatastrophen in Asien und Europa ist, wie sich gezeigt hat, nicht unproblematisch, und am ehesten dürfte diese Vermutung für große Naturkatastrophen zutreffen, vielleicht auch für Hungersnöte, in die aber auf jeden Fall ein organisatorisches (soziales bzw. politisches) Element eingeht. Ein Indikator für unterschiedliche Gefahrenniveaus könnte in der Fluktuation der Bevölkerungsgröße liegen: Volatilere Umweltbedingungen müssten sich in einer größeren demographischen Volatilität niederschlagen. Die folgende Figur kann vielleicht in diesem Sinne interpretiert werden (Abb. 1).

Die höhere Volatilität der chinesischen im Verhältnis zur europäischen Bevölkerung kann zunächst als unmittelbarer Ausdruck unterschiedlicher Gefahreninzidenz interpretiert werden, also als Ergebnis unterschiedlicher Sterblichkeit aufgrund von Extremereignissen. Hinzu kommt aber ein zweites Argument, das auf unterschiedliches Fertilitätsverhalten zielt. In der Biologie unterscheidet man zwischen r- und K-Strategien der Vermehrung. Die erstere Strategie besteht darin, große Mengen von Nachwuchs zu produzieren, ohne sich weiter um deren Schicksal zu kümmern. Hier investiert die Mutter in das einzelne Exemplar nur sehr wenig Ressourcen, was aber durch die große Zahl kompensiert wird. Im Endeffekt reicht es ja, wenn sich nur zwei Nachkommen behaupten können, denn damit ist gewährleistet, dass die Gene, die diese Strategie steuern, auch in der Zukunftspopulation repräsentiert sind. Die K-Strategie dagegen zielt darauf, relativ wenig Nachwuchs zu produzieren, dafür aber in den einzelnen relativ viel (Ressourcen oder Zeit) zu investieren. Hierbei handelt es sich um ein Verhalten, das unsere Spezies *Homo sapiens* (im Gegensatz zu anderen Säugetieren wie etwa Kaninchen oder Mäusen) auszeichnet. Prinzipiell bilden aber beide Grundstrategien die Extreme eines Kontinuums, wobei sich also im Einzelfall eher r- bzw. K-orientierte Varianten unterscheiden lassen.

Es ist nun grundsätzlich möglich, auch bei einzelnen menschlichen Populationen r- bzw. K-Strategien der Vermehrung zu unterscheiden und deren Vorkommen auf unterschiedliche

Europa: Umwelthistorische Determinanten

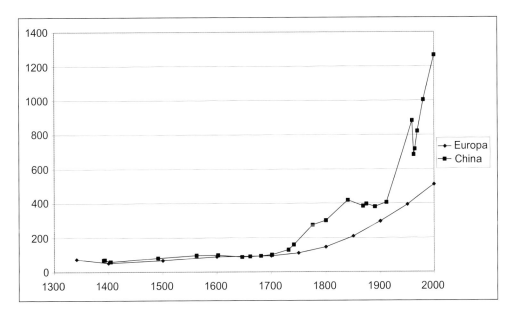

Abb. 1 Bevölkerungsentwicklung in China und Europa (in Mio.). Daten für China bis 1980 nach Zhao WENLIN und Xie SHUJUN, Zhonguo renkoushi. Beijing 1988, danach Official Census, 2000. Ich danke Raimund KOLB für diese Daten. Europa nach LIVI-BACCI 1997.

Umweltbedingungen zu beziehen. Dieses Argument, das sich vor allem bei JONES (1987) findet, hat die folgende Struktur: Wenn die Umweltbedingungen weniger stabil und kalkulierbar sind, hat es nicht viel Sinn, größere Investitionen zu unternehmen, die auf konstante Rahmenbedingungen angewiesen wären. Demographisch hat dies zur Folge, dass unter instabilen Verhältnissen weniger Aufwand für das einzelne Kind getrieben wird, sondern dass man eher viele Kinder in die Welt setzt in der Erwartung, dass schon das eine oder andere überleben wird. Sind dagegen die Erwartungen an die Umwelt stabil, so werden eher weniger Kinder geboren, und es lohnt sich, einen größeren Aufwand für ihre Aufzucht und Ausbildung zu treiben. Hierbei handelt es sich um eine Grundannahme der verbreiteten Theorie des demographischen Übergangs, derzufolge generell die Erwartung einer sinkenden Kindersterblichkeit zu einer Reduktion der Fertilität führt.

JONES zieht aus diesem Zusammenhang sehr weitreichende Schlüsse: Er macht nämlich die Vermutung stark, das „asiatische" (im Gegensatz zum nordwesteuropäischen) Familien- und Heiratsmuster könne als Adaptation an unterschiedliche Umweltfluktuationen und Gefahrenniveaus gedeutet werden. Zu diesem Zweck greift er auf eine idealtypische Unterscheidung zweier Familienformen zurück, denen entgegengesetzte Eigenschaften zugeschrieben werden, wobei er sich vor allem auf Literatur zum nordwesteuropäischen Familienmuster bezieht wie HAJNAL (1965), SMITH (1979) MACFARLANE (1978), GOODY (1983).

(*a.*) **In patrilinearen Großfamilien bzw. Sippenverbänden** gelten relativ viele Menschen als eng miteinander verwandt. In großen Haushalten kann immer noch ein weiteres Familienmitglied durchgefüttert werden, d. h., die Koppelung von Ressourcenverfügung und Familiengröße ist relativ locker. Demographisch kann dies als institutioneller Ausdruck einer r-Strategie interpretiert werden: Man gestattet eine recht hohe Wachstumsrate der Bevölkerung, investiert aber nicht sehr viel in den Einzelnen. Dies kann als Anpassung an stark fluktuierende Umweltbedingungen (Naturkatastrophen, Kriege) verstanden werden, verstärkt aber seinerseits diese Fluktuation, indem immer wieder malthusianische Krisen provoziert werden. Unter risikotheoretischer Perspektive kann das Muster der erweiterten Familie somit als Portfolio-Strategie zur Ruinvermeidung seiner Mitglieder angesichts einer hohen Unsicherheitsinzidenz angesehen werden. Das System sichert seine Mitglieder in mehrerer Hinsicht ab:

– Gegen Naturschwankungen, indem direkte materielle Hilfe geleistet wird, etwa beim Hausbau nach einer Feuersbrunst oder durch Versorgung mit Nahrung nach einer Missernte. Letzteres würde allerdings voraussetzen, dass die Familienmitglieder in ökologisch unterschiedlichen Räumen leben, so dass sie nicht von gleichartigen Umweltfluktuationen betroffen sind.
– Gegen biographische Einbrüche wie Krankheit oder Tod, indem Hinterbliebene (Witwen, Waisen) oder Invalide versorgt werden können.
– Gegen politische Turbulenzen, indem „Rückversicherungen" bei möglichen Machtträgern abgeschlossen werden, wenn verschiedene Angehörige einer Sippe bei unterschiedlichen Parteien vertreten sind – aber das könnte eine Dimension sein, die nur für aristokratische Familien von Bedeutung ist.
– Gegen Übergriffe anderer Individuen, indem kollektive Vergeltung (Blutrache) angedroht wird, deren Glaubwürdigkeit mit der Größe und Schlagkraft des Sippenverbands steigt.
– Öffnung von individuellen Lebenschancen, indem mit Hilfe von „Beziehungen" Positionen verschafft werden, aber auch erwartet wird, dass man bei Bedarf auf Leistungen zurückgreifen kann.

Diese Verstetigungsleistung durch die Sippe hat jedoch ihren Preis. Jedes erfolgreiche Individuum schleppt gewissermassen die gesamte Last der Familie mit sich. Ist jemand reich, so muss er die armen Familienmitglieder ernähren. Ist er mächtig, so muss er sie schützen und protegieren. Ist er stark, muss er für sie kämpfen und ihre Ehre verteidigen. Er verliert also ein großes Stück Autonomie, und eben dies ist der Preis, der für diese Versicherung zu zahlen ist. *Per saldo* könnte dieses Muster auf eine stationäre Fluktuation sozialer Mobilität hinauslaufen, also darauf, dass eine gerichtete Tendenz der Entwicklung eher unterbunden wird: Jeder Erfolg untergräbt seine eigenen Voraussetzungen, da er zu wachsenden Belastungen führt.

(*b.*) Der Gegentypus zur patrilinearen Großfamilie ist die **kognatische Kern- oder Kleinfamilie**, die im Wesentlichen aus den Eltern und deren Nachwuchs, vielleicht noch aus wenigen abhängigen weiteren Familienangehörigen besteht (ledige Geschwister, verwitwete Großeltern). Das Grundprinzip der neolokalen Kleinfamilie lautet, dass mit der Heirat eine neue Familie gegründet wird, das Ehepaar sich also nicht einer bereits bestehenden Familie anschließt. Heirat und die Erzeugung legitimen Nachwuchses sind daher mit der Gründung bzw. Übernahme eines eigenen Hausstands verbunden. Generell sind die folgenden Merkmale zu erwarten:

– späte Heirat (vor allem des Ehemannes), da zur Gründung eines Hausstandes die ökonomische Unabhängigkeit von den Eltern erforderlich ist;
– neolokale Residenz der Verheirateten, folglich höhere räumliche Mobilität;
– Tendenz zu einem höheren Niveau der Haushaltseinkommen, da den Armen die Heirat versagt bleibt.

Die Kleinfamilie hat eine Reihe von Implikationen, die auf Entwicklungstrajektorien verweisen, die im Kontext der erweiterten Familie eher unwahrscheinlich sind:

– Sie schließt den Kreis von Nachwuchs und Ressourcen sehr eng, d. h., sie ist gegenüber Knappheitssituationen hoch sensibel. Daher könnte sie die Funktion haben, das Bevölkerungswachstum enger an die Ressourcenbasis zu binden, als dies andere Familienformen tun. Die Kleinfamilie wirkt damit demographisch stabilisierend.
– Wenn Verehelichung an einen selbständigen Hausstand gebunden ist, bleibt immer ein großer Teil der Bevölkerung unverheiratet. Damit existiert aber eine demographische Reserve, auf die nach erhöhter Sterblichkeit (Krieg, Seuchen) zurückgegriffen werden kann.
– Im Kontext der Kleinfamilie kann im Prinzip jeder zum Familienoberhaupt werden, der dazu ökonomisch qualifiziert ist, sich also einen eigenen Hausstand leisten kann. Dies ist durch Sparsamkeit, Disziplin und Geduld zu erreichen. Umgekehrt gilt nur der Hausvater als vollwertiger Bürger, so dass ein hoher Anreiz dazu besteht, in diese Kategorie aufzusteigen. In diesem Kontext wird also das individuelle Leistungsprinzip begünstigt.
– In der nordwesteuropäischen Kleinfamilie ist es üblich, dass heranwachsende Kinder Gesindedienst in fremden Haushalten leisten, sei es in der Landwirtschaft, sei es im Handwerk (Lehrlinge, Gesellen). Dies hat zur Folge, dass der Ehemann vor der Eheschließung (geographisch und sozial) mobil gewesen ist, also Lehr- und Wanderjahre absolviert hat, bis er sich als Meister etc. niederlassen kann. Dieses Muster der temporären Knechtschaft begünstigt soziale Durchlässigkeit, und die Wanderung von Handwerksgesellen ermöglicht einen raschen Technologietransfer.
– Die Kleinfamilie bringt ökonomische Verpflichtungen nur gegenüber Angehörigen der Kleinfamilie selbst. Unterhaltsverpflichtungen gegenüber entfernteren Verwandten entfallen. Dies bedeutet, dass nicht jeder Reichtum sofort von der gesamten Sippe aufgezehrt wird, sondern dass private Akkumulation möglich wird.
– Umgekehrt bedeutet dies, dass viele Menschen in Gefahr geraten, aus allen Bindungen herauszufallen (Witwen, Waise, Alte). Sie müssen daher aus öffentlichen Mitteln erhalten werden (Kirche, Gemeinde, Sozialstaat). Es entsteht so eine Dualität von Privatsphäre („bürgerliche Gesellschaft") und Öffentlichkeit („Staat"), und öffentlichen Institutionen wachsen soziale Aufgaben zu.

JONES zieht aus dieser Unterscheidung sehr weitreichende Konsequenzen: In Asien habe das höhere objektive Unsicherheitsniveau die Existenz der Großfamilie begünstigt und die Gesellschaft damit auf eine Entwicklungslinie festgelegt, die den Übergang zur Industrialisierung, zu Individualisierung, Initiative, Rechtsstaat und Marktwirtschaft erschwert hat. Die europäische Entwicklung wurde damit nicht zuletzt durch günstigere, weil stabilere Umweltbedingungen geprägt. Auch wenn dies nicht im deterministischen Sinne verstanden wird, hat sich doch eine Trajektorie aufgetan, die schließlich in die Entwicklung des industriellen Kapitalismus in Europa einmündete (so LASLETT 1988). Das auf diese Weise unterschiedene asiatische bzw. europäische Muster der Risikobewältigung kann wie folgt zusammengefasst werden:

– Die Lage in Asien wurde von hoher Unsicherheit, starken Naturschwankungen, sozialpolitischer Anarchie und permanenter Präsenz von Gewaltdrohung, also einer starken objektiven Unsicherheitsinzidenz geprägt. Daher war die Existenz eines Netzwerkes sehr verlässlicher Beziehungen von hohem Nutzen, so dass eine Strategie der Bildung starker

Personenverbände begünstigt wurde, die es ermöglichten, stabile Erwartungen innerhalb einer stark fluktuierenden Umwelt aufzubauen. Wenn diese Verbände durch Verwandtschaft definiert wurden, hatte dies gegenüber kontraktuellen Lösungen den Vorzug der Eindeutigkeit der Zuschreibung. Dies hatte jedoch zur Folge, dass sich der Typus des „unternehmerischen Individuums", der die neuzeitliche Entwicklung Europas prägte, nicht durchsetzen konnte.

– In Europa war das natürliche Gefahrenniveau geringer, es entstanden staatliche Institutionen der Konfliktregulation unabhängig vom individuellen Status (Rechtsstaat, Rechtssicherheit, Gleichheit vor dem Gesetz), schließlich gar öffentliche Institutionen der Daseinsvorsorge (Gemeinden, Sozialstaat). Damit schwand der Nutzen von Personenverbänden, und es bleiben lediglich deren Kosten übrig. In dieser Situation wurde eine Strategie der Individualisierung prämiert, die eine sehr viel vollständigere Zuschreibung von individueller Leistung und individuellem Ertrag ermöglichte. Die erweiterte Familie wurde im Kontext dieser Tendenz zu teuer, zu aufwendig und zu schwerfällig, so dass die individuelle Sezession aus der Familie nicht mehr mit sozialen und ökonomischen Nachteilen sanktioniert wurde. Im Gegenteil: Jetzt wurden die Resultate des individuellen Leistungsprinzips prämiert.

Hierbei handelt es sich um eine grobe idealtypische Unterscheidung, die zwei entgegengesetzte Grundtendenzen fixieren möchte. Wichtig kann aber sein, dass immer dann, wenn sich bestimmte Randbedingungen ändern, eine evolutionäre Prämie darauf liegt, sich ein Stück in die Richtung auf eines der beiden Extreme zu begeben. Wenn also die Sicherheitslage besser wird, können Schritte in Richtung auf Rechtsstaat/Individualisierung eingeschlagen werden, die ihrerseits die Sicherheitslage verbessern. So könnte ein Prozess positiver Rückkoppelung in Gang gekommen sein, der in Europa angesichts einer geringeren objektiven Unsicherheitsinzidenz auf den Weg in den rechtsstaatlich-marktwirtschaftlichen Kapitalismus geführt hat.

Aus dieser Perspektive könnte die europäische Transformation der Neuzeit als Ergebnis einer Ausgangslage mit geringeren Umweltrisiken und höherer Verlässlichkeit der Institutionen verstanden werden, die funktional mit der Kleinfamilie gekoppelt war. Die Kleinfamilie wäre ihrerseits Resultat kontingenter Bedingungen, die erst unter veränderten Umständen eine adaptive Funktion erhielten (OESTERDIEKHOFF 2002). Die europäische Präadaptation könnte auf sehr weit zurückliegende Zeiten verweisen, auf das hohe Mittelalter (MITTERAUER 1997, 2000), das frühe Mittelalter (GOODY 1983), auf das (antike) Christentum (STARK 1996), wenn nicht auf prähistorische Epochen (CLARK und PIGGOTT 1965). In Asien wäre komplementär dazu eine Tradition zu beobachten, die auf eine ältere Adaptation oder auf ein freies kulturelles Muster zurückgeht, das jedoch quer zu den funktionalen Bedingungen der kapitalistisch-rechtsstaatlichen Entwicklung steht.

Die von JONES (1987), LASLETT (1988) oder MACFARLANE (1978) betonten Unterschiede zwischen dem europäischen und dem asiatischen Familienmuster werden in der neueren Literatur allerdings stark relativiert (vgl. GOODY 2000). Dies betrifft einerseits die Allgemeingültigkeit des Musters der Kleinfamilie in Nordwesteuropa selbst, die anhand einzelner Ausnahmefälle bezweifelt wird. Diese Einwände treffen aber das generelle Argument nicht. Gravierender sind dagegen Einwände, die die funktionale Eindeutigkeit des „asiatischen" Musters bestreiten. So wurde darauf hingewiesen, dass große Ähnlichkeiten zwischen dem europäischen und dem japanischen Familienmuster bestehen (MCNEILL 1996). Von prinzipieller Bedeutung sind schließlich die Ergebnisse neuerer Untersuchungen zur chinesischen Familie, die demonstrieren, dass es sich bei der angenommenen Beziehung zwischen patrilinearer erweiterter Familie und einem demographischen Regime, das dem Muster der r-Strategie folgt, offenbar um eine Legende handelt (Eglauer 2001 in Anlehnung an Lee und Wang 1999).

Zwar ist die chinesische Familie durch Patrilinearität geprägt, wozu funktional Tendenzen zur Patrilokalität, zu einem frühen Heiratsalter sowie ein hoher Anteil der Verheirateten an der Gesamtbevölkerung gehören. Es hat sich jedoch nicht bestätigt, dass dies auch zu größeren Familieneinheiten oder einer höheren Fertilität führt. Im spätkaiserlichen China lag die durchschnittliche Haushaltsgröße bei 5,5 Personen, was sich nicht erheblich von den 4,75 Personen im England des 17. und 18. Jahrhunderts unterscheidet. Überraschend ist aber vor allem die Tatsache, dass sich die signifikant frühere Eheschließung in China nicht in deutlich höheren Fertilitätsraten niederschlägt, sondern durch längere Intervalle zwischen den Geburten und eine frühere Beendigung der fruchtbaren Periode der Frauen kompensiert wird, was nur durch innereheliche Prävention erklärt werden kann. Dies bedeutete, dass die tatsächliche Reproduktionsphase der Frauen in Europa trotz des höheren Heiratsalters länger war als in China, mit der Folge, dass dort die Geburtenzahlen bei 5–6 Kindern pro Frau lagen, gegenüber 7–10 Kindern im England des 17. und 18. Jahrhunderts (Eglauer 2001, Wrigley et al. 1997).

Hierbei ist allerdings zu berücksichtigen, dass in China die Praxis des Infantizids, vor allem an Mädchen, weitaus verbreiteter gewesen sein dürfte als in Europa. Dies führt nicht nur zu einer beträchtlichen Dunkelziffer, was die Zahl der Geburten betrifft, da getötete bzw. unmittelbar nach der Geburt gestorbene Kinder in den Rohdaten nicht auftauchen, die den rekonstruierten Bevölkerungsstatistiken zugrundeliegen. Ein weiterer Effekt ist eine Verschiebung der Geschlechtszusammensetzung der Bevölkerung zu einem überproportional hohen Männeranteil, was wiederum die Fertilitätsrate mindert.

Da allerdings Infantizid als eine Form der Fertilitätskontrolle durch kulturell gesteuertes Verhalten angesehen werden kann, handelt es sich hierbei um eine Variante der Prävention, ist also Ausdruck einer K-Strategie der Vermehrung. Damit würde aber ein Eckpfeiler der Argumentation von Jones in sich zusammenbrechen: Die Chinesen hätten gerade nicht als Reaktion auf hochvolatile Umweltbedingungen wahllos Kinder in Welt gesetzt, wobei die umweltbedingte Sterblichkeit schließlich für einen Ausgleich gesorgt hätte. Im Gegenteil: Sie hätten durch gezielte Methoden, von der Steuerung der Geburtenintervalle bis hin zum Infantizid dafür gesorgt, dass genau so viel Nachwuchs geboren wird, wie die Familie unter den gegebenen Bedingungen verkraften kann. Dann bestünde aber kein prinzipieller Unterschied zum europäischen Fertilitätsmuster mehr, sondern wir hätten lediglich unterschiedliche methodische Varianten einer vergleichbaren Problemlösungsstrategie vor uns.

Daraus folgt nicht unbedingt, dass der Unterschied zwischen dem nordwesteuropäischen Muster der kognatischen Kleinfamilie und dem Muster der patrilinearen erweiterten Familie keine wichtigen Konsequenzen haben konnte, die dazu führten, dass die jeweiligen Gesellschaften auf unterschiedliche Entwicklungstrajektorien gerieten. Es bleibt noch immer die Möglichkeit, dass Neolokalität, Gesindedienst und spätes Heiratsalter Verhaltensweisen provozierten, die funktional eng auf die Industrialisierung bezogen waren. Eine direkte Koppelung zu differenzieller Unsicherheitsinzidenz besteht dabei jedoch nicht, d. h., die Gegenüberstellung eines malthusianisch positiven und eines präventiven Fertilitätsregimes als Adaptation an unterschiedliche Umweltbedingungen in Asien und Europa lässt sich nicht aufrechterhalten.

Rolf Peter Sieferle

Literatur

Abel, W.: Agrarkrisen und Agrarkonjunktur. Hamburg, Berlin: Parey 1978
Anderson, J. J., and Jones, E. L.: Natural disasters and the historical response. Australian Economic History Review 28, 3–20 (1988)
Bartlome, N., und Flückiger, E.: Stadtzerstörung und Wiederaufbau in der mittelalterlichen und frühneuzeitlichen Schweiz. In: Körner, M. (Ed.): Stadtzerstörung und Wiederaufbau. Bd. *1*, S. 123–146. Bern: Haupt 1999
Borst, A.: Das Erdbeben von 1348. Historische Zeitschrift *233*, 529–569 (1981)
Boschi, E.: Catalogo dei forti terremoti in Italia dal 461 a. C. al 1980. Bologna: SGA 1995
Botkin, D. B.: Discordant Harmonies. A New Ecology for the 21st Century. New York: Oxford University Press 1990
Braudel, F.: La Méditerranée et le monde Méditerranéen à l'époque de Philippe II. Paris: Colin 1949
Braudel, F.: Capitalism and Material Life. London: Fontana 1974
Braudel, F.: Sozialgeschichte des 15.–18. Jahrhunderts. München: Kindler 1986
Buckle, H. T.: History of Civilisation in England. London: John W. Parker and Son 1857
Clark, G., and Piggott, S.: Prehistoric Societies. London: Hutchinson 1965
Cornell, J.: The Great International Disaster Book. 3rd. Edit. New York: Scribner 1982
Diamond, J. M.: Guns, Germs, and Steel. A Short History of Everybody for the last 13000 years. London: Vintage 1997
Eglauer, M.: Familie und Haushalt im China der späten Kaiserzeit. Stuttgart: Breuninger-Stiftung 2001
Ewald, P.: The Evolution of Infectious Disease. Oxford: Oxford University Press 1994
Glacken, C. J.: Traces on the Rhodian Shore. Nature and Culture in Western Thought from Ancient Times to the End of the 18th Century. Berkeley: Univ. of Calif. Press 1967
Glaser, R.: Klimageschichte Mitteleuropas. 1000 Jahre Wetter, Klima, Katastrophen. Darmstadt: Primus-Verlag 2001
Goody, J.: The Development of the Family and Marriage in Europe. Cambridge: Cambridge University Press 1983
Goody, J.: The European Family. An Historico-anthropological Essay. Oxford: Blackwell 2000
Guidoboni, E.: Catalogue of Ancient Earthquakes in the Mediterranean Area up to the 10th Century. Rom: Istituto Nazionale di Geofisica 1994
Hajnal, H. J.: European marriage patterns in perspective. In: Glass, D. V., and Eversley, D. E. C. (Eds.): Population in History; pp. 101–143. London: Arnold 1965
Helbling, J.: Agriculture, population and state in China in comparison to Europe, 1500–1900. In: Sieferle, R. P., and Breuninger, H. (Eds.): Agriculture, Population and Economic Development in China and Europe; pp. 90–199. Stuttgart: Breuninger-Stiftung 2003
Herlihy, D.: The Black Death and the Transformation of the West. Cambridge: Harvard University Press 1997
Herrmann, A.: Katastrophen. Naturgewalten und Menschenschicksale. Berlin: Schönfeld 1936
Huntington, E.: Civilization and Climate. New Haven: Yale University Press 1915
Jakubowski-Tiessen, M.: Sturmflut 1717. Die Bewältigung einer Naturkatastrophe in der frühen Neuzeit. München: Oldenbourg 1992
Jones, E. L.: The European Miracle. Environments, Economics and Geopolitics in the History of Europe and Asia. Cambridge: Cambridge University Press 1987
Kirch, P. V.: The Evolution of the Polynesian Chiefdoms. Cambridge: Cambridge University Press 1984
Kolb, R.: About Figures and Aggregates: Some Arguments for a More Scrupulous Evaluation of Quantitative Data on the History of Population and Agriculture in China (1644–1949). In: Sieferle, R. P., and Breuninger, H. (Eds.): Agriculture, Population and Economic Development in China and Europe; pp. 200–275. Stuttgart: Breuninger Stiftung 2003
Laslett, P.: The European family and early industrialization. In: Baechler, J., Hall, J. A., and Mann, M. (Eds.): Europe and the Rise of Capitalism; pp. 234–241. Oxford: Blackwell 1988
Latter, J. H.: Natural disasters. The Advancement of Science *25*, 362–380 (1968/1969)
Lee, J. Z., and Wang, F.: One Quarter of Humanity. Malthusian Mythology and Chinese Realities, 1700–2000. Cambridge: Harvard University Press 1999
Livi-Bacci, M.: A Concise History of World Population. Oxford: Blackwell 1997
Macfarlane, A.: The Origins of English Individualism. The Family, Property and Social Transition. Oxford: Blackwell 1978
Marquardt, B.: Die „Europäische Union" des vorindustriellen Zeitalters. Vom Universalreich zum Staatskörper des Jus Publicum Europeum (800–1800). Zürich: Schulthess 2005
McCloskey, D.: English open fields as behavior toward risk. In: Uselding, P. (Ed.): Research in Economic History. Vol. *1*, pp. 124–170. Amsterdam: JAI Press 1976

McNeill, J.: The reserve army of the unmarried in world economic history: flexible fertility regimes and the wealth of nations. In: Aldcroft, D. H., and Catterall, R. E. (Eds.): Rich Nations – Poor Nations. The Long-Run Perspective; pp. 23–38. Cheltenham: Elgar 1996
McNeill, W.: Plagues and Peoples. Garden City: Anchor Press/Doubleday 1976
Mitterauer, M.: Zu mittelalterlichen Grundlagen europäischer Sozialformen. Beiträge zur historischen Sozialkunde 27, 40–46 (1997)
Mitterauer, M.: Die Terminologie der Verwandtschaft. Zu mittelalterlichen Grundlagen von Wandel und Beharrung im europäischen Vergleich. Ethnologia Balkanica 4, 11–44 (2000)
Mitterauer, M.: Roggen, Reis und Zuckerrohr. Drei Agrarrevolutionen des Mittelalters im Vergleich. Saeculum 52, 245–265 (2001)
Mols, R.: Population in Europe, 1500–1700. In: The Fontana Economic History of Europe. Vol. 2, pp. 15–82. Glasgow: Collins/Fontana Books 1974
Montanari, M.: Der Hunger und der Überfluß. Kulturgeschichte der Ernährung in Europa. München: Beck 1993
Oesterdiekhoff, G. W.: Familie, Wirtschaft und Gesellschaft in Europa. Die historische Entwicklung von Familie und Ehe im Kulturvergleich. Stuttgart: Breuninger-Stiftung 2002
Papazachos, B., and Papazachou, C.: The Earthquakes of Greece. Thessaloniki: P. Ziti and Co. 1997
Pfister, C.: Wetternachhersage. 500 Jahre Klimavariationen und Naturkatastrophen, 1496–1995. Bern: Haupt 1999
Popplow, M.: Europa auf Achse. Innovationen des Landtransports im Vorfeld der Industrialisierung. In: Sieferle, R. P. (Ed.): Transportgeschichte. Münster: LIT Verlag 2008
Pryor, F. L.: Climatic Fluctuations as a Cause of the Differential Economic Growth of the Orient and Occident. Journal of Economic History 45, 667–673 (1985)
Quesada, M.-A. L.: Earthquakes in the cities of Andalusia at the beginning of the modern age. In: Körner, M. (Ed.): Stadtzerstörung und Wiederaufbau. Bd. 1, S. 87–103. Bern: Haupt 1999
Shao-Wu, W., and Zong-Ci, Z.: Droughts and floods in China, 1470–1979. In: Wigley, T. M. L., Ingram, M. J., and Farmer, G. (Eds.): Climate and History; pp. 271–288. Cambridge: Cambridge University Press 1981
Sieferle, R. P.: Die Risikospirale. Wissenschaft und Umwelt interdisziplinär 10, 157–166 (2006)
Sieferle, R. P. (Ed.): Transportgeschichte. Münster: LIT Verlag 2008
Smith, R. M.: Some reflections on the evidence for the origins of the 'European marriage pattern' in England. In: Harris, C. (Ed.): The Sociology of the Family; pp. 74–112. Keele: University of Keele 1979
Stark, R.: The Rise of Christianity. Princeton: Princeton University Press 1996
Wallerstein, I.: The Modern World System. Vol. 1: Capitalist Agriculture and the Origins of the European World-Economy in the 16th Century. New York: Academic Press 1974
Webb, W. P.: The Great Frontier. Boston: Houghton Mifflin 1952
Worster, D.: The ecology of order and chaos. Environmental History Review 14, 1–18 (1990)
Wrigley, E. A., Davies, R. S., Oeppen, J. E., and Schofield, R. S.: English Population History from Family Reconstitution, 1580–1837. Cambridge: Cambridge University Press 1997

       Prof. Dr. Rolf Peter Sieferle
       Universität St. Gallen
       Kulturwissenschaftliche Abteilung
       Gatterstrasse 1
       St. Gallen
       Switzerland
       Phone: +41 71 2242730
       Fax: +41 71 2242669
       E-Mail: rolf.sieferle@unisg.ch
       http://www.kwa.unisg.ch/org/kwa/web.nsf/ww

Evolution und Menschwerdung

Vorträge anlässlich der Jahresversammlung vom 7. bis 9. Oktober 2005
zu Halle (Saale)

    Nova Acta Leopoldina N. F., Bd. *93*, Nr. 345
    Herausgegeben von Harald ZUR HAUSEN (Heidelberg)
    (2006, 282 Seiten, 65 Abbildungen, 3 Tabellen, 34,95 Euro,
    ISBN-13: 978-3-8047-2370-2)

Evolution und Menschwerdung gehören noch immer zu den interessantesten Themen, mit denen sich die Naturwissenschaft auseinandersetzt und die die Öffentlichkeit faszinieren. Die Thematik verlangt eine interdisziplinäre Auseinandersetzung, für die eine Akademie wie die Leopoldina prädestiniert ist. Daher griff die Jahresversammlung 2005 verschiedene Aspekte hierzu auf.
Die Schwerpunkte der Tagung spiegeln den enormen Fortschritt der Erkenntnisse über das Evolutionsgeschehen und den veränderten Blickwinkel wider, der sich aufgrund des außerordentlich großen Wissenszuwachses und veränderter Diskussionsebenen in der Forschung, aber auch zwischen Wissenschaft und Gesellschaft ergeben. Die Evolution des Menschen und dessen physische, geistige und kulturelle Entwicklungstendenzen stehen dabei im Zentrum.
Der Band spannt den Bogen vom Urknall und der Bildung der Planetensysteme über die Entstehung des Lebens, die Entwicklung von Prokaryoten und Eukaryoten, die Evolution und das Sterben der Saurier, die Analyse von Insektenstaaten bis hin zu Fragen der Menschwerdung und Formen der menschlichen Kultur. Hier werden unter anderem „Das Sprachmosaik und seine Evolution", die „Evolution durch Schrift", Rituale, Religionen, Gemeinschaftsbildung und sozialer Wandel unter evolutionären Aspekten untersucht, aber auch „Bilder in Evolution und Evolutionstheorie" sowie die „Griechischen Anfänge der Wissenschaft" betrachtet.

Wissenschaftliche Verlagsgesellschaft mbH Stuttgart

Ecological Imperialism, Plants Transfers, and African Environmental History

William Beinart (Oxford)

Abstract

My paper touches on aspects of the history of plant transfers to Africa, especially Southern Africa, and their implications for Alfred Crosby's idea of ecological imperialism. Crosby noted the significance of plant transfers from the 'old' world of Eurasia to the 'new' world of the Americas in facilitating settler colonialism, and the demographic transformation of the Americas. Plants are not as significant in his analysis as disease and domesticated animals; nevertheless, he focused especially on weeds in this unequal exchange. I argue, however, that from the vantage point of Africa, part of the 'old' world, Crosby's discussion of asymmetrical plant exchange is problematic. The most important flow of plants was in the other direction: from rather than to the Americas. Africans welcomed and absorbed many American plants; these included cultivated plants such as maize and useful invaders such as prickly pear. South Africa's declared weeds are disproportionately from the Americas. My analysis suggests the complications in using a notion of botanical imperialism. Useful plants can flow against the routes of power. They can do so without direct colonialism. They don't necessarily facilitate settler colonialism. True, maize and some other American crops because the basis for large white-owned commercial farms in South Africa. But overall American plants been most important for African people and have provided economic opportunity and possibly demographic strength. We need to be cautious in specifying asymmetrical plant flows and also in applying the concept of ecological imperialism with respect to plants.

Zusammenfassung

Der Beitrag befasst sich mit einigen Aspekten der Geschichte des Pflanzentransfers nach Afrika, insbesondere nach Südafrika, und damit, welche Auswirkungen ein solcher Transfer auf die Theorie von Alfred Crosby über den ökologischen Imperialismus hatte. Crosby erwähnte, dass der Pflanzentransfer von der ‚alten' Welt Eurasien in die ‚neue' Welt des amerikanischen Kontinents eine wichtige Rolle spielte, da er den Kolonialismus der Siedler und den demographischen Wandel auf dem amerikanischen Kontinent förderte. Pflanzen sind in seiner Analyse nicht so wichtig wie Krankheiten und Haustiere; nichtsdestotrotz konzentrierte er sich bei diesem ungleichen Austausch insbesondere auf Unkräuter. Der Beitrag dagegen argumentiert, dass vom afrikanischem Standpunkt aus betrachtet – Afrika ist ja ein Teil der ‚alten' Welt –, Crosbys Diskussion des asymmetrischen Pflanzenaustauschs problematisch ist. Der weitaus bedeutendere Austausch von Pflanzen erfolgte in die andere Richtung: vom amerikanischen Kontinent in andere Kontinente anstatt umgekehrt. Die Afrikaner waren froh über amerikanische Pflanzen und integrierten viele davon in ihre Pflanzenwelt, so z. B. Kulturpflanzen wie Mais und nützliche Eindringlinge wie den Feigenkaktus. In Südafrika als Unkräuter bezeichnete Pflanzen wurden überwiegend aus Amerika eingeschleppt. Die Analyse weist darauf hin, dass es zu Komplikationen führt, wenn man den Begriff des botanischen Imperialismus benutzt. Ein Austausch nützlicher Pflanzen kann auch in Gegenrichtung zu den Wegen der Macht erfolgen. Pflanzen können durchaus ohne direkten Kolonialismus ausgetauscht werden. Durch den Pflanzenaustausch wird nicht notwendigerweise der Siedlerkolonialismus gefördert. Es ist zwar wahr, dass Mais und einige andere amerikanische Nutzpflanzen die Basis für große kommerzielle, im Besitz von Weißen befindliche landwirtschaftliche Betriebe in Südafrika bildeten. Insgesamt waren jedoch amerikanische Pflanzen äußerst wichtig für die Afrikaner und haben ökonomische Chancen eröffnet und vermutlich auch zur demographischen Stärke beigetragen. Wir müssen vorsichtig sein, wenn wir den asymmetrischen Pflanzenaustausch beschreiben, ebenso können wir das Konzept des ökologischen Imperialismus nur mit Vorsicht im Hinblick auf Pflanzen anwenden.

William Beinart

1. Crosby and Ecological Imperialism

I have been asked to speak on the environmental history of Africa, but that is so large a task, that I will touch on only a few themes arising out of some recent, and rather specific, research. I will address some aspects of the history of global plant transfers and their implications for Alfred Crosby's idea of *Ecological Imperialism* (Crosby 1986). Crosby did not write much about the environmental history of Africa. But he has some interesting points to make – in particular comparing the relative failure of settler colonialism on this continent to its success in the Americas and Australasia. And we should address his propositions which, along with Jared Diamond's, have been so influential in both academic and popular views of global environment history (Diamond 1998). Both authors have emphasized environmental causation in shaping major historical processes, and that is another reason to develop a careful critique of their work. Historians – certainly those working on Africa – are generally cautious about environmental causation in which they see tendencies towards determinism. So much environmental history that is written by academic historians concentrates on the history of ideas, or on the history of environmental politics, or on the environmental impact of European expansion. We find it difficult to deploy environmental explanations on a major scale, and should give close attention to those who take the risks of doing so. Their work can lead in exciting and important directions; it can also lead to some potentially problematic propositions.

The failure, for the most part, of settler colonialism has been an absolutely central element in Africa's recent history. Perhaps in 1950, at the height of the colonial period, the picture looked different. White settlers were nowhere demographically predominant but they held political power or had significant influence in a number of countries: aside from South Africa/Namibia, also in Zimbabwe, Algeria, Mozambique, Angola, Kenya and Zambia. But some of those white populations had been established for little more than half a century, and most diminished rapidly after the transfer of power. Viewed over the longer term, and even in the context of the twentieth century as a whole, this was a relatively weak historical force. Viewed comparatively, with the partial exception of South Africa, settler colonialism was limited.

In his brief observations, Crosby offered essentially environmental explanations for this failure. His major themes concern an understanding of role of pathogens, plant transfers, and animals in facilitating settler colonialism. He argued convincingly that it is difficult to conceptualize European imperialism adequately without understanding the impact of these environmental influences. In the Americas and Australasia, such forces helped to produce what he called neo-Europes – where Europeans dominated demographically as well as politically. In West Africa, by contrast, the disease environment worked against European settlers; their attempts to use European cultivars failed and their livestock were decimated by trypanosomiasis (Crosby 1986, p. 136). In Southern Africa, Europeans established a firmer foothold but Africans defended some of their territory, and even at the height of settler power, were demographically resurgent (Crosby 1986, p. 146).

Arguments about the impact of disease in African colonization, especially with respect to West Africa, have been extensively debated (Beinart and Hughes 2007, Chapter 2). But less has been written about the role of plants in this context. Plant transfers, as much as domestications, have a central place in world history. They have been fundamental in facilitating major expansions of people, agrarian systems, and of empires. As Crosby notes, agrarian complexes in the Americas and much of the southern hemisphere, which are now amongst the most productive in the world, resulted from the migration or adoption of a wide range of plant species, totally new to these areas. Even where agricultural systems, such as those in the Middle East, China, Africa and the Americas, are still based partly upon plants that were domesticated within the same region (wheat, rice, millet and maize respectively), the regional spread of these crops requires explanation. Non-food species, such as trees and weeds have also spread globally, transforming natural environments.

With respect to plant transfers, Crosby makes two intriguing points. He suggests that unintentional plant transfers, as much as intentional ones, were critical in shaping and benefiting settler colonialism in the Americas. Secondly, an adjunct to his main proposition about the impact of diseases is that exported Eurasian plant species, both domesticated and especially weeds, proved more powerful than those originating in the Americas and Australasia. He distinguishes sharply between the interconnected 'old world', and the isolated 'new world', and he sees a clear flow of plant species from the former to the latter. By contrast he argued that relatively few American species established themselves in the old world. He offers both botanical/ecological and socio-economic reasons for the success of old world weeds: the capacity to reproduce rapidly; similarity in climate; and the degree of social and ecological disturbance in the new world.

Crosby is not entirely consistent; there is a strange disjuncture in his work. While he notes the weakness of new world plants in *Ecological Imperialism*, his earlier book on the *Columbian Exchange* has a fascinating chapter on the impact of American domesticates in the rest of the world, not least Africa (Crosby 1972). He includes Africa firmly in the old world. Crosby also modified his view in the introduction to the second edition (Crosby 1993). Nevertheless we are left with a picture, in *Ecological Imperialism*, of an asymmetry of exchanges between the old and new worlds in respect of plants, as well as diseases, and especially an asymmetry in regard to weeds. Although it is a more muted theme in his text, Diamond reiterates elements of this asymmetry, emphasizing especially the limits of crop domestication in the Americas. He also reasserts Crosby's point that the north/south, rather than east/west, axis of the Americas inhibited the spread of domesticated species, because of climatic variation. This had consequences for later social developments (Diamond 1998).

In general terms Crosby can be criticized for environmental determinism, for his lack of conceptualization of social and ecological hybridity, and perhaps for overestimating the longevity of the neo-Europes that emerged. I nevertheless still find his intervention fruitful and challenging. But is he right about the idea that old world plant species were advantaged? Is the pattern different in relation to cultivated plants and weeds? Karen Middleton and I have tried to explore some of these issues, juxtaposing literature by scientists and historians (Beinart and Middleton 2004). Despite the fact that some scientists also suggest an asymmetry, we called for caution concerning this thesis, and for more clarity about what is being measured. We questioned the value of the old/new world distinction, or of treating these as two single plant power blocs.

I will not repeat all the arguments here. I would like to explore further whether this apparent asymmetry holds with respect to Africa, especially southern Africa. Africa was part of the old world in respect of its disease pathogens. Africans had resistance to smallpox and other infectious diseases that Native Americans lacked. With respect to livestock, no significant American species crossed the Atlantic to Africa. But does the same apply to plants and what is the implication of such an analysis of the concept of ecological imperialism? My interest in these issues was first stimulated by a project on the history of opuntia, or prickly pear, in South Africa. It is a good plant with which to think about these concepts. American in origin, it crossed continents, and crossed boundaries of useful and invasive, of crop and weed.

William Beinart

2. Plant Transfers: American Food Crops in Africa

Certainly, if we take domesticated and useful plants as a baseline, the evidence is of a counter movement or washback between the Americas and Africa. Over the last three centuries sub-Saharan Africa came to depend increasingly on American domesticates: maize, cassava/manioc, sweet potatoes, bean varieties, some gourds, potatoes, tomatoes, tobacco, peanuts, cocoa, avocado, chili, peppers, cinchona, agave, guava, pineapple, as well as prickly pear. True, coffee (African or Middle Eastern in origin), cotton and sugar cane, from the east, are amongst the top agricultural exports from Africa as a whole. Plantains and bananas originally from Asia became major food crops and are exported from some African countries. Plantains played some role in the expansion of Bantu-speaking peoples across the forest zone, into eastern Africa and then southwards. Many other Asian plants have been incorporated in African agriculture. Mangoes were planted along the routes of slave caravans in east Africa. Tea and citrus are, in pockets, major plantation crops. Through careful use of irrigation, and specialist cultivation, South Africa became one of the largest citrus exporters globally, behind only Spain and the United States. Though it accounted for about 2% of global production, it produced 11% of global exports in 2003. But collectively, plants of American origin are probably of greater importance in Africa.

Maize was introduced by Portuguese traders and slavers into Africa soon after it was brought back from the Americas (MIRACLE 1966, CROSBY 1972, McCANN 2005). Its advantages were many. It could serve as a vegetable, eaten fresh after boiling or roasting, as well as a dried, stored and ground grain. Its yields in favorable conditions were relatively high, compared to the well established sorghums and millets. Its covered cob was a protection against voracious birds, and some insects. Maize was rapidly inserted into the agricultural repertoire of the Asante people in West Africa, who later became perhaps the largest and most powerful kingdom in Africa. Its spread was uneven, but relentless. Despite its nutritional disadvantages – a lack of protein compared to sorghum or wheat – it became the major food crop in Africa during the twentieth century. The boundaries of maize production are still expanding. McCANN (2005, p. 7) estimated that in southern Africa maize makes up 50% of calorie intake. South Africa, where – unusually in Africa – the bulk of production is on large commercial farms, has by far the highest output. But the people of three other southern African countries, Lesotho, Malawi and Zambia, consume a higher proportion of maize in their diets than anywhere else in the world. Maize seems to many of its consumers quintessentially African and tied up with long established foodways and cultural expressions. Since the early decades of the twentieth century, South Africa has exported some maize, and other countries sold from time to time into regional markets.

Maize is increasingly grown, on smallholdings as well as commercial farms, as a monocrop which displaces indigenous species, prepares the ground for weeds, can quickly exhaust soil, and precipitate soil erosion. McCANN argues that it is associated with the spread of malaria. But any environmental critique must be tempered by recognition that it is the most important and preferred food source. People need food, and all food production requires ecological disturbance. Techniques of cultivation that reduce soil erosion are certainly available. Cassava, sweet potatoes, bean varieties, tomatoes, chilis, potatoes and peanuts became widespread in different African countries and key components of characteristic African dishes.

CROSBY (1972) notes the coincidence of dramatic African population increase since 1850 and the spread of plants derived from American cultivars. Between 1900 and 2000 global population grew from roughly 1.6 billion to 6 billion (3.75 times). Africa's share grew disproportionately from 8 % to 13 % of the world's total: in absolute numbers from an estimated 130 to 800 million, or about six times. It is worth noting that the period of probably most rapid African population increase, c. 1960–1990, coincided with surging maize production. These decades in Africa may have witnessed one of the most dramatic demographic surges in world history. Some African countries experienced acute food shortages, especially during droughts, and this period is usually associated with increasing food insecurity (Commins et al. 1986, Timberlake 1988). Famine was, however, nowhere a demographic check. Even in Ethiopia, which probably experienced the greatest loss of life in the 1980s, population increased from 23 million in 1960 to 49 million in 1990. While maize took up a lower percentage of cropland in Ethiopia than in most African countries, production there expanded hugely (McCann 2005, p. 220). In Africa as a whole, food production quite nearly kept pace with population growth. African demographic expansion requires complex explanations. It is clearly associated with rurality, poverty and with women's lack of control over their fertility. The role of food production was not necessarily central. But we can note that new plants helped to underpin a degree of food security at various phases of African population expansion. In this sense, maize and its American siblings had a dual environmental impact – that arising indirectly from population increase as well as directly from clearing the land.

American domesticates also became increasingly important as cash crops. Cocoa underpinned the export revenue of Ghana for much of the colonial period and after. At times it brought in 90 % of colonial Gold Coast's export earnings. Ivory Coast production rivaled it, especially after independence, and cocoa exceeded 50 % of that country's export revenue in some years. Peanuts played a similar role in Senegal; tobacco in Zimbabwe and Malawi.

Plants domesticated in Africa, or adopted very early, such as millet, sorghum, teff and a type of red rice, are increasingly marginal in many countries. With relatively few exceptions, African plants have been less successful as colonizers. West African rice had its moment in the coastal wetlands of the Americas (Carney 2001). Sorghum is a minor crop in North America. West African palms have become a major plantation crop, for palm oil, in South East Asia. Flowering plants from South Africa are ubiquitous – often grown as pot plants. But African food and medicinal crops have been less successful than American in the global context; a number of authors note the limited number and variety of domestications.

3. Plant Transfers: American Weeds in South Africa

What about weeds? Crosby cites as evidence for asymmetry that roughly 50 % of farmland weeds in the United States, out of about 500, and 60 % in Canada, were of Eurasian, largely European origin (Crosby 1986, p. 164). But South Africa shows exactly the opposite pattern (Henderson et. al. 1987). Of the 47 main Declared Weeds in South Africa in a 1987 publication, at the time when Crosby wrote, 35 or 74 % were from the Americas, mostly south and central America, and a further four from Australia, making 83 % in all from the new world. Of a further nine plants classified as alien invader plants, all of them trees, seven were hakea and wattle species from Australia, and one, prosopis or mesquite, was from America. Only one, pinus pinaster, or cluster pine, widely used in afforestation, was from Europe. (The current list of Declared Weeds – on the web – is longer, well over 200, with a lower percentage from the Americas; it includes more plants that have been cultivated in gardens or, like jacaranda, used for public avenues.)

William Beinart

The plants from the Americas included some of the most difficult weeds such as burrweed (*Xanthium spinosum*) which was the first to be officially declared 'noxious' in the nineteenth century. It stuck in the wool of sheep, then South Africa's most valuable export. Mexican marigold spread in cultivated land. Lantana, a garden and hedge plant, colonized disturbed urban land. *Opuntia aurantiaca*, the jointed cactus, which reputedly escaped from a mission station rockery in the 1860s, stuck in the wool and hides of livestock, which spread it very widely. Mesquite was deliberately introduced in the late nineteenth century as a fodder and shade plant in the driest pastoral districts. Its value was still praised in the mid-twentieth century, but by the 1990s it had spread rampantly in some areas displacing sparse indigenous vegetation (Hoffman et al. 1999, pp. 143ff.) Some of the most recent trouble makers, such as the pompom weed, similar to a daisy, and first noted in Johannesburg in 1962, are also American. Profligate in seed distribution, it invades grasslands and fields in the higher rainfall districts.

American weeds were highly and disproportionately effective in South Africa, and clearly new world plants could be as powerful as old. Perhaps a closer investigation of European weeds would also indicate a more complex picture. But Crosby's discussion had a further dimension. He suggests that some of the most successful old world invaders were not so much dangerous weeds but valuable self-spreaders, such as clover and Kentucky bluegrass. These provided familiar pasture resources to livestock as they spread out with settlers from the coast. There are a few examples of such plants in South Africa, although they were perhaps not beneficial on the same scale. Black wattle, introduced to Natal for tanning and timber in the 1830s, was later planted by Africans around their homesteads as a quick-growing source of timber and fuel in higher rainfall districts along the east coast. It could be pollarded, and also spread itself, diminishing the need for systematic planting. Similarly, the Australian salt bush, atriplex species, was cultivated as a fodder crop in semi-arid districts in the first half of the twentieth century and spread to a limited degree by itself. It was a valuable resource, alongside karoo bushes. Prosopis was useful in the arid areas. One species of amaranth, a domesticated plant from the Americas, spread partly as an edible weed in cultivated fields in some of the wetter east coast districts of South Africa and was used as a green spinach leaf.

Perhaps the best example of a useful self-spreader in South Africa was *Opuntia ficus indica*, the common or sweet prickly pear. Prickly pear was planted, but unlike maize, it spread largely through non-human agency. It is difficult to classify. In some contexts, and in the eyes of different people, it could be a cultivated plant, a useful self-spreader, a weed, or a serious and damaging invader. Alone of the American self-spreaders, it did become of significant economic value in South Africa (Beinart 2003, Beinart and Wotshela 2003).

Opuntia species originate in the Americas, most in meso-America, and especially in the semi-arid parts of Mexico and the southern USA. They were deliberately brought back from the Americas in the early sixteenth century. By the seventeenth century some species had reached much of the Mediterranean littoral, the Canary Islands, the Cape, and India. One species became widespread in southern Madagascar from the late eighteenth century and another in Australia during the nineteenth century (Middleton 2003). It was probably in these two countries that prickly pear was most successful as a self-spreader; in the first it was a great boon – the foundation of the cattle economy – and in the second a perceived pest.

Prickly pear reproduces most easily from its large fleshy leaves or cladodes, which put out roots when they come into contact with soil. It is easy to plant and also spreads when cladodes fall or are carried by water or other means. While prickly pear tends not to grow from seed when it is deposited directly onto the ground, passage through a digestive tract facilitates germination. Thus the spread of prickly pear in some contexts is closely related to its consumption by animals and birds. Livestock were one agent. In South Africa, monkeys and baboons were especially fond of the fruit and, together with birds, unintentionally sewed the plants in the remotest areas.

At least 10 species became of some significance in South Africa of which the *Opuntia ficus indica* was by far the most useful. In Africa, they are also found in North Africa, in Ethiopia and Eritrea, in parts of Kenya, Tanzania and pockets of Mozambique, and in other semi-arid zones in southern African, notably Botswana. In Israel, Jewish settlers were named Sabra after the prickly pear fruit. In Eritrea, expatriates are called Beles, also after the fruit, because they return in the summer when it is widely harvested and sold. The great value of prickly pear was the multiple purposes that it served. It may not have been the best plant for any of these, but the ease with which it reproduced in certain climatic zones made it especially valued by poor peasant and farm-worker communities. Planted close together, prickly pear forms a dense hedge which can be used for keeping animals away from crops or in corrals. One nineteenth century African chiefdom in South Africa used it for defense around the homesteads.

The spread of prickly pear coincided historically with the spread of pastoralism in areas where water supplies were insecure. Although prickly pear cladodes were too low in nutrients to provide an all round fodder, they were particularly useful in droughts because of their high water content. In southern Madagascar prickly pear became an essential element in the cattle economy; the plants were less significant in South Africa, but widely fed to, or grazed by, ostriches, pigs and sheep as well as cattle. Introduced livestock adapted to prickly pear in the Americas as well. Thorns were treated by chopping or burning. Spineless varieties were propagated and spread in Mexico; by the eighteenth century they had reached South Africa. These plants were particularly valuable as a standing drought fodder because they needed no treatment. However, if spread by seed, many of the thornless plants would revert to the thorny variety. The spineless variety had to be reproduced by cloning from the cladodes and this could not be easily controlled.

The sweet fruits of *Opuntia ficus indica* were equally attractive to people, both white and black. Though difficult to peel, they are reliable, have a long season, and grow with little cultivation. They were also widely hawked and traded – and are still sold in some quantity on the roadsides. Rural African people used the fruit, which does not keep well, to make beer. For many decades in the late nineteenth and the first half of the twentieth century, prickly pear fruit beer was probably the main brew for poor black people in districts of the Eastern Cape where the plants thrived. The plant was also used for yeast, soap, and medicinal purposes.

Prickly pear can also damage livestock, displace other vegetation, and take over the best riverine sites for fields in semi-arid districts. By the early twentieth century, all species were becoming regarded as invasive weeds by agricultural officials and those who were investing in higher value, pedigree animals. In the 1930s, when prickly pear reached its peak in South Africa, dense stands commanded about 1 million hectares. In the 1920s in Australia, 10 million hectares was densely infested. Prickly pears from apparently unpromising 'new' world environments have proved to be highly adaptable. Major state-sponsored biological eradication campaigns were undertaken in both countries, initially with more success in Australia. In Madagascar, the unofficial introduction of cochineal insects decimated the plants, causing a major famine. In South Africa, the eradication campaign proceeded more slowly but by 1980, the main species of prickly pear was reduced by about 90%. The fruits are still gathered, eaten and brewed but on a far smaller scale. Some opuntia species, such as jointed cactus, remain dangerous weeds.

Prickly pear inserted itself into the lives of rural and small town people in the Eastern Cape region of South Africa and remains part of the folklore. It was especially important for poor rural African people and profoundly affected the ecology of many districts in the midland and Eastern Cape for over a century.

4. Conclusion

In conclusion, if Southern Africa is included as part of the old world, there is little basis for CROSBY's model of asymmetrical plant transfer and weed invasion. As he recognized elsewhere, sub-Saharan Africa is hugely dependent on plants originating from American cultivars. South Africa's weeds and invasive plants are largely from the Americas and Australia. This includes some useful weeds, such as prosopis and prickly pear.

I would be cautious, however, about arguing for an inversion of CROSBY's model, and suggesting that American plants are in some ways more powerful. Some scientists, such as WILLIAMSON (1996), are skeptical of attempts to generalize about the botanical characteristics of plant invaders, or about the environments from which they originate, or those that they invade. Much depends on climate, how the plants adapt, whether they are hindered by natural enemies, whether they find suitable ground, disturbed or otherwise, and whether their spread is facilitated by human action, intentional and unintentional. However, even human agency is not necessarily determinant. Long distance plant transfers do usually depend on human intervention, and thus can reflect the routes of trade, empire and settlement at particular moments. But South Africa did not have major trade relations with the Americas, and no settler colonization took place from there.

This brief survey may also signal caution about DIAMOND's views on the significance of early domestications in shaping the global location of civilizations and power. A surprising number of useful plants were at least semi-domesticated in the Americas. While it is true that these did not all spread widely in those continents, lack of domestication was not a primary determinant in shaping later colonization. Similarly, there was probably less diverse domestication of plants in Africa, and certainly in southern Africa. Yet African people were able to resist colonization far more effectively than American. It is striking how quickly plants can spread, and form the basis for new and more effective agrarian complexes. Northern Europe is another case in point. Centrality in early domestication of plants, at least, is perhaps not so central a determinant of later zones of world power as effective plant transfer. African power was in part based on disease resistance and in part on their capacity to command imports and transfers. General analyses of pre-colonial African history, and especially the formation of states in Africa, have focused on control of trade (including the slave trade), labor, weapons, and livestock. ILIFFE (1995) has emphasized the significance of environmental control in African history. But there is as yet a limited literature which explores the role of expanding repertoires of cultivation, and introduced cultivars, in underpinning population expansion and political centralization.

Both the general points that I have raised and the case of prickly pear reveal some of the problems in CROSBY's notion of botanical imperialism. Useful plants, cultivated or not, can flow against the routes of power. They can do so without direct colonialism, although seaborne transport was central to plant transfers in the age of imperialism. And Africans largely welcomed and absorbed many American cultivars, both cultivated plants such as maize and useful invaders such as prickly pear and black wattle. True, maize and some other American crops because the basis for large settler owned commercial farms in South Africa and Zimbabwe. But overall, and in sub-Saharan Africa as a whole, American crops and useful plants advantaged African people and helped to bolster their economic and demographic strength. We need to be cautious not only about asymmetrical flows but also about the concept of ecological imperialism in respect of plants. Of course, it could have been different if African people had been displaced by disease.

References

BEINART, W.: The Rise of Conservation in South Africa: Settlers, Livestock and the Environment. Oxford: Oxford University Press 2003

BEINART, W., and HUGHES, L.: Environment and Empire. Oxford: Oxford University Press 2007

BEINART, W., and MIDDLETON, K.: Plant transfers in historical perspective: A review article. Environment and History *10*, 3–29 (2004)

BEINART, W., and WOTSHELA, L.: Prickly pear in the Eastern Cape since the 1950s – perspectives from interviews. Kronos Journal of Cape History *29*, 191–209 (2003)

CARNEY, J.: Black Rice: The African Origins of Rice Cultivation in the Americas. Cambridge, Mass.: Harvard University Press 2001

COMMINS, S., LOFCHIE, M., and PAYNE, R. (Eds.): Africa's Agrarian Crisis. The Roots of Famine. Boulder: Lynne Rienner Publishers 1986

CROSBY, A. W.: The Columbian Exchange: Biological and Cultural Consequences of 1492. Westport, Connecticut: Greenwood Press 1972

CROSBY, A. W.: Ecological Imperialism: The Biological Expansion of Europe 900–1900. Cambridge: Cambridge University Press 1986; revised: Cambridge: Canto 1993

DIAMOND, J.: Guns, Germs and Steel: A Short History of Everybody for the Last 13,000 Years. London: Vintage 1988

HENDERSON, L., HENDERSON, M., FOURIE, D. M. C., and WELLS, M. J.: Declared weeds and alien invader plants in South Africa. Pretoria, Botanical Research Institute. Bulletin *413* (1987)

HOFFMAN, T., TODD, S., NTSHONA, Z., and TURNER, S.: Land Degradation in South Africa. Pretoria: Department of Environmental Affairs and Tourism 1999

ILIFFE, J.: Africans: The History of a Continent. Cambridge: Cambridge University Press 1995

MCCANN, J. C.: Maize and Grace: Africa's Encounter with a New World Crop 1500–2000. Cambridge, Mass: Harvard University Press 2005

MIDDLETON, K.: The ironies of plant transfer: the case of prickly pear in Madagascar. In: BEINART, W., and MCGREGOR, J. (Eds.): Social History and African Environments; pp. 43–59. Oxford: James Currey 2003

MIRACLE, M.: Maize in Tropical Africa. Madison: University of Wisconsin Press 1966

TIMBERLAKE, L.: Africa in Crisis: the Causes, the Cures of Environmental Bankruptcy. London: Earthscan 1988

WILLIAMSON, M.: Biological Invasions. London: Chapman and Hall 1996

Prof. Dr. William BEINART
University of Oxford
African Studies Centre
92 Woodstock Road
Oxford
OX2 7ND
Great Britain
Phone: +44 1865 613900
Fax: +44 1865 613906
E-Mail: william.beinart@sant.ox.ac.uk

Altern in Deutschland

Die Deutsche Akademie der Naturforscher Leopoldina und die Deutsche Akademie für Technikwissenschaften acatech gründeten im Mai 2005 eine gemeinsame interdisziplinäre Akademiengruppe „Altern in Deutschland", die auf der Grundlage der besten verfügbaren wissenschaftlichen Evidenz öffentliche Empfehlungen erarbeitete, um die Chancen der im letzten Jahrhundert erheblich gestiegenen Lebenserwartung – die „gewonnenen Jahre" – vernünftig zu nutzen und mit den Herausforderungen des demographischen Alterns klug umzugehen.

Nova Acta Leopoldina N. F.

Bd. *99*, Nr. 363 – Altern in Deutschland Band 1
Bilder des Alterns im Wandel
Herausgegeben von Josef EHMER und Otfried HÖFFE unter Mitarbeit von
Dirk BRANTL und Werner LAUSECKER
(2009, 244 Seiten, 32 Abbildungen, 1 Tabelle, 24,00 Euro, ISBN: 978-3-8047-2542-3)

Bd. *100*, Nr. 364 – Altern in Deutschland Band 2
Altern, Bildung und lebenslanges Lernen
Herausgegeben von Ursula M. STAUDINGER und Heike HEIDEMEIER
(2009, 279 Seiten, 35 Abbildungen, 9 Tabellen, 24,00 Euro, ISBN: 978-3-8047-2543-0)

Bd. *101*, Nr. 365 – Altern in Deutschland Band 3
Altern, Arbeit und Betrieb
Herausgegeben von Uschi BACKES-GELLNER und Stephan VEEN
(2009, 157 Seiten, 29 Abbildungen, 20 Tabellen, 24,00 Euro, ISBN: 978-3-8047-2544-7)

Bd. *102*, Nr. 366 – Altern in Deutschland Band 4
Produktivität in alternden Gesellschaften
Herausgegeben von Axel BÖRSCH-SUPAN, Marcel ERLINGHAGEN, Karsten HANK, Hendrik JÜRGES und Gert G. WAGNER
(2009, 157 Seiten, 28 Abbildungen, 2 Tabellen, 24,00 Euro, ISBN: 978-3-8047-2545-4)

Bd. *103*, Nr. 367 – Altern in Deutschland Band 5
Altern in Gemeinde und Region
Stephan BEETZ, Bernhard MÜLLER, Klaus J. BECKMANN und Reinhard F. HÜTTL
(2009, 210 Seiten, 10 Abbildungen, 11 Tabellen, 24,00 Euro, ISBN: 978-3-8047-2546-1)

Bd. *104*, Nr. 368 – Altern in Deutschland Band 6 (in Vorbereitung)
Altern und Technik
Herausgegeben von Ulman LINDENBERGER, Jürgen NEHMER, Elisabeth STEINHAGEN-THIESSEN, Julia DELIUS und Michael SCHELLENBACH

Bd. *105*, Nr. 369 – Altern in Deutschland Band 7
Altern und Gesundheit
Herausgegeben von Kurt KOCHSIEK
(2009, 302 Seiten, 46 Abbildungen, 18 Tabellen, 24,00 Euro, ISBN: 978-3-8047-2548-5)

Bd. *106*, Nr. 370 – Altern in Deutschland Band 8
Altern: Familie, Zivilgesellschaft und Politik
Herausgegeben von Jürgen KOCKA, Martin KOHLI und Wolfgang STREECK unter Mitarbeit von
Kai BRAUER und Anna K. SKARPELIS
(2009, 343 Seiten, 44 Abbildungen, 9 Tabellen, 24,00 Euro, ISBN: 978-3-8047-2549-2)

Bd. *107*, Nr. 371 (2009) – Altern in Deutschland Band 9
Gewonnene Jahre. Empfehlungen der Akademiengruppe Altern in Deutschland
(2009, 102 Seiten, 1 Abbildung, 12,00 Euro, ISBN: 978-3-8047-2550-8)

Wissenschaftliche Verlagsgesellschaft mbH Stuttgart

Nature, Technology and Organization in Late-Imperial China

Mark ELVIN (Canberra)

With 7 Figures

Abstract

Behind present-day China's huge population, environmental degradation, and shortages of resources like water and wood, and prior to the arrival of the new solutions and new problems brought by the introduction of modern technology, lie three thousand years of the almost continuous clearing of forests and other vegetation cover for the purposes of farming, and restructuring of catchments for the purposes of drainage, irrigation and transport. Since the mediaeval economic revolution about a thousand years ago, as good farmland rather than labor-power became the factor of production in shortest supply, these processes were joined by the ever-increasing labor-intensification of farming in a relation of interactive positive feedback with the growing population, and accelerated by the assessment of agricultural taxes on the basis of cultivated land rather than output.

Zusammenfassung

Bevor das heutige China entstand – mit seiner riesigen Bevölkerungsdichte, seiner Umweltzerstörung und seiner Ressourcenknappheit im Bereich Wasser und Holz – und noch bevor neue Lösungen und neue Probleme durch die Einführung moderner Technologie entstanden, wurden bereits dreitausend Jahre lang unablässig Wälder und anderer pflanzlicher Oberflächenbewuchs zum Zwecke der landwirtschaftlichen Nutzung gerodet und Wassereinzugsgebiete zum Zwecke der Trockenlegung, Bewässerung und des Baus von Verkehrswegen neu strukturiert. Seit der wirtschaftlichen Revolution im Mittelalter vor etwa tausend Jahren, als gute landwirtschaftliche Nutzflächen – im Gegensatz zur Arbeitskraft – zu dem Produktionsfaktor wurden, der nicht in ausreichendem Maße zur Verfügung stand, wurden diese Prozesse noch verstärkt durch die ständig steigende Arbeitsintensivierung in der Landwirtschaft, in einem Verhältnis von interaktivem positivem Feedback zur wachsenden Bevölkerung, und sie wurden beschleunigt durch die Festsetzung der Besteuerung im Agrarbereich auf Grundlage des kultivierten Landes statt auf Grundlage des Ertrags.

1. Introduction

The focus of this short survey is on the two main characteristics of the late-imperial premodern Chinese economy that created the strongest pressures on the natural environment, and on the causes and consequences of their development over time. These were the *Gartenbau*, or 'horticultural-style farming', of the relatively lowlying core of China, and the distinctive dual style of agricultural management that combined the farming of the *land* by individual *families* with the *collective* management of *water* (Fig. 1).

Gartenbau was based on an extremely intensive use of labor in terms of hour of work *per hectare* that made possible practices like the construction of intricate local mud-walled ir-

Fig. 1 *Gartenbau* on Yingbing Mountain on the south side of Lake Chao in northern Anhui province. The fields seen here, together with a small sunken stream (on the right but not visible) fill the entire level part of the small valley. Most of the houses are perched on the slope out of sight to the left. Photo: Mark ELVIN 2003

rigation systems with high maintenance requirements, the planting out of the seedbed-grown shoots of some crops such as rice, repeated weedings, the incessant collection and maturing of materials for manure that, combined with the rotation of crops, made the fallowing of the land unnecessary, as well as applications of extra fertilizer during the period of the growth of crops, the intercropping of different crops, and the multiple cropping of a given area of land during each farming year, even if this necessitated the regular rebuilding of fields (as for dryfield wheat following wetfield rice and the reverse). The collective creation and management of all but the smallest water-control systems permitted the building of huge earthworks for defense against floods and the intrusion of the sea, extensive drainage channels for wetlands, and sizeable permanent and temporary storage basins with distribution networks controlled by weirs and dams, as well as long transport canals (Fig. 2). "Collective" mostly meant at least some participation by the *state* authorities, ranging from government entrepreneurial initiative followed by full managerial and legal control, down to mere supervision and intervention only when the system was functioning badly. It could, however, take a variety of forms, such as consortia of landlords, and including, though rarely, purely commercial organizations in which water was treated as a commodity separated from the possession of land.[1]

The present survey also sketches in the pattern through time of the development of labor-intensification, and its environmental effects, and identifies what were the main economic and technical pressures pushing this intensification, since, until the density of population

---

[1] There is a useful summary outline of these points in ELVIN 1975.

Fig. 2 Government-mobilized workforce rebuilding the dam across outflow of Lake Chao in northern Anhui province. The lake itself lies on the far side of the causeway that forms the righthand half of the quasi-horizon. Photo: Mark ELVIN 2003

has become acute (by premodern standards) and additional good potentially arable land has become in very short supply, it does not seem *a priori* economically rational to put more than at most a little additional labor into farming *a given area* with increased labor-intensification as opposed to extending the cultivated area, and farming both in a way that brings optimal returns per hour. After an initial rise, extra labor on a given area brings in declining returns per hour expended, though this simple formulation does not always adequately reflect the complexities of reality.

For example, introducing an additional crop, or even just staying up at night to protect one's ripening harvest from theft, could clearly bring in very large benefits per hour under certain circumstances. It is suggested here, though tentatively for the moment, that the *area-based* system of agricultural taxation that was in force during most of the second millennium CE probably imposed a significant additional pressure in the direction of intensification.

The essentials of this perspective are presented in the four images reproduced in this introduction. They are greatly simplified, especially since the major differences between China north and south of the Yangzi Valley are not analyzed separately. Note, too, that Figure 1 and Figure 2 above are modern photographs, but ones that capture essentially traditional features.

Figure 1 shows minuscule irrigated fields that, with a small watercourse, cover the width of the level land in a tightly restricted hill valley that slopes downhill to the right (approximately the West). In contrast, Figure 2 shows the mobilization of temporary mass labor under government direction rebuilding the cross-dam regulating the outflow from Lake Chao, not many kilometers away from the location of the first scene. The grey sky is the haze due mainly to industrial pollution that now hangs over many parts of the countryside for much of the year.

Mark Elvin

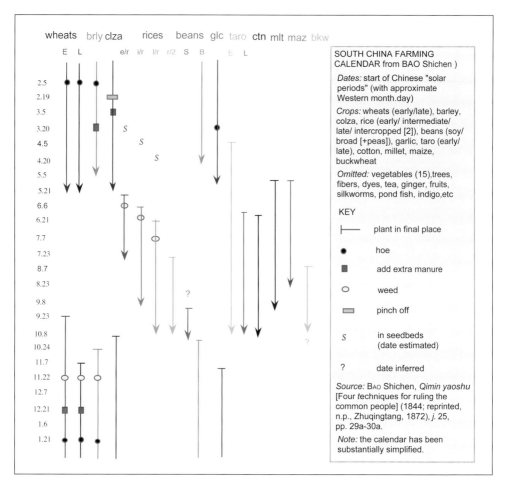

Fig. 3 Farming calendar for south China showing multiple cropping options. From Bao Shichen 1844 (reprinted 1872 by Zhiqingtang [n. p.]. Juan 25, 29a–30a).

Figure 3 shows an annual cropping calendar based on data in a Chinese agricultural encyclopedia from the first half of the nineteenth century. Brief examination makes it clear how the seasonal spacing of the crops would have made it possible to grow many of these harvests in succession on the *same* land, as they commonly were. A number of minor crops have had to be omitted from the diagram in the interests of legibility. Although the full picture would thus often have been more complex than that shown here, we can also assume that not every farm, or even all large ones, would have grown all the items shown or included in the original text.

In contrast, Figure 4 shifts the focus of attention to north China. It shows, for a span of two thousand years, the changing probability in a given year of a breach occurring in the protective dykes that flanked the Yellow River. These breaks were mostly due to the erosion resulting from the clearance and farming of the fragile slopes in the river's middle and upper course. This led to a rapid increase in the load of suspended sediment in the river, and hence a rise in the river bed and the increased risk of an overflow during the period of spate.

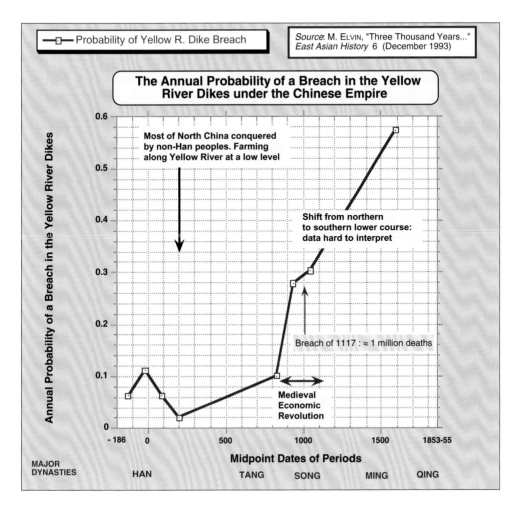

Fig. 4 The annual probability of a breach in the dikes along the Yellow River. Based on data listed in ELVIN 1993a, pp. 30–32. The mean frequencies of dike-breaches recorded for successive periods (not always completely continuous) determine midpoints that are linked to form the graph, with the exception of the final point, which is the late-Qing mean.

There are some complications hidden in the seemingly straightforward upward line after the middle of the first millennium CE. In particular, the complicated shifts southward in the lower course of the Yellow River during the thirteenth and fourteenth centuries make the trend-line close to meaningless during this period. In 1855, after 10 vertical meters of deposited sediment had built up at the southern mouth, it changed back to a northern lower course, where it is today – at least when the flow actually reaches the sea.

Overall, though it is plain that the long-term upward surge in *population* during and after the Song dynasty accompanying and following the medieval economic revolution, which was on the order of a quadrupling in a millennium,[2] was reflected – notwithstanding some shorter-

---

[2] Broadly agreed to be from around 100 million in the early Northern Song to over 400 million in the middle of the nineteenth century, but in a 'China' occupying a considerably larger area.

term irregularities in the graph not captured in this simple diagram – in the increasing environmental pressures that it imposed on the Yellow River. And also, in fact, on most of China.

Note, though, that it should not be assumed that increasing labor-intensification in farming *in itself* (as opposed to the clearing of a greater area of land) would always have worsened erosion. My guess would be that there were some positive as well as negative effects, depending on local circumstances. For example, it is reasonable to imagine that a contour irrigation system, once established, might often have eventually come, after a time, to reduce the rate of *further* losses of soil. See Figure 5 for the illustration of a typical situation.

Population growth did of course tend to promote intensification. By late-imperial times, there were probably no large-scale landscapes in the premodern world that had been so extensively anthropogenically transformed as that of China. The eighteenth-century Jesuits who observed the Chinese countryside felt both surprise and admiration at the extent of its remodeling, a reaction that underlines the contrast with the western Europe of the same date.[3]

2. Environmental Aggression

Widespread aggression towards the natural environment first becomes evident in China around the middle of the second millennium BCE with the formation of early military-agrarian states with urban defensive centers, and the first elements of bureaucratic organization and literacy. These states were in essence systems for mobilizing the efforts of entire populations to support warfare against other similar systems, and other external aggressors; and competitive interaction intensified this character. They needed to maximize manpower for armies and construction works, and the supplies of resources to maintain them. Given these objectives, the promotion of settled farming made practical good sense, even if not always to those mobilized for these ends.[4]

Clearing forests to make possible settled farming was regarded as the foundation of the state and civilization, especially under the Zhou dynasty in the late second millennium BCE and the first half of the first millennium BCE. This emerges clearly from some of its anthems celebrating the destruction of trees.[5] In the more than a thousand years between, very roughly, 500 BCE and 600 CE this government-led clearance of the original vegetation cover continued in some respects (such as in the creation of military colonies), but it was now counterbalanced by the concern of the states and empires of this age to stop their populations deserting settled farming to take refuge in the still little cultivated zones known as 'mountains and marshes', and either hunting and fishing there, or developing semi-independent holdings, or a mix of these. The motives behind this policy were to keep control over manpower and revenue on the one hand, and on the other to profit in monopolistic fashion from the resources of these areas. To some extent, though, it was also a partly intended and partly unintended process of nature conservation. In the long run it proved impossible to enforce; and these controls were largely abandoned and state power extended over the people in these once restricted areas.[6]

---

3 Elvin 2007, Vol. 2, pp. 13–17.
4 Elvin 2004, Chapter 5, "The logic of short-term advantage."
5 Elvin 2004, Chapter 3, "The great deforestation — an overview," and Chapter 4, "The great deforestation – regions and species."
6 Elvin 1993a, pp. 18–21, 25–27.

Fig. 5 Hillslope terracing in northern Anhui. Note that all the forest cover here is either replanting or natural regrowth since the stripping of the hills in the 1970s and 1980s (according to local oral information). Photo: Mark ELVIN 2003

The occupation of most of north China by non-Han peoples from the early fourth century CE (see Fig. 4) led to a migration of many Han into the Yangzi valley and south China. This in turn led to the development of several parts of this previously relatively underdeveloped region, especially by the extension of wetfield rice farming and the building of seawalls and polders. Rice-culture here, in spite of its long local history, was still rather unsophisticated and extensive, sowing seed broadcast rather than transplanting sprouts first grown in special seedbeds; but this area was in later centuries to become the heartland of intensified cultivation techniques. By the end of the tenth century, under the Northern Song dynasty, the economic weight of the lower Yangzi was moving towards parity with that of the northern Chinese plain, and the transition from *labor-power* being the factor of production in short supply to *farmable land* being the critical factor began slowly to spread outwards from the economic centers.

The wide-ranging commercial and technological transformation of the Chinese economy under the late Tang and Song, that is about a thousand years ago,[7] was the starting-point of the long-term sustained shift towards labor-intensive farming, a trend that – it should be stressed – continued long after the moment when the mediaeval economic 'revolution' changed China's economic style. An example of one of the focal points of a later set of changes but in the *same style* was the prefecture of Jiaxing in the lower Yangzi valley. Here in the sixteenth century there was a major upsurge in the growing of supplementary crops like beans, and colza,[8] and various green vegetables, in the context of rice farming.[9]

---

7 ELVIN 1973, Part 2, "The Mediaeval Economic Revolution."
8 Used for oil and oilcake. Similar to 'canola' and 'rapeseed' from *Brassica rapa, Brassica napus*, etc.
9 See KAWAKATSU 1992, pp. 111, 121, 114.

At the same time, there were slow, almost invisible, transforming processes of fundamental importance that should not be forgotten. Two of the major areas of historic China were once upon a time gigantic swamps. There were the north China plain, the historic floodplain of the Yellow River, and the wetlands of the central Yangzi, once famous as the Yunmeng Marshes. A third region, smaller but still sizeable, was the lower Yangzi delta, originally a brackish wetland except for its inland edges. All of these were drained and dried out over time as the result of human endeavors. And all of them have required some degree of maintenance ever since. Wilhelm WAGNER who taught farming in China for more than twenty years, said that his observations of the rare, small, neglected patches of land in north China suggested that most of the region would revert in a century or two to swamps if all human intervention were withdrawn.[10]

Maintenance was also the constant theme of the more immediately visible great water-control systems of imperial times. To a greater or lesser degree, their hydrology was always unstable. Hence an immense commitment of labor, of resources like timber to be cut and stone to be quarried, of money to be levied, and of political and organizational talents for management, was mortgaged almost in perpetuity for a system just to stay in the same place. Even so, not all systems survived. The water supply of the great ZHENG and BAI irrigation canals built under the First Emperor of Qin eventually dug down through the loess too deeply to feed the channels in a satisfactory manner, and proved impossible to resurrect.[11] The Min River system in the western province of Sichuan has survived for close to two-and-a-half millennia, in part thanks to its almost perfect topography for a gravity-driven distribution network, but it has also depended on unfaltering maintenance, some of it dangerous. Each year the barrier that deflects part of the river's flow into the system has to be shifted, so that the mud-clogged channels dry out and can be dredged, and the crumbling channel walls shored up. Then the barrier made of huge bamboo baskets full of heavy stones must be dragged back across the split main course, so that it once again deflects a share of the water back into the capillary-like network of ditches.[12] Occasionally dynamic processes free a system from the need for major maintenance. Thus the great seawall that ran along the southern side of Hangzhou from Tang times on, when joined with lock-gates that held back the outflow of rivers in spate, so that the ebb tides were too weak on their own to wash clear the sediments piling up along the outside, was eventually left like a peaceable but stranded whale in the midst of the land. Inevitably the newly formed land on the outer side of the wall was then reclaimed, desalinated, and a new wall built to protect it from the sea; and this occurred repeatedly.[13]

It is worth noting that in the early days of the creation of irrigation systems in the Yangzi delta area, they were sometimes called 'man-made clouds and rain'. In late-imperial times, engineers even aspired to change the courses of the great rivers. In the course of the 1570s the hydraulic engineer PAN Jixun, aware that the greater the celerity of a flow the greater its carrying capacity, consolidated the multiple channels of the lower course of the Yellow River into a single one. The swollen and accelerated current swept the sediments piled up on the riverbed down to the sea, and the delta grew outwards for 11 years at the rate of 1.5 kilometers a year. In the end it built up a crossbar there that blocked the outflow. Remedial measures had to be taken and PAN lost his job. Coastal currents also carried some of the offshore sediment

---

10 WAGNER 1926, pp. 181, and also 179.
11 WILL 1998, Chapter 9, in ELVIN and LIU 1998.
12 ELVIN 2004, Chapter 6, "Water and the costs of sustainability." On the Min River see also ELVIN 1993a, "Three thousand years of unsustainable growth," pp. 22–23.
13 ELVIN and SU 1995a, b.

down the coast and deposited it on the southern shore of the outer part of Hangzhou Bay, 400 kilometers away, where it formed a massive salient that is still there today. In the eighteenth century efforts were even made (as it was said) to "engineer the sea", by altering the exit taken by the Qiantang River where it enters the outer Hangzhou Bay, but without success.[14]

Chinese poems on hydraulics not infrequently use the metaphors of warfare, and even of assault on the gods. Some verses written on the indomitable Li Benzhong who for thirty-five years in the first half of the nineteenth century spent his own money on employing workers to split and remove the rocks in the Jinshajiang (the upper Yangzi), and elsewhere, that threatened ships, sum up this attitude:[15]

> By heating them with fiery coals he split stubborn rocks to clasts,[16]
> Flame-lights flashing downward to the River God's palaces,
> While the ringing of his hammers filled the precipice-walled valleys.
> ...
> He cleared away the hazards, on both routes, step by step.
> *Once the human will is fixed, then Heaven has no effect!*

How could western scholars in times past have seen this culture as one of eternal standstill, one unable to strive after a rational mastery of nature?

The Chinese also waged war on wild animals, and exterminated many of them, or reduced their numbers. The classics praise the use of fire by early culture heroes to expel dangerous beasts such as tigers, leopards, rhinoceros, and elephants.[17] Elephants were gradually driven out of all but a small corner of southwestern China by peasants who disliked their habit of raiding crops. Elephant habitats were destroyed by removing the shade-cover they needed.[18] Anadromous fishes that come up rivers to reproduce, but spend most of their adult lives at sea, were blocked off from their spawning-grounds by seawalls and lock-gates. Several unidentified species of dog-sized or wolf-sized animals, whose names have survived for a more than a millennium-and-a-half, vanished long ago from real life as their woodland habitats were cleared.[19] Rhinoceros were hunted for their tough hides that soldiers used as armor, and their horns which were thought to protect against most poisons; and so died out.[20] Their long gestation period gave them insufficient demographic resilience. Historical legends from Dali in the far southwest, and not strictly speaking a Chinese area in mediaeval times, also point to the existence of a long-lasting battle between self-sacrificing warriors and the giant pythons whom they allowed to swallow their bodies which they had covered with razor-sharp knives.[21]

3. The Environmental and Economic Periphery

Outside the central areas of China in which labor-intensive farming was possible lay a periphery composed widely varying environments. These included forests in the northeast, steppes

---

14 Both episodes are treated in ELVIN and SU 1998.
15 Poem by LI Zhan in ZHANG 1869, reprinted 1960, p. 13.
16 Fragments of rock.
17 For example LEGGE 1861, III.2.ix (pp. 156–157).
18 ELVIN 2004, Chapter 2, "Humans v. elephants: the three thousand years war."
19 ELVIN 2004, Chapter 10, "Nature as revelation," translation of XIE Lingyun's rhapsody "Living in the hills."
20 WEN 1995, especially pp. 189 and 193.
21 YIN 2002, e.g., pp. I. 68–69, I. 71–72, I. 168, III. 488.

in the north with a pastoral or mixed pastoral and farming economy, semi-deserts dotted with oases in the northwest, the lower slopes and valleys of the high mountains in the west and some of the southwest, and the jungle and semi-jungle in the other parts of the southwest. Some zones, functionally equivalent to peripheries, also lay in inner parts of the heartland, typically on higher ground between the valley floors and moderate slopes.

The aggregation of statistics on the core and the peripheries, which is often done by writers on the economy, tends to blur the distinctive characters of each. An example is figures for the area of farmed land per person, the periphery having only scattered oases of *Gartenbau*, which is highly productive per unit of area, while the farm economy of the heartland is based on this system. The population densities were correspondingly radically different, relatively dense in the core, thin in the outlying areas. The periphery, though it contained varying proportions of Han Chinese in-migrants, was predominantly the home of non-Han peoples, today's 'national minorities'. Many of these had, and sometimes still have, markedly different kinship structures and values from the mainline Chinese, including such institutions in various places as the levirate, cross-generation remarriage, polyandry, and the mother-run households of the Naxi.

In the periphery, in contrast to the rearing of small numbers of larger animals like cows and oxen in the late-imperial heartland, substantial herds of animals were common; and in places, like some parts of the environment shown in Figure 6, they still are. Historically, much of the farming in the periphery was of a shifting nature, and not fully settled. This was especially true after the end of the sixteenth century when crops from the New World like maize and sweet potatoes began to be grown, most notably in many parts of the internal 'periphery' where the soils were quickly exhausted, and the farmers sometimes did not legally own or rent the land they worked, nor normally pay taxes.[22]

Though Chinese attempts at expansion into the semi-desert go back over two thousand years, it was especially during the later empire, with its swelling population numbers, that farmers made repeated efforts to extend the farmed area into places not well suited to it; and were repeatedly defeated by processes such as salinization, or the rapid erosion of fragile subtropical soils with low nutrient content. Some lines from a poem of the eighteenth century sum up an eyewitness account of how the first of these caused the collapse of a farming venture in an arid part of the northwest:

> "After the snowmelt, mud in the sands is all we have to drink.
> Yet, in times gone by, our forebears here – managed both ploughing and ridging.
> Countless hoes chopped at the hills. – *Then the hills arteries changed*,
> And, as the first sprouts opened, so too did saline flakes.
> They tell tales, as well, of mosquito larvae, huge as caterpillars' pupae.
> The state-given hoes were tossed away, as trifles no longer useful."[23]

The defeated farmers, after trying unsuccessfully to rear cattle, were finally reduced to eating boiled grasses. Figure 6 offers a view of the frontier-line between the farmable and the non-farmable caused by the second of these problems, fragile soils, in another area in recent times.

---

22 Illustrative locality studies are of VERMEER 1998 and OSBORNE 1998.
23 ZHANG 1869, pp. 174–175. The poet was ZHOU Xipu.

Nature, Technology and Organization in Late-Imperial China

Fig. 6 Valley on the east side of Lake Erhai in western Yunnan showing scars of unsuccessful attempts at farming on the steeper slopes above the valley floor. Photo: Mark ELVIN 2003

Fig. 7 Peaks on the east side of lake Erhai in western Yunnan: the end of the cultivable periphery. Photo: Mark ELVIN 2003

In a few places, such as Zunhua in the area near the eastern end of the Ming-dynasty Great Wall, a successful mixed economy was developed that drew on a combination of *Gartenbau*, silviculture for nuts, fruits, and timber, and of some herding of animals, plus a little residual hunting. It is interesting, though, that attempts here to build transport canals, and more than

Mark Elvin

the smallest of irrigation systems, were defeated by the excessively sandy nature of many of the soils, and had to be abandoned.[24] The environment proved too strong for the premodern economic culture.

Beyond the periphery lay the wilderness, a glimpse of which is shown here in Figure 7. It was at times of a transcendental beauty, or soul-absorbing fascination, but rarely supportive of human life.

4. The Interactions between Population, Farm Technology, State Taxation, and the Environment

It is difficult to discover much about the dynamics of the Chinese population before the last two centuries of the Empire other than the approximate national and local totals. If we base ourselves on these last two centuries, however, we can make an educated guess at its basic character in late-imperial times, while bearing in mind that there were, demonstrably, substantial and still inadequately explained variations even between units as small as counties in the same region.[25] It combined almost universal marriage for females at an early age, namely about 17, with a very tight distribution about the mean, and – crucially – the ability to limit births *within marriage*. It seems that childbearing started fast, presumably since infant mortality was high (slightly under 300 per thousand in the first year), and establishing a family quickly a major priority; but this eased off once some children, presumably preferably sons, had reached at least the age of 5, by which time they had a much lower mortality than did newborns. The level of female infanticide and infant mortality due to differential neglect is not clear, but a minimum of 11 to 12% over the first year seems likely, and it might have been somewhat higher. Reproduction was thus *flexible*, and we may reasonably guess that people tended to have just slightly more children than was needed for a stationary population. If this is roughly correct, then over-reproduction at a level leading to 'Malthusian' disasters was not a pressing danger, but, rather, there was at most a continual but slow intensification of difficulties allowing sufficient time for them to be, to a significant measure, eased or handled by increased productivity and by out-migration. There were demographic disasters in the late-imperial period, notably the crisis in the middle of the seventeenth century associated with the transition from the Ming to the Qing dynasty, and, in much of the Yangzi valley, that caused by the Taiping Uprising and its suppression in the middle of the nineteenth. The country recovered from both of them, however, in about two generations; and the population resumed its upward path.

The physical means used to achieve the birth limitation within marriage required by the combination of general human reproductive biology and the data as to what probably happened in this specific case are not clear. It is unlikely that the various contraceptive medicines and abortifacients that it has been shown were known to at least some of the population[26] were in wide enough use for so general an effect. The best default hypothesis thus has to be the guess that it was mainly *coitus interruptus*, and that the lack of any known literary Chinese

---

24 ELVIN 2004, Chapter 9, "The riddle of longevity: why Zunhua?" The Zunhua e_0 has now been lowered significantly, using the method described in the source in the next footnote.
25 The following discussion is based on the research by the project on the population dynamics of the lower Yangzi valley run by Dr. Josephine Fox and myself. See the website at http://gis.sinica.edu.tw/QingDemography.
26 BRAY 1997, Part 3. Review by ELVIN 1998.

term for this practice simply reflects the use of euphemisms (as was the case in the UK when I was young) that we have so far failed to identify.²⁷

Conventional Chinese thinking linked a low density of population with slipshod farming. Thus one observer in the twelfth century wrote of the then backward area that is now Hubei province:

> "The land is so sparsely peopled that they do not have to bestow any great effort on their farming. They sow without planting out [seedlings from a seedbed] or weeding. If perchance they have weeded, they do not apply manure, so seeds and other sprouts grow up together. They cultivate vast areas, but have poor harvests."²⁸

In later periods it is very rare to find much emphasis placed in the returns to labor or time in farming, or on seed-to-yield ratios, though both can be found here and there.²⁹ The measure is almost always returns to units of farm land. Why?

Part of the answer is clearly that by about a thousand years ago, in the more advanced parts of China, good land had become the key scarce factor of production, and that this trend gradually became nearly universal in the Chinese economic heartland as the centuries passed. On the other hand, it is easy to show that the opening up of new land continued even beyond the end of the eighteenth century.³⁰ So it is surprising that there was little, if any, discussion in the agricultural handbooks of the relative economic merits of *intensifying* the cultivation on the land that a farmer already owned and of opening or buying *new* land, assuming that his family did the farmwork rather than renting it to others.

The answer is almost certainly that after about 1142 the yearly tax on agriculture was assessed in terms of the *cultivated area*, though in principle this was divided into nine grades by estimated quality, which would have made the levy more nearly equitable.³¹ It was not linked to annual output. Under the Ming it was ordained by the founder of the dynasty that ordinary commoners' fields had to pay each year 5 *sheng* and 3 *ge* of grain per unit area of 1 *mu*. This is about 5.3 liters of grain per 7% of a hectare.³² This is a low level, but it seems likely that it bore little relation to reality. Thus the local history for Yongjia county in the middle of the sixteenth century says:

> "In recent years, the land taxes and obligatory labour-services [which latter were also imposed according to the amount of land owned] have become multiple and burdensome. They are five times what they used to be in the past. Mountains and wetland have also been measured in *mu* units by pacing out, and then taxed. The other miscellaneous taxes that have been levied in addition [presumably also with the quantities linked to the area land owned] cannot be counted off one by one."³³

More than two hundred years later a passage from the *Veritable Records* for 1806 notes that:

---

27 Drawing on the ideas of SANTOW 1993, 1995.
28 ELVIN 1973, pp. 120–121.
29 On labor, see ELVIN 1982, pp. 29–30. On seed-to-yield ratios, see ELVIN 2004, pp. 208–209.
30 This remark is based on the relevant sections of *Nankai University Department of History* 1959.
31 SUDÔ 1962, especially pp. 474–479, on the process of the crystallization of the system.
32 WADA 1957, p. 151.
33 KIYOMIZU 1968, pp. 504–505.

> "When the tribute grain is collected, the local officials in the provinces collect more than the amount sanctioned by law. They make arrangements to have gentry of bad character act as their agents [...]. They bribe them in advance, granting them the right to contract for a certain portion of the tribute grain. The rustics and the poor have a redoubled burden because these persons can levy an excess amount from them just as they please."[34]

A substantial number of other items from the same source confirm that this kind of malpractice was not uncommon. An entry from 1735 reads: "In one place after another reports of tax grain resulting from the opening of new land have added nothing to the state's [actual] quota of taxes. Not only has it not proved possible to do the land-surveying [required], but the clauses in the demand for the first reporting [of liability] have been the occasion for *intimidatory and fraudulent assignations* [to pay], the high officials [seeking] a reputation for zeal for the public good [by claiming to be raising more taxes] and the lesser officials having the additional motive of seeking for profits."[35] A number of other passages from official sources show that taxes and labor-services (which were, as already noted, based on the acreage owned) were seen as a serious burden in various places at various times because of official malpractice. As early as 1668 the Emperor KANGXI thought that official cheating of the taxpayers had been rife.[36]

There has been a general assumption among many historians that the Chinese land tax did not constitute a serious burden, and if this were true then the argument put forward here would carry little weight. Although the issue needs more investigation, the sample of evidence given above, which is a small part of what is available, suggests that this view probably needs revising.

Assuming, then, for the moment, that the burden was not trivial, the economic logic for an independent peasant was to go much further in trying to grow more on his existing holding, before adding to his acreage, rather than farming more land less labor-intensively than he would have done in a tax-free situation, even if the latter course gave better *pre-tax* returns to his hours of work. The crux was that in the first case, unlike the second, there was no requirement that he share any of the increase from his additional efforts with the state. The widely varying situations of tenant-farmers are harder to evaluate in such a simple fashion, though the logic would have been almost identical in cases where there was a stable or permanent tenancy if there was also a fixed monetary rent. In general, landlords seem to have taken only a share of the *main* crop. This was better adapted than area-based tax to variations in weather, crop diseases, and pests, but, again, the tenant would have kept for himself the increase from at least the supplementary crops, which as partially shown in Figure 3, were not an inconsiderable part of the total harvest. In a more general way, tax pressures on landlords would also have been transmitted to their tenants through the terms of rental contracts.

The hypothesis suggested here is therefore that a slow but persistent upward population pressure interacted in mutually self-reinforcing fashion with the increased labor-intensification of farming that was needed to support growing numbers, given the incentive for farmers to avoid as far as possible extensification because of the penalties imposed by the area-based tax assessment system. This is of course only a hypothesis at present, and needs more thor-

---

34 ELVIN 1973, p. 267.
35 *Nankai Department of History* 1959, p. 75.
36 *Nankai Department of History* 1959, p. 664.

ough testing and sharpening. It goes some way, however, to explaining China's large late-imperial population, which has been the root of many of her later environmental problems.

The system that resulted had a distinctive mixture of strengths and weaknesses. For example, in late-imperial southern China, with its adequate days of sunshine and days free of frost, plus a widespread use of irrigation water, it was possible to grow 6 cereal crops in every 3 years on a unit of land, plus a wide range of subsidiary crops, in contrast with the traditional medieval European system that yielded 2 crops every 3 years when a field lay fallow every third year. Northern China was not so fortunate, but at least did not need to fallow. Multiple cropping also provided some increased resilience when faced with periods of unsuitable weather. In the other pan of the balance the Chinese farmer had to devote much larger efforts to fertility maintenance, having very few large animals to provide manure and no fallow fields for them to graze on, and to maintaining the channels and dams of water-control systems which were mostly made in premodern times only out of sun-hardened mud. In a wider perspective, the demands of the dense heartland population led by the end of late-imperial times to severe shortages not only of good land, but also of wood, both for fuel and for building, in many areas, and likewise shortages of water.

The introduction in recent times of inputs based on modern science, such as chemical fertilizers, concrete, pesticides, as well as an astonishing array of different kinds of machinery driven by steam, gasoline or electricity, plus modern transport, brought at least temporary relief from this situation, as well as their own problems, but that is another story. Anyone who travels these days in China in that half or two-thirds of the country that is away from the spectacular economic growth of the large cities – whose mastery of many modern skills in some cases goes back to the late nineteenth century[37] – cannot fail but to be aware that in many respects the economic and environmental past has barely begun to release its grip on the present.

References

BAO, S.: Qimin sishu [Four techniques for ruling the common people]. Reprinted 1872 by Zhiqingtang [n.p.]. 1844
BRAY, F.: Technology and Gender: Fabrics of Power in Late Imperial China. Berkeley, Los Angeles: University of California Press 1997
ELVIN, M.: The Pattern of the Chinese Past. Stanford: Stanford University Press 1973
ELVIN, M.: On water control and management during the Ming and Ch'ing periods. *Ch'ing-shih wen-t'i* III.3. (Review article of the *Shindai suiri-shi kenkyū* [Researches on the history of water control during the Qing dynasty] by MORITA Akira.) 1975
ELVIN, M.: The technology of farming in late-traditional China. In: BARKER, R., and SINHA, R. (Eds.): The Chinese Agricultural Economy. Boulder CO: Westview Press 1982
ELVIN, M.: Three thousand years of unsustainable growth: China's environment from archaic times to the present. East Asian History 6 (1993a)
ELVIN, M.: Le transfert des technologies en Chine avant la Seconde Guerre Mondiale. Nouveaux Mondes *2* (Genève: CRES) 1993b
ELVIN, M., and SU, N.: Man against the sea: Natural and anthropogenic factors in the changing morphology of Harngzhou Bay, circa 1000–1800. Environment and History 1. 1. 1995a
ELVIN, M., and SU, S.: Engineering the sea: Hydraulic systems and pre-modern technological lock-in in the Hangzhou Bay area circa 1000–1800. In: ITÔ, S., and YASUDA, Y. (Eds.): Nature and Humankind in the Age of Environmental Crisis. Kyoto: International Research Center for Japanese Studies 1995b

---

37 ELVIN 1993b.

Mark Elvin

Elvin, M.: Review article on Francesca Bray (1997), Technology and Gender: Fabrics of Power in Late Imperial China. Berkeley and Los Angeles: University of California Press. Journal of Social History *32.2*. (1998)

Elvin, M., and Liu, T. (Eds.): Sediments of Time. Environment and Society in Chinese History. Cambridge, New York: Cambridge University Press 1998

Elvin, M., and Su, N.: Action at a distance: The influence of the Yellow River in Hangzhou Bay since A.D. 1000. In: Elvin, M., and Liu, T. (Eds.): Sediments of Time. Environment and Society in Chinese History; pp. 344–410. Cambridge, New York: Cambridge University Press 1998

Elvin, M.: The Retreat of the Elephants. An Environmental History of China. New Haven, London: Yale University Press 2004

Elvin, M.: Economic pressures on the environment in China during the eighteenth century seen from a contemporary European perspective: Insights from the Jesuit *Mémoires*. *Tōyō Bunko hachijū-nen shi* [Eighty-year history of the Tōyō Bunko]. General editor: Shiba Yoshinobu. Tokyo: Tōyō Bunko. 3 Vol. 2007

Elvin, M., Fox, J., and Wen, T.-H.: Qing Demography Project website: http://idv.sinica.edu.tw/wenthung/Demography/ 2007

Legge, J.: The Works of Mencius. Vol. 2. In: The Chinese Classics with a Translation, Critical and Exegetical Notes, Prolegomena, and Copious Indexes. London: Trübner 1861

Kawakatsu, M.: Min-Shin Kōnan keizai-shi kenkyū [Researches on the farm economy of Ming and Qing Jiangnan]. Tokyo: Tōkyō Daigaku shuppankai 1992

Kiyomizu, T.: Mindai tochi seido-shi kenkyū [Researches on the history of the land system under the Ming dynasty]. Tokyo: Daion 1968

Nankai University Department of History (Ed.): Qing shilu jingji ziliao jiyao [Digest of materials on economics from the Qing-dynasty 'Veritable Records'. Beijing: Zhonghua shuju 1959

Osborne, A.: Highlands and Lowlands: Economic and ecological interactions in the lower Yangzi region under the Qing. In: Elvin, M., and Liu, T. (Eds.): Sediments of Time. Environment and Society in Chinese History; pp. 203–234. Cambridge, New York: Cambridge University Press 1998

Santow, G.: Coitus interruptus in the twentieth century. Population and Development Review. 19. 4. 1993

Santow, G.: Coitus interruptus and the control of natural fertility. Population Studies *49* (1995)

Sudō, Y.: Sōdai keizai-shi kenkyū [Studies on the economic history of the Song dynasty]. Tokyo: Tōkyō Daigaku shuppankai 1962

Vermeer, E.: Population and ecology along the frontier in Qing China. In: Elvin, M., and Liu, T. (Eds.): Sediments of Time. Environment and Society in Chinese History; pp. 235–282. Cambridge, New York: Cambridge University Press 1998

Wada, S.: Meishi shokkashi yakuchū. [Annotated translation of: The Economic Monograph of the 'Ming History']. 2 Vol. Tokyo: Tōkyō Bunko 1957

Wagner, W.: Die chinesische Landwirtschaft. Berlin: Parey 1926

Wen, H.: Zhongguo lishi shiqi zhiwu yu dongwu bianqian yanjiu [Studies on changes in plants and animals in China during historical times]. Chongqing: Chongqing chubanshe 1995

Will, P.-É.: Clear waters *versus* muddy Waters. The Zheng-Bai irrigation system of Shaanxi province in the late-imperial period. In: Elvin, M., and Liu, T. (Eds.): Sediments of Time. Environment and Society in Chinese History; pp. 283–343. Cambridge, New York: Cambridge University Press 1998

Yin, M. (Ed.): Dali gu yishu chao [Transcription of ancient lost books from Dali]. Sponsored by the Literary Association of Dali. Kunming: Yunnan xinhua chubanshe 2002

Zhang, Y. (Ed.): Qing shiduo [The bell of poesy of the Qing dynasty]. Reprinted 1960. Beijing: Zhonghua shuju 1869

    Prof. J. Mark D. Elvin, PhD
    Emeritus Professor and Visiting Fellow
    Division of Pacific and Asian History
    Research School of Pacific and Asian Studies
    Australian National University
    Canberra
    Australian Capital Territory
    Australia 0200
    Phone: +44 1993 880 197
    E-Mail: elvin@sino.uni-heidelberg.de

The Iconic Quality of Land in Australia and Oceania

Thomas Bargatzky (Bayreuth)

With 8 Figures

Abstract

Humans create order and meaning in their world views by transforming randomness into patterns of stability. The dichotomy between nature and culture has become an integral part of Western science and philosophy, but it has to be replaced with regard to pre-modern societies by models emphasizing the polarity of chaos and order. In both Australia and Oceania, for example, land has been of paramount practical and symbolic importance to the aboriginal people. Land is accorded *iconical* quality in both areas. An *icon* is the form in which incidents in the visible world are interpreted as occurrences in the world of the gods (J. Assmann). In Australia, the iconic status of land is bestowed *in illo tempore* by the gods who have been converted in landforms and continue to be effective through *real presence* (Realpräsenz) until the present time. In Oceania, however, land is transformed into iconic landscape through institutions like the kinship group and the sacred community center in Polynesia (*marae/malae*).

Zusammenfassung

Um sich in ihrem Dasein in Natur und Gesellschaft zurechtzufinden, benötigen Menschen Orientierungsrahmen, die das Zufällige zugunsten von Stablilitätsmustern reduzieren. Der Gegensatz von Natur und Kultur hat in dieser Hinsicht im Abendland lange als Grundlage von Weltdeutungen gedient, er versagt aber, wenn wir uns mit vormodernen Kulturen und ihrem Umwelthandeln befassen. Zur Illustration dieser Thesen werden in diesem Aufsatz anhand ausgewählter Beispiele die *ikonischen* Vergegenwärtigungen von Land in Australien und Ozeanien vorgeführt. ‚Ikon' ist die Form, in der nach dem Prinzip der sakramentalen Ausdeutung ein Vorgang in der sichtbaren Wirklichkeit als ein Ereignis in der Götterwelt ausgelegt wird (J. Assmann). In Australien geschieht diese Ausdeutung unmittelbar durch die Realpräsenz numinoser Wesen in den Landschaftsformen selber; in Ozeanien treten dagegen Institutionen zwischen Land und Mensch. In Polynesien sind dies Familie und Clan, aber vor allem Zeremonialplätze (*marae/malae*), in denen die ikonische Qualität von Land mittelbar wirksam wird.

1. Introduction

At a glance, Australia and Oceania do not have many things in common: with the exception of New Guinea, the small or tiny Pacific islands cannot offer very much when compared to the huge land mass of the Australian continent. Water, the sea, one may surmise, must be of paramount practical and symbolic importance to the Pacific islanders and plays a major role in their world view. Yet without land humans cannot survive. Land, therefore, is paramount not only to the aboriginal people of Australia, but of Oceania, too.

Humans encounter land not only in the form of gardens, as a means for subsistence, but also as landscape. The landscape as an ensemble of material facts has been claimed by cultural geographers as their arena of research. Myth has conventionally been the domain of anthropologists who have examined myths to understand social organization among small-scale, pre-modern communities. Myth and landscape, however, are "distinct but articulated signifying systems through which social relations among individuals and groups and human relations with the physical world are reproduced and represented" (COSGROVE 1993, p. 281).

In the historically strict sense, the term 'landscape' was coined in an emergent capitalist world to evoke a particular set of elite experiences represented in literature, painting, and music.[1] The term 'landscape', however, has a double meaning. On the one hand, a landscape is a perceived and selected set of localized and specific physical features visible on the ground; on the other hand, there are "archetypal landscapes imaginatively constituted from human experiences in the material world" (COSGROVE 1993, p. 281). Myths may shape both kinds of landscape and be shaped by them.

A powerful interpretive code for representing landscape in Europe has been the idea of the four elements. The Classical Greek doctrine of the four elements Earth, Water, Air, and Fire probably dates from pre-Socratic times (WINDELBAND 1957, pp. 28–41). It has exercised a profound influence on European culture and science. Parallel philosophical systems exist in Hindu, Japanese and Chinese thought (MALL 1995, CHANTEPIE DE LA SAUSSAYE 1925). Hence, they all belong to a class of sophisticated philosophical and theological reflections which occur in societies of the 'Early State' level of political organization (CLAESSEN and SKALNÍK 1978). One would not expect to encounter similar interpretive systems among the pre-modern people of Australia and Oceania, since philosophy presupposes the existence of a degree of abstraction in the social relations that comes into existence on the basis of social relations necessitated by commodity exchange only (SOHN-RETHEL 1989).

Given the great cultural, linguistic, and environmental diversity of the people and cultures of Australia and Oceania, it is not easy to work out traits shared by all their different cultures in relation to attitudes focused on features of nature, such as, for example, the four elements.[2] This can only be done at the most general level, but not by way of induction, but deductively, starting from premises concerning the most general features of their world view. These features can be comprehended with the help of the term 'iconicity' (German: *Ikonizität*) created by Jan ASSMANN (1984, pp. 135–138) with regard to the religion of Ancient Egypt. To put it plainly: in ASSMANN's reconstruction of the term, an 'icon' is the form in which incidents in the temporal, visible world can be explained as occurrences in the timeless world of the gods. In its most general meaning, a Christian sacrament would belong to the class of icons, since 'sacrament' refers to material means, through which God touches us; a sacrament being a means of communication with God, the material form in which God is present in the visible world. Baptism, water, wine, bread, oil, incense, candles, altars, matrimony etc. are sacraments in the diverse Christian belief systems. I hold that in Australia and Oceania, land has iconic status; land is to the indigenous people what a sacrament is to a Christian believer.

To render the concept of iconicity intelligible, an outline of the ontology of myth will be given in section 2. In section 3, it will be demonstrated that land is perceived as an icon in significantly different ways in Australia and Oceania. A model of human interaction with

---

[1] Cf. BENDER 1993, RITTER 1974, pp. 141–163. A critical examination of the historical veracity of PETRARCA's alleged ascent of Mont Vernoux, taken by RITTER as a stating point of his essay on landscape, is given by GROH and GROH 1996, pp. 17–82.

[2] For comparative purposes, BELLWOOD 1978, though somewhat dated, is still a good start, as is the monumental work on the German South Pacific, edited by HIERY 2001.

'nature' under the premises of an ontology of myth will be outlined in section 4. Finally, in section 5, suggestions are made for further research on the iconic status of land in Oceania and its different local expressions.

2. The Ontology of Myth

Anthropologists studying non-Western, or pre-modern cultures in general, mostly agree that 'nature' and 'culture' are conceived in ways which are very different from the corresponding terms in Western culture (BARGATZKY and KUSCHEL 1994). The modern Western concept of nature was developed during the age of Enlightenment and came into being essentially about the turn of the 18th century (HORIGAN 1988, BECKER 1994, COSGROVE 1993, THEOBALD 2003). It has its roots in European civil society and was fully developed during the decades following the industrial revolution (GLACKEN 1985). In modern times, nature has become both a quarry of economic resources, open for exploitation, and a repository of aesthetic and moral values; a haven offering peace and consolation for estranged and alienated romantic individuals (LEPENIES 1969). Hence, the specific contents of this modern Western concept of nature, as well as the concomitant rigid opposition between nature and culture, is not the product of enquiry but rather the condition of enquiry, as indicated by HORIGAN (1988, p. 6). It originated from the specific and singular events of European history leading to the emergence of civil society and its social and cultural repercussions. It would be naïve to assume that people with a very different historical background, and a pre-modern mode of production under conditions outside the laws dictated by commodity exchange (SOHN-RETHEL 1989), would share with modern Westerners a concept which is not older than some 300-odd years.

It is not the nature/culture dichotomy, but the ontology of myth, which serves as a prerequisite to our understanding of meaningful human action under non-modern settings. Myth makes no distinction between mind and matter, the part and the whole, the inside and the outside, nature and culture. All these distinctions lie behind the scientific mode of explanation, they follow from René DESCARTES' fundamental dichotomy of *res cogitans* and *res extensa* (HÜBNER 1983, chapter 15, HÜBNER 1985).[3] The history of ideas teaches us that the precise concept of nature seems to be a derivate of mechanical thinking (SOHN-RETHEL 1989, pp 66–67).[4] Hence, in myth, there is no nature in the modern Western sense. Any so-called 'natural' object can be considered alive under specific conditions, and be the potential or actual epiphany of a god. Since many gods active within 'nature' are also active within the human domain, there is no clear-cut distinction between the spheres of the gods and humans. Take *Hephaistos* for an example, the ancient Greek god of metalworking, or the pre-Christian mythology of the ancient Hawaiians (CHARLOT 1983). According to Valerio VALERI, in

---

[3] A systematic treatise on the concept of the ontology of myth and its value for cross-cultural interpretation is given by BARGATZKY 2007. – The Melanesian person has been described as a multiple person, a 'dividual', in contrast to the autonomous, indivisible, and self-animated Western 'individual' (LIPUMA 1998, STRATHERN 1988). Dividual persons "contain a generalized sociality within" (STRATHERN 1988, p. 13). What is more, the concept of the Melanesian dividual proclaims a unity between person and place (HESS 2006). Hence, this concept bears the unmistakable marks of the ontology of myth. Alas, contemporary anthropology is not yet ready to benefit from the explanatory power of HÜBNER's generalized theory (BARGATZKY 2003).

[4] SOHN-RETHEL 1989 refers to similar conclusions reached by CASSIRER 1910, pp. 155–158.

Hawai'i so-called 'natural' phenomena, closely linked to a god, signify certain predicates of the human species (VALERI 1985, p. 31).

The mythical conceptions of quality and causality are very different from their counterparts in science. Mythical qualities have their own specific kinds of efficacy which are defined through particular tales or 'stories' which I call *archái* (sg. *arché*), following Kurt HÜBNER (1983, pp. 236–239). An *arché* is a sacred event, a generic deed, the story of a god or holy person. Each of these *archái* is an individual story with a beginning and an end. They constitute time in the mythical sense, as a primordial *structure*, not as a continuum of individual events. HÜBNER likens the mythical flow of time to the turning of pages of a book: "each of which is new, in the book of these cosmic tales, until we finally reach the point where they begin to repeat themselves cyclically over and over again" (HÜBNER 1983, p. 236). Mythical time is cyclical insofar the *archái* incorporate an *identical* recurrence of the events described in it. Take, for example, the eternal rhythm of the departure and return of Persephone in Greek mythology, which constitutes the mythical interpretation of the succession of the seasons.

In myth, we are always dealing with the recurrence of the same events. The *very same* sacred generic deed or event is repeated over and over again. This is valid not only for *archái* related to 'natural' events, but also to so-called historical *archái*. A historical *arché* functions as a time-structure whenever some holy person bestows or establishes something for the first time. In Greek mythology, *Athena* bestows the olive tree and establishes the art of weaving; *Apollo* establishes the order of the state and bestows that of music; *Hermes* is the author of business, barter, and trade. "Since all of these are both stories and sequences of events, both of which belong to the mythical quality and substance of the divinity in question, these substances are also active whenever and wherever people plant olive trees or use the loom, make music or engage in business, etc."[5] Whenever such things occur, the same *arché* repeats itself, and the corresponding holy person is present and is called upon or entreated. Hence, it is appropriate to speak of the *real presence* (German: *Realpräsenz*) of the holy person, or of the *mythical substance* of the ritual or of everyday acts of humans when they follow the guidelines established by the corresponding gods in the *arché*. To put it differently, *archái* are sequences of events involving mythic substances that recur in identical form in the actions which are based upon them. In philosophy, the mythic substance in the ontology of myth can be likened to the Platonic idea, a substantial essence that remains identical in everything that partakes of it (HÜBNER 1985, pp. 36–38). Take the majesty of king *Agamemnon* in the Iliad as an illustration of the concept of mythic substance. This majesty "is a mythic substance invested by Zeus and passed down through his line like some hereditary trait. It also survives in such attributes of power as the king's sceptre, which is far more than a mere interchangeable symbol […] Agamemnon inherited the sceptre from his ancestors, one of whom had received it from Zeus. When oaths are sworn by Agamemnon's sceptre, the point is always made that this is a sceptre which can never be replaced and which partakes of the numinous substance of royal dignity."[6]

From a comparative point of view, rituals like the Christian Eucharist and the Samoan Kava Ceremony are iconic ritual commemorations which bring to mind *archái*. The sequences of events in theses rituals involve mythic substances which recur *sacramental*, as *real presence*, that is, as *real symbols*, in identical form, in the actions based upon them

---

5 HÜBNER 1983, p. 239.
6 BORCHMEYER 1992, pp. 33–34.

(BARGATZKY 1997, pp. 90–91). According to Mircea ELIADE (1986), they would belong to a class of polyvalent, universal hierarchies which can be translated into many diverse local cultural expressions. In the following section, the different ways in which *land* is attributed iconic quality in the world views of the indigenous people of Australia and Oceania is dealt with. Given the vastness of the area under consideration, this topic can be dealt with only by way of select ethnographical examples.

3. Land and Iconicity in Australia and Oceania

Basing our considerations on the ontology of myth as the point of departure for the interpretation of the relation between humans and land in Australia and Oceania, we find out that land, as an icon, can become a real symbol of the generic deeds of gods, *more-han-human-persons*. They created it *in illo tempore*, together with the institutions that shall help humans to subsist and survive and to maintain relations with the gods.[7] Land is converted into an icon in two different ways: either directly, in aboriginal Australia, or indirectly, through the mediation of human activities and institutions.

3.1 Australia

For the first Australians, the mythical substance of *archái* is active in human action, especially in ritual action. Yet in Australia, there is a specific way aboriginal holy persons make their presence a *real presence* in the way outlined in the preceding section. In Australian aboriginal mythology, 'dreamtime' is the era of creation when the great Spirit Ancestors walked on earth (FLOOD 1983). These ancestors created the world as it is today, both its 'natural' features and the human institutions. Locations and landforms, mountains, rock formations, plants, animals as well as certain human-made features, engravings or bull-roarers, are sacred since they are visible, ever-present and highly charged physical manifestations of events that occurred during 'dreamtime'. They are reference points that literally ground the culture to the land. Specific locations are the visible part of holy persons, in a manner analogous to the way bread and wine as the visible parts of the body of Christ in Roman catholic Eucharist, for example. Dreamtime geography is sacred geography, since by tying specific locations or features of the physical environment to dreamtime events, those events are made real in the sense of *real presence*. Thus, they provide both cultural meaning and personal orientation for life. Aboriginal 'landscape', therefore, is not conceived as a 'natural' or 'historical' construct, it is a template for the memory of generic dreamtime events (KÜCHLER 1993). Links between the present and the ancestral past are continuously being recreated by human action, but also by spirit conception which is "part of the way in which continuity is established between ancestral beings, social groups and land. Thus features of the landscape are signs both of people and of the embodiment of spiritual forces" (MORPHY 1993, p. 232). To put it briefly, land is not pure materiality, pure 'nature' in the modern Western sense of the term, but land sometimes *is* a superhuman person, or the visible part of such a person. Hence, in Australia,

---

[7] *In illo tempore* is the formula persistently used by Mircea ELIADE in his numerous writings to refer to the generic actions of gods in the beginning and which *ab origine* continue to be operative in the present (cf. ELIADE 1984, p. 32).

the sacral iconic quality of land is mediated through the real presence of holy persons and the vestiges of their *archái*, or generic deeds, in the land itself.

Take the bandicoot myth of the Northern Aranda as an example (STREHLOW 1947, pp. 6–10). The myth tells the story of the bandicoot ancestor *Karora*, who lived at a place now known as the Ilbalintja Soak. It is given here in a very reduced form only, since it is only the aspect of 'land as a superhuman person' which is under consideration here. – *Karora* is father of many sons who emerged from underneath his arm-pits. Soon, father and sons have killed and eaten all the bandicoots which had originally also sprung from *Karora's* body. In their search for food, the sons encounter an animal they take to be a sandhill wallaby and break its leg. This, however, is a holy person by the name of *Tjenterama* and who tells them 'I am a man as you are' and limps away. The sons return to their father. Then, the great flood of sweet honey comes from the east and engulfes them all. The father, *Karora*, is swirled into the Ilbalintja Soak and remains there; but the sons are carried by the flood under the ground to a spot three miles further on. Here they rejoin *Tjenterama* who is now their new chief. *Karora* is lying in eternal sleep at the bottom of the soak.

In STREHLOW's time, the people would point out the rocks and stones at a ceremonial ground which represent the undying bodies of the brothers which lie on top of a round stone which is said to be the body of *Tjenterama*. He is represented in all present-day bandicoot ceremonies as the great bandicoot chief of Ilbalintja. Still at the present time, the explanatory power of the 'story' is a trait in the belief system of aboriginal Australians that is interwoven with Christian teachings (KRINES 2001).

3.2 Polynesia

In a famous book titled *Vikings of the Sunrise* (1954), the New Zealand scholar Sir Peter BUCK[8] portrayed the Polynesians as a daring people of seafarers who settled the Pacific islands, likening them to the Vikings of a later era who settled Iceland and Greenland and crossed the northern Seas in courageous voyages of discovery and conquest. For many readers, the Viking-metaphor was decisive in creating an imagination of the Polynesians as a people whose most important bond was and still is to the sea. Yet everyone who has a personal acquaintance with Polynesians knows that for them land is of utmost importance. Canoe building and navigation were highly respected arts and crafts for specialists, yet garden produce was the staple food, and everyone was a gardener, even the chiefs.

In Polynesia, too, certain natural features like rocks are sometimes pointed out as the remains of legendary holy persons *in illo tempore*, but the character of such tales is rather folkloristic. Other than in Australia, landforms in Polynesia are not represented as *ritually important and meaningful* visible physical manifestations of the bodies or body parts of holy persons. Land is translated into *sacral iconic landscape* through the social organization of kinship groups, institutions of order, rank, and government, which have their origins in the *archái* of holy persons. In present times, these foundations of social order are validated by Atua, the God of Christianity.

The sacral quality of land is translated into the social organization to become an icon in two ways: (*i*) through the union of land and descent group in language, sentiment, and mean-

---

[8] Sir Peter's Maori name is TE RANGI HIROA. *Vikings of the Sunrise* was republished under the title *Vikings of the Pacific* (1959).

ing; (*ii*) through the chiefly and kingly names of the kinship groups who live on the land. Chiefly names and titles, as well as the sacred places (*malae* in Western Polynesia; *marae* in Eastern Polynesia) of the political community, which reflect the polities' ranking system (BELLWOOD 1978, pp. 339–370), carry the mythical substance of the land, the individuality of territory, and render it visible through human action.

3.2.1 The Descent Group (**kaainga*) and its Chiefly Names and Titles as Mythical Substance of the Sacral Quality of Land

Language is one key to our understanding of indigenous peoples' attitudes concerning the meaning of features such as land in their world-view. Since the languages of Polynesia belong to one language family, it is easier to work out common features than in the case of Micronesian or Melanesian languages which are rather more diverse (BELLWOOD 1978, pp. 117–134).

Studies in Oceanic linguistics and anthropology have succeeded in reconstructing at least two Proto-Polynesian (PPN) terms for social groups in Ancestral Polynesia: **kaainga* and **kainanga*.[9] In modern Polynesia, variants of **kaainga* have two significations: (*i*) the land on which food is grown; (*ii*) a social group based on kinship relations (BARTHEL 1961, p. 270; MILKE 1982). Hence, the term **kaainga* has to do with both land *and* people.[10] Associated meanings are 'place of residence, home, people of the place'. The core set of denotata shared by all of the principal branches of the Polynesian phylogeny include "a primary reference to land, and more specifically to an estate, but also to a social group that controlled rights to the estate. Moreover, it seems certain that the estate included a principal dwelling or house site."[11]

Modern Polynesian glosses of **kaainga* exhibit a different range of meanings from those associated with **kainanga*. The PPN term **kaainga* "referred to a minimal descent group or extended household, together with lands occupied and cultivated by that group. Probably, residence was a criterion for group membership, which would not have been the case with the larger and more inclusive **kainanga*" (KIRCH 1984, p. 66).

The core denotata of **kainanga* are: (*i*) a descent group, tracing descent back to a founding ancestor; (*ii*) unilineality of descent; (*iii*) exogamy; and (*iv*) control over land. Hence, **kainanga* indexes a 'land-holding or controlling group tracing to a common ancestor'. Moreover, **kainanga* were larger than minimal residential groups – the **kaainga* groups – and incorporated several such smaller groups. A priest-chief – the **qariki* – was the leader or titular head of the **kainanga*.

Summing up, we can state that in PPN **kaainga*, family group or minimal descent group and land are blended in emotion, meaning, and language. This finding is the basis for an understanding of the deep symbolic significance of garden produce (e.g., root crops and tubers such as taro) across Polynesia (Fig. 1). Shared meals of tubers which are cultivated on the family lands bestows communion not only among the living members of a descent group, but also with the ancestors buried on the family land.

---

9   The doubling of the 'a' is used here to mark vowel length. The ' marks the glottal stop.
10 For a concise summary of the linguistic research concerning the **kaainga-*kainanga* problem, cf. KIRCH 1984, pp. 65–66; KIRCH and GREEN 2001, pp. 207–218.
11 KIRCH and GREEN 2001, p. 215.

Thomas Bargatzky

Fig. 1 Taro garden near Luatuanu'u, island of Upolu, Samoa, 1981. Photo: T. BARGATZKY

In this connection it is striking that a PPN word meaning 'land' is **fanua*. This word is a retention from the Proto-Oceanic term **panua*, 'land, inhabited land or territory'. Pacific linguistics shows that **fanua* had a double meaning to the PPN speakers: 'land in the most general sense', but also 'placenta, afterbirth'! "This linkage of the two meanings metaphorically suggests the intimate linkage between people and their natal land."[12]

Cook Island Maori conceptions are an example of the intimate connection between humans and land, mediated by *'enua*, the modern derivate of **fanua*. Moreover, Cook Island Maori conceptions relative to land show that between descent groups (*ngati*), its names and titles, the land and its *marae*, there is a bond created in the mythical past through the *archái* of the founding heroes *Karika* and *Tangiia* which cannot be dissolved. It continues to be effective and meaningful even in modern times, when new concepts of land use, introduced foreign crops and new agricultural methods, have brought change to the indigenous Polynesians during the past 180 years (HARTAN 2002, PASCHT 2007). The 'mystical' quality of the relationship between humans, the land, and the realm of the *more-than-human-persons* manifests itself strikingly in the proverb: "If you have tradition, if you have land, and if you have a god, you are a man", communicated by Nihi VINI (1987).[13]

Ethnohistorical studies on the lifeways of the Cook Islanders show that love for the land (*'enua*) is deeply rooted in the mythical past, the founding era of the *archái* of the epic deeds of *Karika* and *Tangiia*, who allegedly settled the land, and to whom the present-day family groups (*ngati*) trace their descent. Arno PASCHT shows that in the past, when a Cook Islands Maori had been asked for his name, his answer would not consist only of his name proper, but he would formally recite the history and traditions of his family who owns this name,

---

12 KIRCH and GREEN 2001, p. 105; cf. table 4.1 on pp. 103–103 ibid.
13 I would like to thank Arno PASCHT who has drawn my attention to the proverb and to the article by VINI which has not been accessible to me while writing this article.

the land his name and family belong to, and the god or gods associated with the land of his kinship group.

In the Cook Islands, the relationship between humans and land is the basis for social stability, emotional security and personal identity. Pascht (2006, 2007) has demonstrated that this has not been so in the past only, but is also valid for the Cook Island Maori in present times. Rituals underline the intimate connection between humans and their land. The placenta was ritually buried on the family land, as is the dead person's body. Given the double meaning for PPN *fanua* ('land', 'placenta'), it is no surprise that in the Rarotangan variety of the Cook Islands' dialect, the term *'enua* also carries this double meaning of 'land' and 'placenta'. The placenta "was in ancient days considered highly sacred, and was buried with formal ritualistic ceremonial. A tree, generally a coconut, was afterwards planted over the spot" (Savage 1980, p. 63). The burying of the placenta, however, is not practiced very often today, but the ritual is orally transmitted and kept alive in the memory of the people as a powerful reminder of their relationship with the land.[14]

According to the Cook Islands' agricultural census of 1996, as rendered by Marianne Hartan (2005), 76% of all the households in the islands were engaged in agricultural production. This figure includes households which were active in subsistence agriculture only. Hence, for at least three quarters of the Cook Island Polynesians, land, and working the land, was of paramount importance. On the basis of her investigations concerning modern Cook Island Maori land tenure and agricultural practice, Hartan draws the conclusion, that it is land, not the sea, which carries a major impact on everyday life.

The example of the Cook Island Maori shows that in Eastern Polynesia, conceptions of the sacred quality of the land find their expression in the idea of an inextinguishable connection of the people and its lands. In Samoa, this relationship comes most clearly to the fore in the idea of the Samoan *malae* as the sacred center of the community.

3.2.2 The Samoan Malae as Political and Ceremonial Center of the Community

In the Samoan language, too, the term *fanua* has the double meaning of 'land' and 'placenta' (Milner 1966, p. 58). As in the other Polynesian societies, the iconic quality of land is mediated not only through kinship groups such as the **kaainga*, but also through the sacred chiefly names and titles of the kinship groups which live on this land. In Polynesia, the *malae or marae* is the visible center of political organization, and it is through the association with such a center that chiefly names and titles derive status, meaning, and prestige. The *malae/marae* is defined through its chiefly names and titles, and these are acknowledged through their connection to such a center. Landscapes are famous on account of the *malae/marae* of its communities, the families, chiefly titles and myths connected to them (Fig. 2).

This connection between land, the Samoan political 'family', and the *malae* will be illustrated by the example of the political and ceremonial structure of the Samoan subdistrict Safata.

Samoan spatial and social orientation is based on two kinds of dimensions (Shore 1982, pp. 48–51). The first is the pattern familiar to Polynesian directional orientation: seaward (*tai*) and landward (*uta*). *Tai* and *uta* may be conceived both as places, or relative directions.

---

14 The dictionary by Buse and Raututi 1995, p. 100, corroborates the double meaning of *'enua* as 'land, country, territory', and 'afterbirth, placenta'. Strangely, Savage 1980, p. 63, gives only the meaning 'placenta' for the term *'enua*.

Thomas Bargatzky

Fig. 2 The *malae* of the community Mauga, island of Savai'i, Gaga'emaunga district, Samoa, with chiefs' guest houses in the background, 1981. Photo: T. BARGATZKY

Symbolically, however, *i tai* ('toward the sea') suggests the more populated and ordered arenas of Samoan life. On the other hand, *i uta* means 'toward the rear part of the community', or 'toward the bush'. – The second distinction divides a community into concentric zones of center and periphery. The ideal community is conceived of as a circular structure, with the *malae* as the sacred political ground and center of dignified activity in the center (Fig. 3).

Around the *malae* there is a secondary circle made up of the chiefly guest houses or meeting houses closer to the center, and the smaller living quarters, huts, and cookhouses radiating outward. This is the design of an ideal community, however.[15] At the periphery of such a community, there are the boundaries with other communities, the sea, and the bush. These areas are outside the control of the chiefs of the community.

The seaward/landward distinction and the center/periphery distinction allow for intricate patterns of cross-cutting, highly symbolic and practical orientations which Samoans learn to master. A more detailed description of this dimension of Samoan spatial and social orientation is beyond the scope of this article, however.

The traditional Samoan religious rituals were inseparably connected with political organization and the Samoan *malae*, like its eastern Polynesian counterpart, the *marae*, was a community centre of major religious importance. This is still so today, despite the successful Christianization of Samoa. The Kava Ceremony, the most important public ceremony performed in the chiefly residences surrounding the *malae*, is a key to our understanding of the significance of the *malae* in relation to the connection between religion, politics, and landscape (BARGATZKY 1997).

---

15 Today, communities tend to be linear rather than circular, straddling both sides of the main government road. The circular model, however, is still a very powerful ideal model of the Samoan politically independent community (*nu'u*).

The Iconic Quality of Land in Australia and Oceania

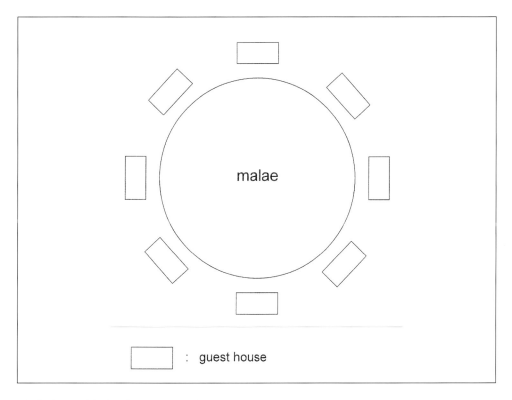

Fig. 3 Ideal model of the Samoan *malae* with guest houses

In pre-Christian traditions, the Gods were believed to assemble in the Ninth Heaven, the dwelling place of *Tangaloa*, the most important one of the old Samoan deities. When the Gods made the decrees concerning the formation of the Samoan universe, kava was served and the Kava Ceremony established. The Kava Ceremony and the Eucharist are homologous in structure, they belong to a class of polyvalent, universal hierophanies which can be translated into many diverse local cultural expressions. In traditional Samoa, this hierophany found its expression in the Kava Ceremony performed on the *malae* which is a sacred (*sa*; *pa'ia*) place. Hence, the Samoan chiefs (*matai*) officiating in the Kava Ceremony were the original priesthood of traditional Samoa. In the Christian era, they consecrated the Kava Ceremony to Jehova, thus it became possible for them to continue to officiate in a traditional politico-religious ritual, and to be Christians at the same time.

A *malae* is intimately connected to a political community and the region in which this community is located. A region is identified by the *malae* of the communities located there. Its essence, its 'personality', is encoded in the *archái* connected with the founding of the communities and the political family groups ('*aaiga*) linked to it. The example of Safata may serve as an illustration.

Safata is situated in the southwestern part of Tuamasanga on the island Upolu (Fig. 4).

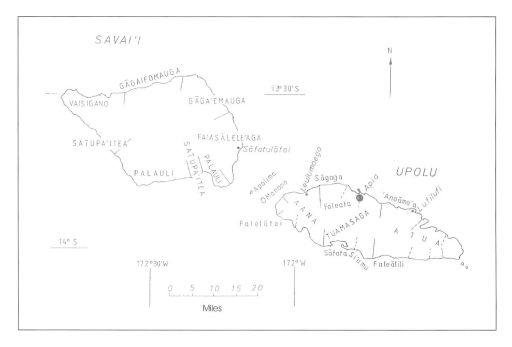

Fig. 4 Map: The Independent State of Samoa

While Tuamasanga is normally called a 'district' of Upolu, Safata is labeled a 'subdistrict'. From east to west, Safata comprises nine politically independent communities (Sam. *nu'u*): Mulivai, Tafitoala, Fausanga, Fusi, Vaie'e, Niusuatia, Lotofangá, Sataoa, and Sa'anapu. Each community has its own *malae*, and each one is governed by its own council of the holders of chiefly family titles (*matai*). These nine *nu'u* can be conceived as foci of an intricate traditional network of ceremonial relations between the *matai*-titles of the respective communities, and diverse social and ceremonial corporations (Fig. 5).[16]

The *malae* of Sataoa in the east and Vaie'e in the west are the paramount *malae* of Safata. Sataoa's *malae* is known as Siulepa, and the *malae* of Vaie'e goes by the name Tongamau. As *malae taaua*, they embody the political-ceremonial structure of Safata. George B. MILNER (1966, p. 247) renders the Samoan word *taaua* into English as "(Be) dear, precious, valuable", and "(Be) important, essential". They are said to be of equal rank.[17] Sataoa and Vaie'e per-

---

16 The Occasional Ceremonial Hierarchy of Safata has been described in detail elsewhere (BARGATZKY 1988, 1990). Ethnological fieldwork in Samoa was conducted in 1980–1981, 1985 and 1995 with financial support from the *Deutsche Forschungsgemeinschaft* (DFG). My Dr. phil. habil dissertation (BARGATZKY 1987) is based to a large degree on private information obtained from Samoans and remains unpublished, for the time being. It can be consulted, however, by investigators on mutual agreement. – I am deeply obliged to VASA Komiti and his brother AUSEUNGAEFAA Laupolatasi, *matai* from Fusi, who welcomed me in their '*aaiga* and shared their profound knowledge of Safata with me. I would also like to express my gratitude to high chief NONUMALO Sofara for his invaluable information and advice concerning all matters Samoan.

17 It is not easy to achieve unanimity among Samoans on any statement concerning traditional social organization and cultural meaning. Some of my informants would dispute the equality of rank of Sataoa and Vaie'e. I follow the judgment of VASA Komiti, however, who, as one of the traditional orators (*tulafale*) of Fusi, had immediate access to knowledge concerning the functioning of Safata's political-ceremonial organization.

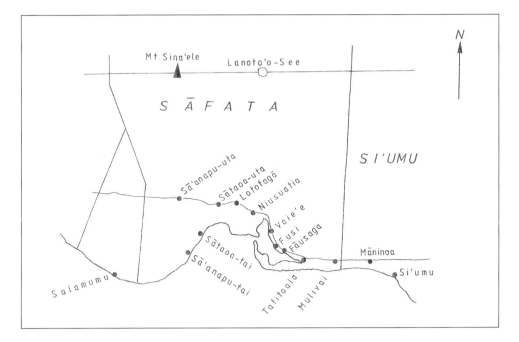

Fig. 5 Map: The subdistrict of Safata, Upolu, Samoa

form complementary functions in the operation of Safata's political-ceremonial organization, with the *malae* of the other communities in a supporting role.[18]

It is through the *malae* of a region, that this region's character and importance is ultimately identified in relation to the complicated network of political-ceremonial ties which unite all the different politically independent *nu'u* of Samoa. The significance of a region, as translated into its *malae*, is given through the *archái*. One might say that the land is 'baptized' through the *malae* which enshrine the *archái* that define what Samoan social, political and cultural identity is to its people.

3.3 Micronesia

In Micronesia, where land is not available in sufficient quality and quantity for human subsistence, humans create land. A good example is agriculture on the Pacific atoll. An atoll is a chain of sandy islets built on a reef surrounding a lagoon. The soil is made almost entirely of calcium carbonate. Ground-water is salty or brackish, except for a fresh-water lens in hydrostatic equilibrium in the ground (FREVERT 1987, p. 35, BARRAU 1965, p. 335).

A coral sand islet constitutes an environment not particularly suitable for horticulture. Yet, on many of these atolls, *Cyrtosperma chamissonis* (swamp taro) is one of the staple food plants. To grow swamp taro on atolls, the islanders dug a pit in the central part of the coral islet down to the level of the fresh water lens. Bottomless baskets made of woven twigs, are

---

[18] A detailed description of the role of Safata's *malae* in the enactment of political-ceremonial events is given elsewhere (BARGATZKY 1987, pp. 69–106).

placed on the muddy pit floor, filled with all available organic material of vegetable origin, including fallen leaves. Cuttings of *Cyrtosperma* are planted in these baskets floating on the fresh water mud. "These pits almost perfectly recreate the natural habitat of *Cyrtosperma*, a striking adaptation to an environment where both [...] the essential fresh water, and the organic matter of the tropical forest are lacking."[19]

The languages spoken in Micronesia are members of the Austronesian language family; the cultures of Micronesia, however, are much more diverse than the Polynesian ones. There is a strong matrilineal emphasis and social hierarchy, even stratification, yet it is more difficult to extract traits common to all of the Micronesian societies, which go beyond very general propositions. As far as land is concerned, however, it is safe to say that it is considered to be more than a deposit of resources for every-day life. It represents the history of the people and is part of their identity (KÄSER and STEIMLE 2001, pp. 475–479).

Peoples' attitude to food is a good indicator of the practical importance and symbolic meaning of land. Referring to Truk and Ponape Districts in the Eastern Carolines, for example, John L. FISCHER states that while food is relatively abundant most of the time on most islands and should not seem to be a matter of great concern from the point of view of supply, it nevertheless has important social and ceremonial values. "By gifts of food people acknowledge their respect for their chiefs and the old people of their family. Welcome to visitors is expressed first of all by gifts of food. On Ponape the ability to grow large yams is a mark of industriousness and respectability [...] On all islands, high and low, the breadfruit is important and the yearly variations in the harvest have been traditionally considered signs of supernatural favor or displeasure; ceremonies were conducted to ensure a bountiful crop."[20] Hence, food seems to be a valid indicator for the bestowal of an iconic status to land.

Despite the importance of land as the basic prerequisite for the survival of the people of Micronesia in general, however, there is a marked duality in their world view. Land and sea are perceived to be complemental, as are the northern and the southern part of an island, the east coast and the west coast, the left side of a village and its east side, brother and sister, husband and wife, men's work and females' work in the division of labor (KÄSER and STEIMLE 2001, p. 484). More research specifically directed to the topic of the symbolic value of land in the traditional mythic world view of the people of Micronesia is needed, however.

3.4 Melanesia

We have given, so far, examples of highly complex, ranked or even stratified Polynesian and Micronesian societies. Once we turn to politically egalitarian Oceanic societies, however, we find the same general properties of a world model which ranks everything according to the polarity of order/chaos; a model which cuts across our classification of nature according to the four elements. The distinctions between the elements earth, water, and air are metaphorically superseded by the overarching polarity of order and chaos. In Melanesia, for example, "Taro and yam, irrigation and drainage, the permanently humid environment of the tropical rain forest and the periodically (and less) humid environment of the monsoon forest, wet and dry [...] these contrasts seem to me of considerable importance in the history of Indo-Pacific agriculture."[21]

---

19 BARRAU 1965, p. 335.
20 FISCHER 1970, p. 233.
21 BARRAU 1965, p. 343.

For the Melanesians of New Caledonia, the garden is the safe domain of the daily life where everything is known, foreseeable, rational. Taro and yams are vegetatively propagated, edible perennials, always reproducing identical to themselves. It comes as no surprise that their clones are the symbols of a comforting and civilized stability, the symbols of the established and accepted social order. In comparison with the garden, the bush fallow or forest is the domain of the unknown, "haunted by ghosts of the dead; everything there is unforeseen, accidental, abnormal; escaped cultivated plants return to the wild state; wild yams are bitter; wild taros hurt the mouth. In the popular nomenclature of the Melanesian gardeners of New Caledonia, wild, bitter and toxic plants are opposed to domesticated, sweet, edible plants [...] In one word, the garden is culture and civilization, while the bush is uncivilized wilderness."[22]

Thorolf LIPP, in his recent fieldwork among the relatively isolated east-coast villages on Pentecost, Vanuatu, has shown that there is a general three-fold classification of order, intermediate realm, and chaos. A person entering a village of the Sa-people of Pentecost comes from the bush (*pane*) and has to enter an intermediate buffer zone (*pone*) demarcated by logs before he gets into the village proper (*lone*), the realm of humans. The village represents order, the forest chaos, but there is the intermediary realm of ambiguity represented by the *pone* where pigs roam about freely. Humans have to pass by the pigpens before entering the sphere of order and culture, represented by the houses. As in Polynesia, the gardens, however, are in the bush and thus represent a pole of order in the middle of the wilderness (LIPP 2008).

The ethnographic examples given so far call for a generalization. In the following section, therefore, the outline of a model of human interaction with the realm called 'nature' in Western thought will be presented. The model will be based on the premises of the ontology of myth as demonstrated in section 2, above.

4. Cosmic Order through Procreation

The opposition between nature and culture has become an integral part of Western science and philosophy. We conceive nature as something external, and independent of ourselves. According to contemporary economic theory, nature is everything which cannot be attributed to human work and achievement. Western culture theory is to a large extent the hostage of the nature/culture dichotomy which serves as a handy template for the explanation of the ways humans reproduce the organizational structure of their social and cultural relations with the realm called 'nature'. Generations of leading cultural anthropologists, e.g., Franz BOAS, Alfred Louis KROEBER, Claude LEVI-STRAUSS, Leslie A. WHITE, Marshall SAHLINS have cast their accounts of pre-modern lifeways into the mould furnished by the nature/culture dichotomy. The tenacity of this interpretive model has hampered our understanding of pre-modern lifeways.

What is at issue here is not the conjecture that binary oppositions are vital to thought. Humans need to create order and meaning in a seemingly chaotic world by transforming randomness into patterns of assumed stability. Binary oppositions are apt to achieve this. We should rather be concerned with the allegedly universal meanings given to nouns in binary oppositions such as nature/culture, since neither the concept of nature nor that of culture can be free from the biases and traditions of the culture in which they were constructed.

---

22 BARRAU 1965, p. 344.

The nature/culture dichotomy needs to be supplanted by a model which takes into account that in myth, there is no 'nature' in the modern Western sense of the term. The modern term for nature is derived from *natura*, which is the Latin translation of the Greek term *physis*.[23] There is a fundamental difference, however, between *physis* and nature. *Physis* (Engl. *growth*; Germ. *Wuchs*) means all that grows, emerges and disappears again. Contrary to modern understanding, however, growth, in myth, is not only a quantitative concept, it has qualitative connotations as well. Mind and matter are within *physis*. In *physis*, there is 'nature', but also human thought and institutions! PICHT shows that everything which modern philosophy has transferred into the Subject originally belonged to the innermost domain of what the Greeks called *physis*: logic, and the paramount principles of epistemology. Hence, the very structure of mind and meaning, which Greek philosophy assigned to *physis*, was projected into human consciousness by modern philosophy.

It is obvious that the concept of *physis* explodes the familiar modern dichotomies mind/matter, nature/culture, *res cogitans/res extensa*, subject/object. I suggest that *physis* is a concept suitable for cross-cultural comparison, because similar conceptions exist in other pre-modern cultural settings. For the Pawnee Indians, for example, *Tirawahat* was the Universe-and-Everything-Inside (O'BRIEN 1994, pp. 137–138); and for the Crow Indians, Mother Earth was the source of life and increase (VOGET 1994, p. 130). The ideas making up these concepts seem to be akin to those informing the concept of *physis*, since they also emphasize entirety, the unity of mind and matter. This is the great design informing the world views of pre-modern cultures, and the first cultures of Australia and Oceania are no exception in this regard.

The Samoans, too, for example, do not distinguish between nature and culture. The modern Samoan term for nature, *natura*, is a European loan-word, borrowed together with the specific European connotation "natural environment" (CAIN 1986, p. 122). Rather, genealogical metaphor is used to account for the origin of things we would classify as either natural or cultural. The most important knowledge of this kind is guarded by the chiefly members of certain ceremonial 'families' (*'aaiga*) who are commonly lumped together with descent groups in the anthropological literature, but who are something different altogether. The possession of the genealogical code thus enables these chiefs (*ali'i*) to 'procreate' metaphorically social and cultural order in its most dignified and sacred aspects (BARGATZKY 1988). Samoan *ali'i*, as members of a ceremonial family, have not only the power of physical procreation – as animals and ordinary human beings do. Through access to sacral genealogical knowledge they also have the power to maintain and continue the social and cultural order; that is, to guarantee its perpetuation through the creation of a king, a *tama'a'aaiga* (BARGATZKY 1996).

In Samoa, a king embodies everything considered dignified. It is the king's duty not to wield power in the sense of politics, but to promote the honor system which is strengthened by his virtuous standing. A king embodies what a Samoan would designate as culture. The origin of the first kings is explained through cosmogonic genealogies of a kind we find in all those non-modern communities where the *nexus rerum* is not created by commodity exchange. Material and immaterial things, plants, animals, mythical beings, the first humans and their institutions originate in successive genealogical steps out of some chaotic 'primary matter'. At the end of the cosmogonic genealogies, we find the first kings, the procreators and 'sons' of their family groups (KOSKINEN 1972), the procreators of chiefs and ordinary humans. The chiefs, in turn, elect the kings. Thus, they are the metaphorical creators of the

---

23 I follow the derivation of the concept *physis* given by PICHT 1989, pp. 54–57, 110–113.

kings and the iconic quality of the land is reflected in the chiefly titles, since they partake of the mythical substance of the *arché* of their original creation.

This distinctive characteristic of productivity links the concept of *physis* to the cosmologies of non-Western pre-modern cultures. As PICHT (1989, p. 113 and passim) has demonstrated, the Greek concept of *poiesis*, Lat. *productio*, refers to the innate creative ability of *physis*. The capacity to produce is an attribute of both 'nature' *and* of the human species. Both are parts of *physis*, hence, humans partake in *physis'* ability to produce. The supreme product of humans is the re-creation of the *archái* during the perennial effort to transform the cosmos from chaos to order. Hence, the binary opposition between nature and culture (Fig. 6) needs to be replaced by the opposition between chaos and order. The processes of *physis*, or all that which grows, emerges and disappears again, moves between the two poles of chaos, instability, bushland/wilderness on the one side, and order, stability, the polis or political community, or the garden, at the other side. Order stands for the *archái*, and the increase of order is inversely related to the decrease of chaos, and *vice versa* (Fig. 7). Everyday life is more prone to chaos and instability than ritual and when ritual takes over, there is the highest degree of order. In this ideal typical model, 'nature' and 'culture' are dissolved in the qualitatively distinct polarity of chaos and order; they do not exist any more as distinct qualities (Fig. 8). In the ideal typical polis and garden ('culture'), the 'elements' air, water, and earth are integrated into the real presence of an encompassing and ordered *gestalt* of the *archái*. The 'elements' are untamed, chaotic and threatening where the power of the *archái* is at its lowest ('nature'). 'Pure nature' does not carry positive connotations. 'Chaos' and 'order' replace 'nature' and 'culture'; they denote different states of aggregation of the 'elements' *and* of the state of human affairs! The categories 'nature' and 'culture' become submerged in a set of concepts which reflect the ontology of myth.

In this polarity of chaos and order, in which the modern Western dichotomy of nature and culture is dissolved, a general model of Oceanic classification is embedded. It is a model, which I will not hesitate to call a universally pre-modern cosmic model which exists as a general template outside Oceania, too.

5. Conservation, the Andesite-Line, and Deforestation in Oceania: Suggestions for Further Research

Summing up, I would like to point to two general conclusions which can be drawn from this review of the evidence concerning land and iconicity in Australia and Oceania. Let us begin with the peculiar Australian way of iconicity, e.g., the conception of the *real presence* of holy persons and the vestiges of their *archái* in the land itself. Seen from a comparative perspective, one may conclude that this way of representing the generic deeds of a particular local cosmology is not limited to the First Australians. It is striking that the North American Navajo Indians, for example, hold beliefs in relation to the iconic quality of land, and 'natural' features in general, which are very similar to those shared by the First Australians (CARMEAN 2002). This may be a trait common to the religious culture of nomadic, or semi-nomadic, hunter-gatherers. Further research, however, would be needed to clarify this issue.

Next, the growing anxiety about the state of the earth and the disillusionment with the role science plays in the process of environmental degradation has given rise to a rethinking of our moral attitudes and philosophical concepts. The study of myth has gained momentum, both in

Thomas Bargatzky

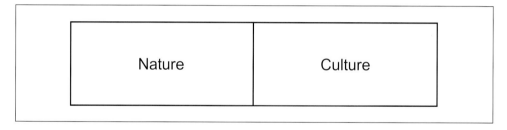

Fig. 6 The Western nature-culture dichotomy

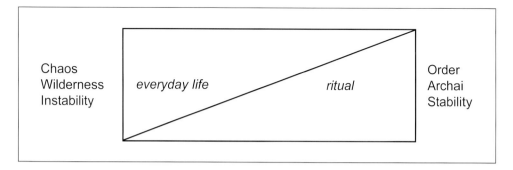

Fig. 7 The inverse relation between chaos and order

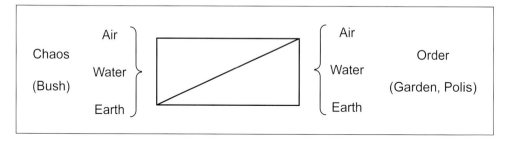

Fig. 8 The elements in relation to chaos and order: beyond 'nature' and 'culture'

the works of philosophers and anthropologists, and in the popular press. The philosopher Kurt HÜBNER published a book on the *Truth of Myth* (1985), and the philosopher Georg PICHT holds that natural science destroys nature and that such a science can not be a true science of nature (PICHT 1989, pp. 13–15). As a consequence of this disillusionment, there is a tendency to portray so-called 'primitive' people as the true paragons and saints of ecological salvation. North American Indians, in particular, have been allocated the role of fervent worshippers of 'Mother Earth'.[24] If we turn to their mythical world-view, it is claimed, all will be well, our natural environment and our souls will be saved.

---

24 For a critique of this concept, cf. GILL 1987.

There is a hidden cynicism in our turning to the world-views of people whose life styles have been thoroughly transformed or even destroyed by our civilization and expecting them to give us the recipes for the treatment of the deceases caused by European greed and expansion. In this attitude, the typical 'throwing-the-baby-out-with-the-bath-water' attitude manifests itself. Science is identified with rationality and both science and rationality are rejected. Western rationality is conceived as the source of evil, including environmental destruction. Myth is said to be 'irrational', and it is this alleged irrationality which attracts Westerners to become engaged in the so-called 'Umweltbewegung', longing for spiritual reorientation in a new 'environmental religion' (THEOBALD 2003). Yet myth has its own rational epistemology, as has been demonstrated convincingly by HÜBNER (1983, 1985) and PICHT (1986); and the so-called 'primitive people' (German: *Naturvölker*) have never been conservationists (BARGATZKY 1986, REDMAN 1999). This rule, for example, holds good for the horse-raiding and buffalo-hunting Indians of the North American Great Plains (VOGET 1994), as well as for the first Polynesian inhabitants of the Cook Islands who heavily taxed the original environment through introducing plants and animals for use (e.g., taro, banana, coconut, pig) which continue to play a major role in the islanders' diet. Since the beginning of the contacts with Europeans, the number of cultivated plants has risen steadily (HARTAN 2002).

Given the polarity of chaos and order inherent to a mythical world-view as delineated in section 4, is very unlikely that the people of Australia and Oceania were in a position to practice a philosophy of conservation, since the very idea of conservation would require the notion of nature as a being separate from the realm of humans, and for such a notion, there is no room within the framework of an ontology of myth, with the all-encompassing concept of *physis*, or its related concepts, as the blueprint for indigenous cosmologies. Yet our model of the iconic quality of land in Oceania is only a very general one. Further research is needed to work out the different ways this model is translated into the reality of the particular Pacific cultures, by taking into account the different environmental conditions under which these cultures have to operate. The Pacific Ocean, for example, is divided by a line famous to geologists and known as the andesite line (BELLWOOD 1978, pp. 20–21). It runs to the east of New Zealand, Tonga, and Fiji, round to the north of the Solomons, Bismarcks and New Guinea, and up again to the east of Yap and the Marianas in Micronesia. "In the Southwest Pacific on the Asian side of that line, a line of volcanoes of the explosive ash type blow out ash that may be wind-carried for hundreds of miles and that maintains the fertility even of islands like New Caledonia that have no volcanoes of their own. In the central and eastern Pacific beyond the Andesite Line, the main aerial input of nutrients to renew soil fertility is instead in dust carried high in the atmosphere by winds from the steppes of Central Asia. Hence islands east of the Andesite Line, and far from Asia's dust plume, ended up more deforested than islands within the Andesite Line or nearer Asia."[25]

The cultural differences between Western Polynesia on the one side, and Central and Eastern Polynesia on the other side, are well-known to anthropologists (e.g., BURROWS 1938, FREEMAN 1983). Central and Eastern Polynesia, to the east of the andesite line, are the areas of the *marae*, the big sacred temple compounds housing the idols. Easter Island, in the eastern point of the Polynesian triangle, and farthest removed from the andesite line, is probably the place of the most spectacular cultural efflorescence in ancient Polynesia, resulting in the most astonishing architectural features (SAHLINS 1955, DIAMOND 2005, pp. 79–119). It is also a

---

25 DIAMOND 2005, p. 117.

place with a very dramatic environmental degradation on account of the deforestation caused by the descendants of the Polynesian immigrants.[26]

In Western Polynesia, ceremonial life in lush and fertile Samoa has been very intense, and competitive, too. Yet the mediation between humans and the gods through the iconic nature of the *'aainga*, its lands and chiefly titles, and the *malae* has taken a different course. The *malae*, too, has been – and still is – a sacred ground, but its sacredness is not emphasized through rich architecture, as in Central and Eastern Polynesia. On the face of it, the Samoan *malae* is a rather inconspicuous ground, demarcated by the circle of the chiefs' guest houses only (cf. Fig. 2, above). Hence, it has often been mistaken for a simple 'village ground, without religious connotations'.[27] It is as if Samoans did not see the necessity to convince themselves of the enduring quality of their relationship with the gods through the idea of permanence, guaranteed by fixed iconic stone structures. The iconicity of the *marae* in Central and Eastern Polynesia was farther removed from the land than in the case of the Western Polynesia *malae*. In Samoa, the iconic character of the land translates itself *directly* into the *ceremonies* conducted on the *malae*. There is no visible and permanent intermediate structure like the temple compounds of Central and Eastern Polynesia, which seem to be most conspicuous where the environment is more vulnerable. This reminds one of a place such as Chaco Canyon in the northwestern corner of New Mexico, where the prehistoric Anasazi built one of the most remarkable and conspicuous ceremonial complexes in North America in one of the dryest, remotest, and most isolated parts of the continent (FRAZIER 1986). The more vulnerable the environment and the more endangered the survival of the society, the more conspicuous the iconic structures which invite the gods to take their seat among humans, in order to bestow their life-giving blessings. In Oceania, the Samoans could – and did – trust their rituals; the inhabitants of Central and Eastern Polynesia had to be on the safe side by constructing their *marae* in the form of elaborate structures. In both ways, however, land is transformed into an icon through which the generic deeds of holy people *in illo tempore* become a sacramental *real presence*.

Acknowledgement

Thanks are due to Marianne HARTAN, Thorolf LIPP and Arno PASCHT for discussing specific aspects of land tenure and symbolism in the Cook Islands and Pentecost with me. I would also like to thank Sandra JACOB for her help in preparing the drawings.

References

ASSMANN, J.: Ägypten: Theologie und Frömmigkeit einer frühen Hochkultur. Stuttgart: Kohlhammer 1984
BARGATZKY, T.: Einführung in die Kulturökologie. Umwelt, Kultur und Gesellschaft. Berlin: Dietrich Reimer 1986

---

[26] The connection between esoteric efflorescence, the intense competition in the building of temple mounds and erecting statues, and environmental degradation on Easter Island has recently been popularized by DIAMOND 2005. The very gist of his argument, however, has been published exactly 50 years before by Marshall SAHLINS 1955. DIAMOND, however, gives no credit to SAHLINS. I am stating the bare facts, however, and do not want to intimate that DIAMOND deliberately fails to mention SAHLINS' contribution.

[27] This erronous viewpoint is refuted by BARGATZKY 1997.

Bargatzky, T.: Die Söhne Tunumafonos. Deszendenz, Metapher und Territorialität am Beispiel der traditionellen politischen Organisation Westsamoas. (MS, unpublished Dr. phil. habil. Dissertation). München: Ludwig-Maximilians-Universität 1987
Bargatzky, T.: Evolution, sequential hierarchy, and areal integration: the case of traditional Samoan society. In: Gledhill, J., Bender, B., and Larsen, M. T. (Eds.): State and Society. The Emergence and Development of Societal Hierarchy and Political Centralization; pp. 43–56. London: Unwin Hyman 1988
Bargatzky, T.: "Only a Name". Family and Territory in Samoa. In: Illius, B., and Laubscher, M. (Eds.): Circumpacifica. Festschrift für Thomas S. Barthel (Vol. II), pp. 21–37. Frankfurt (Main): Peter Lang 1990
Bargatzky, T.: Embodied Ideas. An Essay on Ritual and Politics in Pre-Capitalist Society. In: Claessen, H. J. M., and Oosten, J. G. (Eds.): Ideology and the Formation of Early States; pp. 298–320. Leiden: Brill 1996
Bargatzky, T.: "The Kava Ceremony is a Prophecy": An Interpretation of the Transition to Christianity in Samoa. In: Hiery, H., and MacKenzie, J. M. (Eds.): European Impact and Pacific Influence. British and German Colonial Policy and the Indigenous Response; pp. 82–99. London, New York: I. B. Tauris 1997
Bargatzky, T.: "Mythos semper vivus". Eine Replik. Erwägen Wissen Ethik *14*, 61–73 (2003)
Bargatzky, T.: Mythos, Weg und Welthaus. Erfahrungsreligion als Kultus und Alltag. Münster: Lit 2007
Bargatzky, T., and Kuschel, R. (Eds.): The Invention of Nature. Frankfurt (Main): Peter Lang 1994
Barrau, J.: L'humide et le sec. An Essay on Ethnobiological Adaptation to Contrastive Environments in the Indo-Pacific Area. The Journal of the Polynesian Society *74*, 329–346 (1965)
Barthel, T. S.: Zu einigen gesellschaftlichen Termini der Polynesier. Zeitschrift für Ethnologie *86*, 256–275 (1961)
Becker, G. K.: The Divinization of Nature in Early Modern Thought. In: Bargatzky, T., and Kuschel, R. (Eds.): The Invention of Nature; pp. 47–61. Frankfurt (Main): Peter Lang 1994
Bellwood, P.: Man's Conquest of the Pacific. The Prehistory of Southeast Asia and Oceania. Auckland, Sydney, London: Collins 1978
Bender, B. (Ed.): Landscape: Politics and Perspectives. Providence, Oxford: Berg Publishers 1993
Borchmeyer, D.: On the Uses and Disadvantages of History for the Music Drama: From History to Myth in Wagner's Operatic Development (Transl. Stewart Spencer). Programmheft VI, Bayreuther Festspiele, 27–42 (1992)
Buck, Sir P. (Te Rangi Hiroa): Vikings of the Sunrise. Christchurch: Whitcombe and Tombs 1954
Buck, Sir P. (Te Rangi Hiroa): Vikings of the Pacific. Chicago: University of Chicago Press 1959
Burrows, E. G.: Western Polynesia: A Study in Cultural Differentiation. Etnologiska Studier *7*, 1–192 (1938)
Buse, J., and Raututi, T.: Cook Islands Maori Dictionary, by J. Buse with Raututi Taringa edited by B. Biggs and Rangi Moeka'a. Rarotonga: Ministry of Education, Government of the Cook Islands 1995
Cain, H.: A Lexicon of Foreign Loan-Words in the Samoan Language. Köln, Wien: Böhlau 1986
Carmean, K.: Spider Woman Walks this Land: Traditional Cultural Properties and the Navajo Nation. Walnut Creek: Altamira Press 2002
Cassirer, E.: Substanzbegriff und Funktionsbegriff. Berlin: Bruno Cassirer 1910
Chantepie De La Saussaye, P. D.: Lehrbuch der Religionsgeschichte. (Zwei Bände). Vierte, vollständig neubearbeitete Auflage. Hrsg. von A. Bertholet und E. Lehmann. Tübingen: J. C. B. Mohr (Paul Siebeck) 1925
Charlot, J.: Chanting the Universe. Hawaiian Religious Culture. Honolulu: Emphasis International 1983
Churchward, C. M.: Tongan Dictionary. Tonga: At the Government Press 1959
Claessen, H. J. M., and Skalník, P. (Eds.): The Early State. The Hague: Mouton 1978
Cosgrove, D.: Landscapes and Myths, Gods and Humans. In: Bender, B. (Ed.): Landscape: Politics and Perspectives; pp. 281–305. Providence, Oxford: Berg Publishers 1993
Diamond, J.: Collapse. How Societies Choose to Fall or Succeed. New York: Penguin Books 2005
Eliade, M.: Kosmos und Geschichte: Der Mythos der ewigen Wiederkehr. Frankfurt (Main): Insel 1984
Eliade, M.: Die Religionen und das Heilige. Elemente der Religionsgeschichte. Frankfurt (Main): Insel 1986
Fischer, J. L.: The Eastern Carolines (3rd Edition). New Haven: Human Relations Area Files 1970
Flood, J.: Archaeology of the Dreamtime. A Story of Prehistoric Australia and Her People. Honolulu: University of Hawaii Press 1983
Frazier, K.: People of Chaco: A Canyon and Its Culture. New York, London: W. W. Norton 1986
Freeman, D.: Margaret Mead and Samoa. The Making and Unmaking of an Anthropological Myth. Cambridge (Mass.): Harvard University Press 1983
Frevert, T.: Atolle – Süßwasseroasen im tropischen Ozean. Geographische Rundschau, Themenheft Tropische Inseln *39*/1, 32–38 (1987)
Gill, S. D: Mother Earth. An American Story. Chicago, London: The University of Chicago Press 1987
Glacken, C. J.: Culture and Environment in Western Civilization during the Nineteenth Century. In: Kendall, E. B. (Ed.): Environmental History. Critical Issues in Comparative Perspective; pp. 46–57. Lanham: University Press of America 1985

GOLDMAN, I.: Ancient Polynesian Society. Chicago: University of Chicago Press 1970
GROH, R., and GROH, D.: Die Außenwelt der Innenwelt: Zur Kulturgeschichte der Natur 2. Frankfurt (Main): Suhrkamp 1996
HANDY, E. S. C., and PUKUI, M. K.: The Polynesian Family System in Ka-'u, Hawai'i. Wellington: The Polynesian Society 1958
HARTAN, M.: Zum gesellschaftlichen Umgang mit (Lokal-)Wissen auf den Cookinseln, Polynesien. Entwicklungsethnologie *11*/1, 13–33 (2002)
HARTAN, M.: Landwirtschaft in Polynesien zwischen Tradition und Moderne. Aspekte des lokalen Wissens zum Anbau von Nutzpflanzen auf den Cookinseln. MS, Dr. phil. Dissertation, University of Bayreuth 2005
HESS, S.: Strathern's Melanesian 'dividual' and the Christian individual: a Perspective from Vanua Lava, Vanuatu. Oceania *76*, 285–296 (2006)
HIERY, H. J. (Ed.): Die deutsche Südsee 1884–1914. Ein Handbuch. Paderborn: Ferdinand Schöningh 2001
HORIGAN, S.: Nature and Culture in Western Discourses. London, New York: Routledge 1988
HÜBNER, K.: Critique of Scientific Reason (Translated by P. R. DIXON, Jr., and H. M. DIXON). Chicago: The University of Chicago Press 1983
HÜBNER, K.: Die Wahrheit des Mythos. München: Beck 1985
KÄSER, L., und STEIMLE, P.: Grundzüge des Weltbilds in Gesellschaften Mikronesiens. In: HIERY, H. J. (Ed.): Die deutsche Südsee 1884–1914. S. 475–507. Paderborn: Schöningh 2001
KIRCH, P. V.: The Evolution of the Polynesian Chiefdoms. Cambridge: Cambridge University Press 1984
KIRCH, P. V., and GREEN, R. C.: Hawaiki, Ancestral Polynesia. An Essay in Historical Anthropology. Cambridge: Cambridge University Press 2001
KOSKINEN, A. A.: *Ariki* The First-Born. An Analysis of a Polynesian Chieftain Title. FF (Folklore Fellows) Communications No. *181* (second printing). Helsinki: Suomalainen Tiedeakatemia 1972
KRINES, S.: Rezente christliche Einflüsse in der Traumzeitvorstellung der australischen Aborigines. Anthropos *96*, 157–168 (2001)
KÜCHLER, S.: Landscape as memory: The mapping of process and its representation in a Melanesian Society. In: BENDER, B. (Ed.): Landscape: Politics and Perspectives; pp. 85–106. Providence, Oxford: Berg Publishers 1993
LEPENIES, W.: Melancholie und Gesellschaft. Frankfurt (Main): Suhrkamp 1969
LIPP, T.: Gol – das Turmspringen auf der Insel Pentecost in Vanuatu. Beschreibung und Analyse eines riskanten Spektakels. Münster: Lit 2008
LIPUMA, E.: Modernity and forms of personhood in Melanesia. In: LAMBEK, M., and STRATHERN, A. (Eds.): Bodies and Persons. Comparative Perspectives from Africa ans Melanesia; pp. 53–79. Cambridge: Cambridge University Press 1998
MALL, R. A.: Philosophie im Vergleich der Kulturen. Interkulturelle Philosophie – Eine neue Orientierung. Darmstadt: Wissenschaftliche Buchgesellschaft 1995
MILKE, W.: Der Begriff Kainanga in Polynesian und Mikronesien. Zeitschrift für Ethnologie *107*, 207–217 (1982)
MILNER, G. B.: Samoan Dictionary. Oxford: Oxford University Press 1966
MORPHY, H.: Colonialism, history and the construction of place: The politics of landscape in Northern Australia. In: BENDER, B. (Ed.): Landscape: Politics and Perspectives; pp. 205–243. Providence, Oxford: Berg Publishers 1993
O'BRIEN, P. J.: Pawnee views of nature in the Central Plains: The historic and prehistoric data. In: BARGATZKY, T., and KUSCHEL, R. (Eds.): The Invention of Nature; pp. 137–158. Frankfurt (Main): Peter Lang 1994
PASCHT, A.: Das Erbe von Tangiia und Karika. Landrechte auf Rarotonga. MS, Dr. phil. Dissertation, University of Bayreuth 2006
PASCHT, A.: Die Macht der Traditionen – Maori Customs und Landrechte auf den Cookinseln. Zeitschrift für Ethnologie *132*, 59–76 (2007)
PICHT, G.: Kunst und Mythos. Stuttgart: Klett-Cotta 1986
PICHT, G.: Der Begriff der Natur und seine Geschichte. Stuttgart: Klett-Cotta 1989
REDMAN, C. L.: Human Impact on Ancient Environments. Tucson: The University of Arizona Press 1999
RITTER, J.: Subjektivität. Frankfurt (Main): Suhrkamp 1974
SAVAGE, S.: A Dictionary of the Maori Language of Rarotonga. Suva: Institute of Pacific Studies, University of the South Pacific; in Association with the Ministry of Education, Government of the Cook Islands 1980
SHORE, B. S.: A Samoan Mystery. New York: Columbia University Press 1982
SAHLINS, M. D.: Esoteric Efflorescence in Easter Island. American Anthropologist *57*, 1047–1052 (1955)
SOHN-RETHEL, A.: Geistige und körperliche Arbeit. Zur Epistemologie der abendländischen Geschichte. (Revidierte und ergänzte Neuauflage.) Weinheim: VCH – Acta Humaniora 1989
STRATHERN, M.: The Gender of the Gift. Problems with Women and Problems with Society in Melanesia. Berkeley etc.: University of California Press 1988

STREHLOW, T. G. H.: Aranda Traditions. Melbourne: Melbourne University Press 1947
THEOBALD, W.: Mythos Natur. Die geistigen Grundlagen der Umweltbewegung. Darmstadt: Wissenschaftliche Buchgesellschaft 2003
VALERI, V.: Kingship and Sacrifice. Ritual and Society in Ancient Hawaii. Chicago: The University of Chicago Press 1985
VINI, N.: Outer Islanders on Rarotonga. In: MASON, L., and HERENIKO, P. (Eds.): In Search of a Home; pp.103–109. Suva, Fiji: Institute of Pacific Studies, University of the South Pacific 1987
VOGET, F. W.: Were the Crow Indians conservationists? In: BARGATZKY, T., and KUSCHEL, R. (Eds.): The Invention of Nature; pp. 125–136. Frankfurt (Main): Peter Lang 1994
WINDELBAND, W.: Lehrbuch der Geschichte der Philosophie. (Mit einem Schlusskapitel Die Philosophie im 20. Jahrhundert und einer Übersicht über den Stand der philosophiegeschichtlichen Forschung. Hrsg. von H. HEIMSOETH). Fünfzehnte, durchgesehene und ergänzte Auflage. Tübingen: J. C. B. Mohr (Paul Siebeck) 1957

       Prof. Dr. Thomas BARGATZKY
       Universität Bayreuth
       Facheinheit Ethnologie
       GWII Universitätsstraße 30
       95440 Bayreuth
       Germany
       Phone: +49 921 554137
       Fax: +49 921 554136
       E-Mail: Thomas.Bargatzky@uni-bayreuth.de

Wüsten –
natürlicher und kultureller Wandel in Raum und Zeit

Leopoldina-Meeting
Deutsche Akademie der Naturforscher Leopoldina in Zusammenarbeit mit der Gesellschaft für Erd- und Völkerkunde zu Stuttgart e. V.

am 2. und 3. Mai 2008 in Stuttgart

> Nova Acta Leopoldina N. F. Bd. *108*, Nr. 373
> Herausgegeben von Wolf Dieter BLÜMEL (Stuttgart)
> (2009, 259 Seiten, 141 Abbildungen, 7 Tabellen, 24,95 Euro,
> ISBN: 978-3-8047-2680-2)

Wüsten üben eine eigenwillige Faszination aus: Sie sind heute einerseits attraktive, mystifizierte, abenteuerträchtige Reiseziele, andererseits aber noch immer extrem lebensfeindliche Naturräume. Die aktuelle Diskussion um den globalen Klimawandel und seine möglichen Folgen wirft ein Schlaglicht auf die lebensarmen Wüsten der Erde. Im vorliegenden Band werden vielfältige Aspekte des Lebens- und Wirtschaftsraumes „Wüste" anhand von Beispielen aus der Sahara und der Namib-Wüste in Afrika, der Atacama in Südamerika und den Wüstengebieten Zentralasiens thematisiert, z. B. die Rekonstruktion der klimatischen und landschaftlichen Geschichte, die kulturelle und kulturgeschichtliche Bedeutung, der aktuelle Wandel und die zukünftige Entwicklung dieser Regionen. In diesen Kontext ordnen sich auch archäologische Forschungsbefunde ein und liefern erstaunliche Erkenntnisse über frühere Kulturmilieus. Aber auch Fragen des Wüstentourismus in der Gegenwart werden kritisch beleuchtet. Die Beiträge zum sozialen, wirtschaftlichen und politischen Wandel in wüstenartigen Gebieten zeigen, welche – teils unerwartete – Rolle solchen Grenzräumen der Ökumene zukommt. Es wird deutlich, wie verletzlich diese Naturräume sind, welche – teils verderbliche – Rolle der Mensch in vielen dieser Ökosysteme spielt und welches gesellschaftlich-politische Konfliktpotenzial sich darin verbirgt. Klimatologische Modellierungsansätze werfen einen Blick in die mögliche zukünftige Entwicklung der Wüsten vor dem Hintergrund des aktuellen Klimawandels in einer stark anthropogen beanspruchten und veränderten Welt.

Wissenschaftliche Verlagsgesellschaft mbH Stuttgart

Environmental History in the Americas: The Two Great Invasions

John McNeill (Washington)

Abstract

American ecosystems have been buffeted by two main human invasions. This contribution aims to provide an overview of the environmental consequences of these twin invasions in the history of the Americas. It will give occasional attention to the themes of earth, air, water, and fire, which are the warp and weft of this book. It will also give intermittent attention to other parts of the world in order to show in which respects the American experience was extraordinary, and in which respects commonplace.

Zusammenfassung

Die amerikanischen Ökosysteme sind durch zwei entscheidende Eingriffe des Menschen erschüttert worden. Der Beitrag soll einen Überblick über diese doppelten Eingriffe in der Geschichte der amerikanischen Kontinents geben. Gelegentlich wird auch auf die Themen Erde, Luft, Wasser und Feuer eingegangen, die dieses Buch bevorzugt behandelt. Der Beitrag geht auch auf andere Teile der Welt ein, um aufzuzeigen, in welcher Hinsicht die amerikanische Erfahrung durchaus außergewöhnlich und in welcher Hinsicht sie geradezu alltäglich war.

The environmental history of the Americas since the first human occupation has been one of unusual tumult. The main reason for this is the short history of humankind on the American continents: American ecosystems have been buffeted by two main human invasions, with little time to adjust. The first invasion began perhaps 15,000 years ago when wanderers crossed the land bridge connecting Siberia to Alaska; the second began just over 500 years ago, when Christopher Columbus arrived in American waters, soon to be followed by millions of fellow Europeans, enslaved Africans, and eventually a growing stream from much of the rest of the world.

This chapter aims to provide an overview of the environmental consequences of these twin invasions in the history of the Americas. It will give occasional attention to the themes of earth, air, water, and fire, which are the warp and weft of this book. It will also give intermittent attention to other parts of the world in order to show in which respects the American experience was extraordinary, and in which respects commonplace.

John McNeill

1. The First Invasion

We humans tend to regard other creatures – water hyacinth, zebra mussels, grey squirrels, starlings, rabbits – as invasive species. But by far the most successful invasive species of the past 100,000 years is our own. We have expanded our niche from eastern and southern Africa to include all the continents and thousands of islands, while expanding our population by four or five orders of magnitude. If reigning interpretations of mitochondrial DNA evidence are correct, about 75,000 years ago our numbers amounted only to about 10,000, and we are now more than six billion. Few if any other species can match this record.

1.1 Occupation of the Americas

Our invasion of the American continents probably began only about 15,000 years ago. Archaeologists disagree vehemently about the dating and interpretation of evidence of the first occupation, with some claiming that the first arrivals came more than 40,000 years ago. But the more conventional position, obviously provisional and subject to revision, is a mere 14,000 or 15,000 "Before Present" (BP) (GILBERT et al. 2008, GOEBEL et al. 2008, MANN 2005, MITHEN 2004, pp. 210–245). If this date is anywhere close to correct, then the Americas are easily the last of the continents (barring Antarctica) subjected to human invasion, with Australia a distant second (at perhaps 40–60,000 years BP). The date, however, corresponds roughly with the re-occupation of northern Europe at the end of the last Ice Age, so in that sense the experience of human settlement of the Americas is not so eccentric.

It is likely, of course, that more than one group of invaders crossed Beringia, chasing herds of caribou (reindeer) during the intervals when sea levels were low yet ice-free corridors existed on land. When they arrived they seem to have spread out rather quickly, reaching Chile and easternmost Brazil within a couple of thousand years, if the entry date is indeed close to 15,000 BP. This may indicate boat travel, but it is quite within the realm of possibility that the first few dozen generations of Americans walked to Chile – southward progress of 10 kilometers per year would have brought them there in a little over a thousand years.

The American continents were a giant cul-de-sac. Soon, after the ice melted back and sea level rose, they were cut off from Asia except for sporadic boat traffic. For all intents and purposes, human history in the Americas between its inception and 1492 took place in isolation from the rest of the world, a fact which would powerfully shape events after COLUMBUS.

1.2 Extinctions

The first invaders of the Americas came with late Paleolithic technologies and skills. Presumably, while following herds across Beringia they refined their big-game hunting abilities, and thus once in America enjoyed a brief bonanza, killing and eating a dozen species of large mammals, none of which had any experience of brainy, upright apes armed with spears, bows, language, and fire. Within a few thousand years of the arrival of the first Americans, many species of megafauna went extinct. This too is the subject of great scholarly controversy (MANN 2005); for some it is obviously a result of human overhunting of 'naïve' species, but for others that interpretation is unproven and climate change or epizootics (or both) seem more plausible explanations. There is even evidence of a comet impact that correlates broadly in time with these die-offs (FIRESTONE et al. 2007). The end of the last Ice Age was a climatically chaotic time, and in North America featured many massive floods that could have drowned millions of creatures. Thus there is some evidence of climate change; but there is none for epizootics. That of course does not rule these explanations out. Moreover, there is no reason to suppose each species was necessarily wiped out by the same process: one could have succumbed to overhunting, and another to habitat loss. To me, it seems suspicious that approximately the same thing happened, widespread megafauna extinctions, not long after the first humans arrived in Australia, New Zealand, Madagascar, and many smaller islands. And it seems suggestive that American mammals had survived previous eras of chaotic climate change. This contentious matter will probably never be resolved.

However caused, the late Pleistocene megafauna extinctions in the Americas were comprehensive and consequential. The first Americans encountered smilodons (wildcats with teeth the size of kitchen knives), glyptodons (giant armadillos), ground sloths the size of hippos, beavers the size of bears, as well as mammoths, mastodons, camels, and five species of horses. In North America, 36 species of megafauna (70% of what existed) went extinct at this time, and in South America 46 (80%) (MITHEN 2004, pp. 237–238; FLANNERY 2001, pp. 155–169). These extinctions left the Americas with an impoverished fauna, and broader niches for some of the survivors, such as deer and bison.

1.3 Fire

The first Americans, like most Paleolithic people, were arsonists. They used fire to try to shape landscapes to their advantage, for example to drive game toward hunters, or to create fresh grass and prevent the expansion of forests, in order to provide more habitats for tasty herbivores. With their arrival and spread throughout the Americas, biological selection for compatibility with fire grew somewhat stronger. In the drier parts of the Americas, where natural fire was already frequent, compatibility with fire had long been a useful trait in plants; now it became more useful more widely, and those species that flourished in often-burned landscapes did better than ever before. Soils changed too under more frequent burning. In these respects, the Americas were conforming to the broader pattern where human fire had long been part of the ecological equation: the pre-human Americas were eccentric, the post-human Americas more normal (PYNE 2001, GOUDSBLOM 1992, MANNION 2006, pp. 133–136).

1.4 Agriculture

The dates of the earliest transitions to agriculture in the Americas are just as controversial as the date of the first migrations or the causes of the Late Pleistocene extinctions. It seems likely that in at least three different times and places, groups of early Americans learned to produce rather than merely gather food (PIPERNO and PEARSALL 1998, MANNION 2006, p. 142; BELLWOOD 2005, pp. 146–179). These were, first, Mesoamerica, where maize, squash and (rather later) beans emerged as food crops perhaps beginning as long as 10,000 years ago (some would say 6,500 years ago). Secondly, in the northern Andes and adjacent Pacific coastlands, peppers and squashes and (much later) potatoes were raised from perhaps as far back as 6,000 years ago. Thirdly, in the eastern woodlands of North America another, apparently independent, transition to horticulture and then agriculture took place, featuring goosefoot, sumpweed, sunflower, and squashes, perhaps around 4,000 years ago. These dates are all rough, and controversial. In general, it seems that unlike the situation in southwest Asia, the first food crops in the Americas were snack foods rather than staples, and the great majority of people were not full-time farmers until about 3,000 years ago. Ultimately, in terms of quantities, maize and potatoes dominated the American agricultural landscapes, and the American diet, with the exception of the poor rain forest soils of Amazonia, where cassava became the staple food. Maize eventually reached as far as Quebec (by around 800 or 1000 A. D.)

1.5 Sedentism, Intensification, and Population

From multiple points of origin in the Americas, horticulture and agriculture spread, haltingly, to most suitable regions. The lands most appropriate for potatoes and maize eventually came to support the densest populations and the most complex, differentiated, and hierarchical societies. But settled societies were not necessarily agricultural. Unusually rich fishing grounds could also sustain sedentary peoples. The two best examples of this occurred along the Pacific coasts of Peru, where the upwelling of the Humboldt Current brought (and brings) oxygen-rich waters to the sea surface, forming the basis for a dense food web that includes vast shoals of anchovies; and the salmon-filled rivers of what is now Northern California northward to Alaska, which supported villages and chiefdoms with minimal need for agriculture. A third example occurred along the tropical Gulf of Mexico coast, where the Olmecs and surrounding cultures relied on shellfish, manatees, fish and turtles to form a large part of their diet, although they also ate maize and other crops from at least 4500 BP (Day et al. 2007). Sedentary societies based on marine food resources existed in only a few locations, and could not spread widely nor grow in scale beyond narrow limits. The biggest societies and states were, as elsewhere in the world, agricultural.

Over the centuries, these ways of life spread to suitable geographic regions and grew more refined and productive. Various ingenious forms of irrigation were invented in drier landscapes from Bolivia to New Mexico, raising yields and population densities. Nowhere did domesticated animals play a large role in sustaining human populations, because the best candidates for domestication had gone extinct, leaving only less rewarding ones such as turkeys and guinea pigs, which although domesticable did not yield great quantities of meat. Hunting and fishing continued to figure in the food economy of most parts of the Americas, providing much of what animal protein American diets afforded.

Despite the isolation of the Americas and Afro-Eurasia from one another, in broad strokes their pre-Columbian histories are remarkably similar. In both regions transitions to agriculture took place shortly after the end of the Last Ice Age; agriculture brought ceremonial centers and polities; states relied on religion and organized violence to buttress their rule; states and markets divided between them the job of distributing goods and services within societies; the scale of trade networks and of empires tended to grow over time. And, broadly speaking, the capacity to alter the natural environment grew as well, although that generalization is only a first approximation of the truth.

It is sometimes argued that pre-Columbian American populations, and American Indians in general, pursued lifestyles informed by spiritual connection with their ecosystems that prevented them from inflicting great damage on land, water, and wildlife, an issue discussed in Krech (1998) and Kirch (2005). An early variant of this view appears in the writings of the Dominican missionary Jean-Baptiste Du Tertre, who spent eighteen years in the West Indies (Du Tertre 1667–1671). Similar arguments exist in reference to Buddhists, Hindus, and Daoists, but have not fared well in light of research into Chinese and Indian environmental history (ELVIN 2004, TUAN 1968). If true of Amerindians, this would have made them eccentric. No doubt at times Amerindians did their best to preserve wildlife (in many cases an important part of their food supplies) and their religions involved various forms of nature worship. But the evidence of the Inca state or the Maya polities strongly suggests that religion and ethics did not seriously constrain these peoples, or their rulers at least, in their quests for material well-being and abundance. They sculpted their hills and mountains into terraces; they burned and cleared swathes of forest; they rerouted waterways for irrigation (BEACH et al. 2008). The archeological record shows cases in which pre-Columbian farmers drastically impoverished their soils, and others in which they maintained and resurrected soil fertility over many centuries (SANDOR 1992). In short, like people everywhere, they manipulated their environments to the best of their abilities in routine struggles for security and abundance, constrained by their numbers, their technologies, but less by their religions and ideologies (DENEVAN 1992, KIRCH 2005).

Yet another contentious issue concerning the pre-Columbian Americas is the size of the population (ALCHON 2003, COOK 1998). The arguments are, as usual, fraught with political overtones. Scholars have offered figures for the total population of the hemisphere in 1492 ranging from 5 million to 112 million. Lately, informed opinion has settled around 30–60 million, although some vigorous dissents have been registered. The great centers of population – about this there is little controversy – were the maize heartlands of Central Mexico and the potato highlands of the Andes, with the addition, before 900 Christian Era (C. E.), of the Maya lowlands. Skeletal remains seem to imply that American populations were, by the standards of the rest of the world, unusually healthy. They suffered from deficiency diseases, from violence, but very little from infectious disease. If true, this probably reflects the paucity of domesticated herd animals, the source of most of the 'crowd' diseases that have afflicted most of the human race in recent millennia (STECKEL and ROSE 2002).

2. The Second Invasion

In the wake of COLUMBUS a second great invasion of the Americas followed. It came mainly from the Atlantic coastlands of Europe and Africa, and involved people, plants, animals, and pathogens. For the first 400 years, until the 1880s, the majority of the people coming to the Americas were Africans, not Europeans. Indeed until as late as 1820, some 80% of transatlantic migrants were Africans (ELTIS 2000). Almost all the Africans came as slaves, chiefly from West Africa and Angola. About 11 million survived the crossing to begin what were often short lives in the Americas. European migrants, sparse until the 19th century, came mainly from Iberia and the British Isles until an avalanche of immigrants (over 50 million c. 1880–1914) arrived, including many from Italy and Eastern Europe.

2.1 The Columbian Exchange

COLUMBUS and his followers accidentally brought the commonplace crowd diseases of Eurasia and Africa with them. These scythed down most of the 30–60 million Amerindians within four or five generations in wave after wave of devastating epidemics (COOK 1998, ALCHON 2003, GUERRA 1989, KELTON 2007, RAMENOFSKY 1987, but see HENIGE 1998). Infections spread far in advance of the human invaders, easing their conquests and settlements.

John McNeill

Indeed they often moved into depopulated landscapes – widowed lands – and mistook these for forests primeval. In Panama, for example, in the lee of epidemics and abandonment of villages and fields, huge forests grew up c. 1510 –1690, which led Scottish colonists at Darien in 1698 –1699 to imagine they were settling a pristine wilderness (Borland 1715). Wildlife populations probably grew as well, at least those well adapted to forests. In northern North America it may be no exaggeration to say that in the 17th century, and perhaps before, the beaver, rather than humankind, played the largest role of any creature in shaping the landscape. If so *Castor canadiensis*' reign as environmental monarch was a brief one, as it would soon be driven to the brink of extinction by fur trappers. In general, the catastrophic loss of human population meant less burning and far less agriculture and hunting for a century or two before human numbers recovered. This was a unique moment in environmental history, for never before or since have human numbers dropped so far so fast over so broad a territory as in the Americas after 1492.

The pathogens brought to the Americas after 1492 formed one component of what Alfred Crosby christened the Columbian Exchange (Crosby 1972). Several important animal species crossed the Atlantic, including horses, pigs, cattle, sheep and goats. They provided Amerindians with new sources of food, especially of animal proteins, and of hides and wool. They also disrupted landscapes, rooting up crops, nibbling away on vegetation, and heightening soil erosion (Melville 1994). The horse especially altered ways of life fundamentally, allowing a new nomadism in the Americas, and changing military balances profoundly. This inevitably had ecological effects. In the Canadian prairies in the 18th century, for example, horseborne warriors drove off earlier Amerindian populations, who had carefully preserved beaver numbers because they needed beaver ponds as a hedge against drought. The newcomers, knowing nothing of the cycles of drought, but politically powerful enough to do as they pleased, encouraged beaver trapping without restraint, and thereby made themselves more vulnerable to drought (James Daschuk, pers. comm.)

The Columbian Exchange brought new plants as well, helpful crops as well as irksome weeds. Wheat, rye, barley, oats and other cereal grains added dramatically to the agricultural possibilities in the Americas, and in time revolutionized whole regions such as the Argentine Pampas or the North American plains. Bananas and citrus fruits enriched diets in tropical landscapes, and eventually more broadly. Sugarcane, cotton, and coffee lent themselves to plantation modes of production, conspicuously changing both social and ecological realities in Brazil, the Caribbean, and the US South. Against this roster of important crops, the introduced weeds, even the iconic tumbleweed, which Americans regard as a symbol of the US West (but which is a recent immigrant from Ukraine), pale in significance.

An under-recognized feature of the Columbian Exchange is the degree to which it involved Africa. This was an issue that Crosby (1972) sidestepped. However, in the context of the slave trade from about 1550 to 1850, not only did 11 millions African arrive alive in the Americas, deeply affecting their demography, but so did numerous plants, animals, and diseases. Useful food crops included okra, yams, black-eyed peas, millets, sorghum, sesame, and above all African rice. African rice became a major crop in some of the swampy coastlands of places such as Surinam and South Carolina. Coffee too came from Africa originally, although its route to the Americas did not follow the slave trade. Two of the most dangerous diseases to arrive in the Americas after 1492, yellow fever and falciparum malaria, presumably crossed the ocean from West Africa on slave ships, contributing to the catastrophic series of epidemics among Amerindian peoples, at least in the warm lowlands around the Caribbean and northern

Brazil. Because of their high lethality, they also kept populations from high latitudes small in the Caribbean. People born and raised amid yellow fever and malaria in West Africa (or in the Caribbean itself), and thereby resistant to both diseases, enjoyed better prospects for survival (Kiple 1984). Those born elsewhere, such as at high latitudes, faced very short life expectancies in the lowland American tropics. No African (non-human) animals had much impact on the Americas, except perhaps for mosquitoes as disease vectors.

The Columbian Exchange fundamentally transformed the environment of the Americas, far more so than the export of American plants, animals, and diseases transformed Afro-Eurasia. American food crops – potatoes, maize, peanuts, tomatoes – of course had profound consequences in places such as Ireland, Rumania, southern Africa and China. But American animals and pathogens had much more restricted impacts. Crosby (1986) explained this in terms of the teamwork of ecological imperialism, according to which idea the political (human) imperialism in the Americas, undertaken by Europeans, aided and abetted the spread of elements of the Afro-Eurasian biota (and *vice versa*). The depopulation created newly destabilized landscapes, which opportunistic species of plants and animals readily exploited. Taking a longer time perspective, it might also be valid to suggest that since human settlement in the Americas was only 15,000 years old, American ecological systems were not so much destabilized after Columbus, but never stable in the first place, at least not since the first human invasion.

2.2 Plantation Ecology

The Columbian Exchange provided the essential ingredients for a new feature of the American economy, the plantation. In most cases this involved an Afro-Eurasian crop such as sugar, cotton, rice or coffee, and African (or more rarely European) labor. Tobacco was the only major plantation crop of American origin. Plantations brought their own suite of ecological changes. In the Caribbean, where the plantation complex was most completely installed, sugar was usually the main crop, and sugar revolutionized ecosystems. On an island such as Barbados, the first to undergo a sugar revolution (starting in the 1640s), the profitability of sugar inspired planters to (order their slaves to) burn and clear the island's forests and plant oceans of sugarcane, to the point where Barbados had to import food and fuel from afar by the 1650s (Ligon 1657). Monocultures of sugarcane invited rat and other pest infestations, which planters tried to counter by introducing snakes and mongooses. Plantations of every sort proved hard on soils, because they drew on the same sets of nutrients year after year. Thus sugar, cotton, and tobacco especially, but coffee as well, acquired deserved reputations for exhausting soil (Nelson 2007). Brazilian planters concluded that coffee needed fresh forest soils, and rapidly burned off the southern reaches of the giant Atlantic coastal forest to find fresh soil (Dean 1995). Plantations also promoted soil erosion, because economies of scale encouraged large field size, and the seasonal rhythms of a single crop meant that huge expanses of soil might be bare, and thus maximally vulnerable to wind and water erosion, at the same time. From Brazil to the Chesapeake, from 1600 to 1850, the economically dominant agro-ecosystems on the Atlantic shores of the Americas were plantation ecosystems.

The labor of creating and maintaining plantations was overwhelmingly slaves'. Slaves had no interest in preserving the investment of their masters, and many of them may indeed have preferred to see resources wasted, soils lost, crops burned. Ecological destruction might have appealed to slaves as a form of resistance against their oppressors. On the other hand, they

John McNeill

knew that they would be the ones doing the hard labor if anything had to be rebuilt. CARNEY (2001) gives examples of slaves taking great care with their masters' dikes on South Carolinian rice plantations. At the moment, the environmental history of slavery in the Americas is an unexplored topic. Indeed, the environmental history of slavery and plantations in other settings, from Mauritius to Morocco to Malaya awaits its historians.

2.3 Mining

Another major feature of the colonial economy, aside from plantation agriculture, was mining. By and large, mining took place far from the Atlantic coasts, either deep in the interior of the continents, or along the Pacific rim. The great centers before the 19th-century gold rushes, were in Potosí (in today's Bolivia), and in central and northern New Spain, now Mexico. Between 1570 and 1820, 80% of the world's silver came from mines in the Spanish Empire. Every of kilogram smelted in New Spain required biomass burning equivalent to a soccer pitch worth of oak forest. Over the centuries, this amounted to about 32,000 km^2, an area about the size of Belgium, Taiwan, or Maryland. Coppicing could have reduced the rate of deforestation, but was sparingly practiced. At Potosí the scale of everything was larger still. This was the world's most productive mine for 150 years after 1590. It was joined in the 1690s by gold and (later) diamond mines in Brazil (BAKEWELL 1997, CRAMAUSSEL 1999, DORE 2001, ROCHA MONROY 2002). Everywhere mining brought deforestation and devegetation, as fuel was required at several stages in the mining and smelting processes. Beyond that, pit props could only be made of timber. Huge quantities of waste rock, soil, and slurry had to be dumped wherever possible, often in ravines and streams. Silver amalgamation involved the liberal use of mercury, which inevitably spilled into the environment, and routinely was dumped into waterways.

Although the great colonial mines eventually played out, mining, like plantation agriculture, survived and indeed flourished in the 19th century. New strikes carried miners to new lands, such as the copper zones of Chile and Montana, or the goldfields of Colorado, California, and Alaska. The environmental fallout from mining followed as well, often on even larger scales, and complicated by new techniques such as hydraulic mining, in which high-pressure hoses washed away soils in the search for gold (FOLCHI DONOSO 2001, ISENBERG 2005).

2.4 Hunting and Trapping

Plantations and mines formed the backbones of the colonial economy in the Americas, especially in Spanish and Portuguese America. But there were other branches of the economy with deep environmental implications, such as hunting and trapping. Amerindians had hunted for millennia before the colonial era, but the commercialization of the business, and the connections to wealthy overseas markets, brought hunting and trapping to a new scale after 1600. The most famous case is that of the North American beaver, which had the misfortune to provide pelts for hats that were the height of fashion in the 18th century. By 1800, beaver trappers had pushed west to the Rocky Mountains, and by 1860 had trouble finding beaver anywhere. The near-removal of the beaver from northern North America changed the continent's hydrology, because millions of beaver dams fell to ruin, ponds disappeared, and the species that depended on beaver hydraulic engineering suffered from habitat loss. Large-scale commercial hunting also focused on the deer populations of southeastern North America in the 18th century (SILVER

1990), and on the wild cattle and horses of the Pampas in the 18th and 19th (BRAILOVSKY and FOGUELMAN 1991). Perhaps the most dramatic case in which commercial hunting changed environments was the near-extinction of the bison between 1840 and 1882. On the great sea of grass between Manitoba and Chihuahua, the huge heartland of North America, some 30 million buffalo roamed in the 1830s. By 1882, only about 2,000 remained. It could be that some animal disease such as brucellosis was involved, but without doubt commercial hunting was the main engine behind these events. Bison hide made excellent leather belts for use in the textile mills of the industrializing eastern Unites States (ISENBERG 2001, FLORES 2001).

2.5 The Grasslands Frontiers

The elimination of the vast herds of bison, and of the Plains Indians who depended on them, opened the great sea of grass to agricultural settlement (CUNFER 2005, BINNEMA 2001). Iron plowshares and barbed wire helped of course, and by 1930 cultivation on the grasslands had approached its apogee. Similar processes prevailed on the Argentine Pampas in the 18th and 19th centuries, where domesticated cattle and wheat cultivation gradually pushed aside the feral herds (and gauchos) that had dominated the landscape since the 16th century (BRAILOVSKY and FOGUELMAN 1991, GARAVIGLIA 1990). Eventually prairie fires were prevented, broad spaces fenced and gridded by roads, and almost all life forms aside from wheat and cattle eliminated.

The conversion of the American grasslands was part of a worldwide movement extending to the Russian and Ukrainian steppe (SUNDERLAND 2004, KHODARKOVSKY 2002), the Mongol and Manchurian steppe (REARDON-ANDERSON 2005), the high veld of South Africa, and the plains of New South Wales. Everywhere this was both a political and an ecological process. The former inhabitants of the seas of grass, usually mobile pastoralists, but in the North American case bison hunters, first had to be pushed off the land, which involved military defeat. In the course of the 18th and 19th centuries, China, Russia, the USA, and Argentina all succeeded in destroying the military and political power of the mobile peoples, and replaced them with farmers, crops, and sedentary livestock – creating breadbaskets that would feed a large proportion of the world after 1880.

3. Industrial America

In the 19th century pockets of the Americas underwent the transformation known as industrialization. In this it followed in roughly the same path as had Great Britain, except a generation or two later, on about the same schedule as Germany.

3.1 Water and Coal in Early Industrialization

Among the first manifestations was mechanization in textile manufacturing, using water to power mechanical looms (STEINBERG 1991). By 1860 most of the sizeable North American rivers flowing into the Atlantic had acquired dams, millponds, and mills. Their waters were increasingly polluted with dyes and other effluents, and became progressively more hostile to fish. The rivers themselves were increasingly canalized, to regularize their flows and make them more suitable for navigation. Similar alteration of waterways took place in Europe, probably most comprehensively in Britain and Germany, in the 19th century. Little of this took place in 19th-century South America, where industrialization started later and proceeded more slowly, nor did water power matter much in North America to the West of the Appalachians. But in Eastern Canada and the USA, a prosperous textile industry grew up along the banks of suitable rivers, forming one of the most dynamic sectors of the economy before 1860. Water power had its limits and was of little use in many other economic sectors, such as metallurgy, for which vast quantities of heat energy were needed.

Coal provided the answer in America, as it had in Britain. North America had plenty of it, both in Canada and the USA, and at fairly accessible locations. Latin America, as it happened, had rather little coal, a handicap to its industrialization efforts. After the 1860s coal-fired industries proliferated in North America: iron, steel, bricks, glass, and a hundred other kinds of manufacturing came to rely on coal, and by the 1890s the USA was the world's largest single industrial producer, outstripping Germany and Britain.

The coal mines of 19th century North America, from Nova Scotia to Alabama, featured some of the same forms of environmental distress as the silver mines of Mexico or the copper mines of Chile: deforestation, mountains of slag, acidification. The burning of coal, however, added a whole new dimension to environmental history in America, as in Europe and Japan: acute urban air pollution. Pittsburgh, a center of the US steel industry, appeared to a visitor in 1866 as "Hell with the lid taken off" (DAVIDSON 1979, p. 1037, quoting James PARSONS). Rudyard KIPLING wrote of Chicago in the 1890s that "its air is dirt" (quoted in CRONON 1991, p. 392). For a century or more, dozens of North American cities filled their air with coal smoke, soot, and sulfur dioxide, emitted both through factory smokestacks and household chimneys. Human health suffered, particularly in the form of lung diseases. Plant life withered. Thousands of local laws and regulations tried to restrict the air pollution brought on by coal combustion, but their total effect remained modest until the 1950s. In these respects, the urban air pollution history of North America was broadly similar to that of Britain, Germany, and Japan, where industry also ran on coal.

In some respects, however, the North American experience differed. Distances were greater and population densities lower than in Europe or Japan. A polluted city, such as St. Louis, might be the only island of acute air pollution for 300 km in any direction. A fresh breeze could quickly relieve its problems, if only temporarily. In Europe and Japan, by 1920 whole regions (the Midlands, Ruhr, Hanshin) hosted such concentrations of coal-fired industry and close-packed cities that air pollution levels more frequently stayed high. But in general, the effects of coal in North America were very much like those effects elsewhere.

3.2 Oil and Environment in the Americas

The experience of the Americas with oil was more distinctive. The USA pioneered hardrock oil drilling in 1859, and was the first society to adapt to oil as a fuel. By the 1920s the USA and Canada were mass producing automobiles, trucks, and tractors, and installing the infrastructure associated with motorized transport and farming. In this respect they were about a generation ahead of Europe and Japan, where the transition to oil as the dominant fuel came only in the 1950s or 1960s. Both the USA and Canada had plentiful supplies of oil, and easy access to the further supplies of Mexico and Venezuela, both important producers as early as 1915. North American oil production reached its peak in the 1960s, after which imported oil played an every larger role. But between the first great gushers in 1901 and the tripling of world oil prices in 1973, the USA and Canada enjoyed seven decades of abundant and cheap oil, during which time they thoroughly adapted their economies and societies to petroleum's potentials.

A lifetime of cheap and abundant oil was long enough for North Americans to rebuild their cities and suburbs around cars and roads, creating the urban sprawl that reached its apogee in southern California. It was also long enough for them to create an agricultural system predicated on oil, needed for farm machinery and petroleum-based pesticides and fertilizers. Americans even generated a large share of their electricity by burning oil. Using oil rather than coal in power generation and in many industries reduced overall air pollution, and changed its chemical composition too. But car exhausts added new pollution to the atmospheric mix, which by the 1940s was occasionally intense enough in Los Angeles to deceive locals into thinking their city was under chemical attack by the Japanese navy.

Although most Latin American countries had little oil of their own, in the course of the 20th century almost all of them shifted to an oil-based economic and ecological system. The advantages of motorized transport over the huge distances of Brazil or Argentina were too great to resist. Moreover, since all fossil fuels had to be imported, and oil was much easier to transport than coal (via pipelines and tankers, for example) after 1920 oil increasingly dominated the energy mix in Latin America. Thus similar air pollution problems emerged, most conspicuously in Mexico City, which by the 1970s suffered from air as filthy as anywhere's. Like Santiago and a few other major Latin American cities, it combined copious sunshine and often stagnant air masses, which together with automobile exhausts made the perfect recipe for photochemical smog.

From the 1930s until the 1980s, many Latin American countries tried hard to create their own industrial sectors using fossil fuels. The most successful were the bigger countries, Argentina, Brazil, and Mexico. They inadvertently created pockets of acute industrial pollution, the most notorious of which was Cubatão in southeastern Brazil. By the 1980s, Cubatão's air and water were a byword for environmental degradation. The small city hosted 40% of Brazil's steel and fertilizer production, and was a 'sacrifice zone' for the national economy, earning the nickname "Valley of Death" from Brazilians. Its infant mortality rates were 10 times those of the rest of São Paulo state, it had no birds, its trees died off, and landslides imperiled its denuded slopes. Even rats had difficulty surviving conditions in Cubatão (BOHM et al. 1989). Happily in the late 1980s, after civilian rule returned to Brazil and it became safe to question the state's model of industrial development, citizen and media pressure led to substantial improvement in the industrial pollution levels in Cubatão.

The experience of Cubatão was extreme, but it encapsulates the pattern of pollution history in the era of fossil fuels. In the interest of national economic development, or of private profit, tremendous environmental sacrifices were imposed on industrial regions, which citizens bore, for a century in Pittsburgh, for a generation in Cubatão, at considerable cost to their health. Eventually combinations of citizen pressure, crusading journalism, reforming politicians – and the relocation of industries to different landscapes – reduced these sacrifices. Today Pittsburgh is no more polluted than any North American city its size. But the steelmills are gone, relocated to South Korea or China.

3.3 Forests and Land Use in the 20th Century Americas

In North America the 19th century witnessed a tremendous wave of deforestation and agricultural expansion, especially in the US Midwest (CRONON 1991) and Ontario. For most of the century wood provided the most common fuel, and for all of it wood was the primary construction material. Its abundance led to wasteful use that would have caused distress to any German forester. But timber production in the USA peaked in 1907. Forest area reached its nadir in 1929 (these dates came a little later for Canada). Since 1929, forest area in the USA has grown. Huge areas of the eastern USA, used as farmland or pasture in the 18th and 19th centuries, returned to forest spontaneously because better agricultural land to the west could, with railroads and trucking, feed the eastern cities more cheaply than could the farms of the east. Thus once again, as after the depopulation of Amerindian North America c. 1500–1700, another century of forest resurgence occurred in eastern North America. With it came resurgent populations of deer, bears, beavers, forest birds and so forth. In 1850 Connecticut and Massachusetts were a patchwork of farms, pastures, towns and woodlots. Today, seen from the air, southern New England is almost a single carpet of forest, with islands of human habitation. Thousands of kilometers of stone walls, once used to mark boundaries of fields and meadows, now lie beneath the forest canopy (RUSSELL 1998, WHITNEY 1996).

The reforestation of eastern North America occurred despite continued population growth because of two main things. First, agriculture grew far more efficient. Second, a more integrated international economy made it possible to import timber products from distant lands. After the 1920s, almost no one plowed up new land in North America. But every decade farmers harvested more food. They adopted high-energy farming, with fertilizers, pesticides, and machinery, and they adopted new high-yield seeds. The results were stunning, tripling or quadrupling the amount of wheat or maize that a hectare could give. Thus Americans and Canadians could and did abandon their less rich farmlands and pastures, mainly in the east, as noted above, but eventually also in the Great Lakes region and southern USA.

Their timber and forest products industries grew more efficient too. The prodigal waste of the 19th century could easily be mitigated. Cheap transport meant land could be used for what it did best, so for example the Pacific Northwest (British Columbia, Washington, Oregon, Northern California), very good at growing tall trees, became the most important logging region of North America, succeeded by pine plantations of the American South in the later 20th century (WILLIAMS 1992). The pressures to clear forests all over the continent slackened.

In South America, however, different pressures produced different results. Sustained rapid population growth throughout the 20th century, combined with, in many countries, profoundly unequal land tenure systems, led landless peasants into the forests in their struggles for land and sustenance (FOWERAKER 1981). The great Brazilian Atlantic forest almost vanished entirely, followed after 1960 by about 15% of the world's largest single forest zone, that of Amazonia (DEAN 1995). While the timber trade and other factors played some role in this, by far the largest driving force was poor peasants' quests for land. This process continues to this day, although at slower rates since the 1980s. It remains to be seen whether the forest history of South America will one day show a reversal of the modern trend, that is emulate the North American pattern in which forest expansion outpaces forest contraction. Should that one day happen, it will be, by all indications, decades in the future.

4. Megalopoli

Some time about the year 2000, for the first time in history, more than half of the human race lived in cities. The Americas, North and South, reached this watershed about 1940, and today are both more than 75 % urban in their population. They were in the vanguard of a global trend in which, increasingly, the typical human environment is the built environment of the city. Within that is another trend towards giant cities, megalopolis, with more than 10 million people. Today there are 20 such in the world, and six of them are in the Americas, including two of the biggest. Mexico City and São Paulo each have about 20 million residents, while the conurbations of New York, Los Angeles, Buenos Aires, and Rio de Janeiro all top 10 million, and Lima is not far behind. By and large the Latin American megalopoli include broad rings of shantytowns around their urban cores, home to millions of former peasants hoping to change their luck in the big city. North America's big cities by and large include impoverished cores, surrounded by prosperous suburbs, although this generalization fits Canadian cities less well than US ones.

These patterns represent an extraordinary environmental evolution. It is widely shared in the world, and now feels normal to most of us as citizens of the contemporary age. In some respects it is a triumph of engineering and public health that so many millions can live so tightly packed without sparking murderous epidemics. In other respects it may seem a dismal failure, in that urban transport is so inefficient, and so many opportunities for prudent planning are squandered. In any case, for better and for worse, the human environment in the Americas, more so than in most of the world, is the urban environment, and so it will likely remain, barring events as jarring as those surrounding the first two human invasions of the Americas.

5. Conclusion

Seen as a whole, the environmental history of the Americas since the first human occupation some 15,000 years ago contains two epochs of especially tumultuous change. These come towards the beginning and the end of this period, and correspond to the two great human invasions of the Americas. The first, mainly biological in character, resulted from the arrival of the first Americans towards the end of the last Ice Age. The second, more comprehensive in character, and more visibly inscribed in the earth's surfaces (BEACH 1994, WALTER and MERRITTS 2008), came with the arrival of Europeans and Africans after 1492, and, reached its most acute phase with the arrival of cheap energy in the 20th century. These two great invasions unleashed twin tsunamis of environmental change, close enough together in time (less than 15,000 years) so that, if seen from Olympian heights, it seems the Americas never had time enough to adjust to human occupation.

This is an environmental history quite different from that of Africa, where human presence is oldest, and from Eurasia, where it is old enough and where modern times contained no great external invasions. It resembles to some degree the story of Australia, where human presence dates back perhaps 40,000–60,000 years, and where after 1788 a sudden invasion of exotic species washed over the land. But if one descends from Olympian heights for a moment, sharp differences appear. Aboriginal Australia never had more than a million people, perhaps 2 % of what the pre-Columbian Americas hosted. Its population density at the time of contact (1788 vs. 1492) was roughly one order of magnitude lower than that of the Americas. Aboriginal Australians did not practice agriculture (as that term is conventionally understood).

The Americas are, it seems, a case apart. US historians before the 1970s used to argue for 'American exceptionalism,' by which phrase they meant that the USA did not share the same course of development (and as many saw it, the same pathologies) as Europe. Their successors have by and large abandoned this view, or at least lost interest in it. Environmental history can resurrect this claim, and broaden it to extend to the Americas as a whole: the Americas are exceptional in their environmental history, because of the brevity of their human occupation, and the pervasiveness of the perturbations brought by the two great invasions.

Acknowledgements

Thanks to my Georgetown University Colleague Tim BEACH for suggestions on the text.

References

ALCHON, S. A.: A Pest in the Land: New World Epidemics on Global Perspective. Albuquerque: University of New Mexico Press 2003
BAKEWELL, P. (Ed.): Mines of Silver and Gold in the Americas. Aldershot: Variorum 1997
BEACH, T., LUZZADDER-BEACH, S., DUNNING, N., and COOK, D.: Human and natural impacts on fluvial and karst depressions of the Maya Lowlands. Geomorphology *101*/1–2, 308–331 (2008)
BEACH, T.: The fate of eroded soil: Sediment sinks and sediment budgets of agrarian landscapes in Southern Minnesota, 1851–1988. Annals of the Association of American Geographers *84*, 5–28 (1994)
BELLWOOD, P.: The First Farmers: The Origins of Agricultural Societies. Oxford: Blackwell 2005
BINNEMA, T.: Common and Contested Ground: A Human and Environmental History of the Northwestern Plains. Norman: University of Oklahoma Press 2001
BOHM, G., SALDIVA, P., PASQUALUCCI, C., MASSAD, E., MARTINS, M., ZIN, W., CARDOSO, W., CRIADO, P., KOMATSUZAKI, M., and SAKAE, R.: Biological effects of air pollution in Sao Paulo and Cubatao. Environmental Research *49*, 208–216 (1989)
BORLAND, F.: Memoirs of Darien, Giving a Short Description of that Countrey, with An Accounts of the Attempts of the Company of Scotland to Settle A Colonie in that Place… Glasgow: Hugh Brown 1715
BRAILOVSKY, A., and FOGUELMAN, D.: Memoria verde. História ecológica de Argentina. Buenos Aires: Editorial SudAmerica 1991
COOK, N. D.: Born To Die: Disease and New World Conquest, 1492–1650. New York: Cambridge University Press 1998
CRAMAUSSEL, C.: Sociedad colonial y depredación ecológica: Parral en el siglo XVII. In: GARCÍA MARTÍNEZ, B., and GONZÁLEZ JÁCOME, A. (Eds.): Estudios sobre historia y ambiente en América; pp. 93–107. Mexico City: Instituto Panamericano de Geografía e Historia 1999
CRONON, W.: Nature's Metropolis. New York: Norton 1991
CROSBY, A.: The Columbian Exchange: Biological and Cultural Consequences of 1492. Westport, CT: Greenwood. 1972
CROSBY, A.: Ecological Imperialism: The Biological Expansion of Europe 900–1900. Cambridge: Cambridge University Press 1986
CUNFER, G.: On the Great Plains: Agriculture and Environment. College Station, Texas: A and M Press 2005
DAVIDSON, C.: Air pollution in Pittsburgh: A historical perspective. Journal of the Air Pollution Control Association *29*, 1035–1041 (1979)
DAY, H. W., GUNN, J. D., FOLAN, W. J., YÁÑEZ-ARANCIBIA, A., and HORTON, B. P.: Emergence of complex societies after sea level stabilized. Eos. Transactions of the American Geophysical Union *88*, 169–176 (2007)
DEAN, W.: With Broadax and Firebrand: The Destruction of the Brazilian Atlantic Forest. Berkeley: University of California Press 1995
DENEVAN, W.: The Pristine Myth: The landscape of the Americas in 1492. Annals of the Association of American Geographers *82*, 369–385 (1992)
DORE, E.: Environment and society: Long-term trends in Latin American mining. Environment and History *6*, 1–29 (2000)
DU TERTRE, J.-B.: L'Histoire générale des Antilles habitées par les Français. Paris: Iolly 1667–1671
ELTIS, D.: The Rise of African Slavery in the Americas. New York: Cambridge University Press 2000
ELVIN, M.: Retreat of the Elephants: An Environmental History of China. New Haven: Yale University Press 2004
FIRESTONE, R. B., WEST, A., KENNETT, J. P., BECKER, L., BUNCH, T. E., REVAY, Z. S., SCHULTZ, P. H., BELGYA, T., KENNETT, D. J., ERLANDSON, J. M., DICKENSON, O. J., GOODYEAR, A. C., HARRIS, R. S., HOWARD, G. A.,

Kloosterman, J. B., Lechler, P., Mayewski, P. A., Montgomery, J., Poreda, R., Darrah, T., Que Hee, S. S., Smith, A. R., Stich, A. , Topping, W., Wittke, J. H., and Wolbach, W. S.: Evidence for an extraterrestrial impact 12,900 years ago that contributed to the megafaunal extinctions and the Younger Dryas cooling. Proceedings of the National Academy of Sciences USA *104*, 16016 –16021 (2007)

Flannery, T.: The Eternal Frontier: An Ecological History of North America and Its Peoples. New York: Grove Press 2001

Flores, D.: The Natural West. Environmental History in the Great Plains and Rocky Mountains. Norman: University of Oklahoma Press 2001

Folchi Donoso, M.: La insustenibilidad de la industria del cobre en Chile: Los hornos y los bosques durante el siglo XIX. Revista Mapocho *49*, 149 –175 (2001)

Foweraker, J.: The Struggle for Land: A Political Economy of the Pioneers in Brazil from 1930 to the Present Day. Cambridge: Cambridge University Press 1981

Garavaglia, J. C.: Pastores y labradores de Buenos Aires: Una historia agraria de la campaña bonaerense (1700 –1830). Buenos Aires: Editorial de la Flor 1990

Gilbert, M. T. P., Jenkins, D. L., Götherstrom, A., Naveran, N., Sanchez, J. J., Hofreiter, M., Thomsen, P. F., Binladen, J., Higham, T. F. G., Yohe, R. M. II, Parr, R., Cummings, L. S., and Willerslevet, E.: DNA from pre-clovis human coprolites in Oregon, North America. Science *320*, 786 –789 (2008) [Science online (3 April 2008) http://www.sciencemag.org/cgi/content/abstract/1154116v1 (2008)]

Goebel, T., Waters, M., and O'Rourke, D.: The late Pleistocene dispersal of modern humans in the Americas. Science *319*, 1497–1502 (2008)

Goudsblom, J.: Fire and Civilization. New York: Penguin 1992

Guerra, F.: Epidemiología Americana y Filipina, 1492–1898. Madrid: Ministerio de Sanidad y Consumo 1989

Henige, D.: Numbers from Nowhere: The American Indian Contact Population Debate. Norman: University of Oklahoma Press 1998

Isenberg, A.: The Destruction of the Bison. New York: Cambridge University Press 2001

Isenberg, A.: Mining California. New York: Hill and Wang 2005

Kelton, P.: Epidemics and Enslavement: Biological Catastrophe in the Native Southeast, 1492–1715. Lincoln: University of Nebraska Press 2007

Khodarkovsky, M.: Russia's Steppe Frontier: The Making of a Colonial Empire, 1500 –1800. Bloomington: Indiana University Press 2002

Kiple, K. N.: The Caribbean Slave: A Biological History. New York: Cambridge University Press 1984

Kirch, P. V.: Archaeology and Global Change: The Holocene Record. Annual Review of Environment and Resources *30*, 409 – 440 (2005)

Krech, S.: The Ecological Indian: Myth and History. New York: Norton 1999

Ligon, R.: A True and Exact History of the Island of Barbadoes. London: Parker 1657 [reprinted by Frank Cass Publishers, London 1970]

Mann, C. C.: 1491: New Revelations of the Americas before Columbus. New York: Knopf 2005

Mannion, A. M.: Carbon and its Domestication. Dordrecht: Springer 2006

Melville, E.: A Plague of Sheep: Environmental Consequences of the Conquest of Mexico. New York: Cambridge University Press 1994

Mithen, S.: After the Ice: A Global Human History, 20,000 –5,000 BC. Cambridge: Harvard University Press 2004

Nelson, L.: Pharsalia: An Environmental Biography of a Southern Plantation, 1780 –1880. Athens: University of Georgia Press 2007

Piperno, D., and Pearsall, D.: The Origins of Agriculture in the Lowland Neotropics. San Diego: Academic Press 1998

Pyne, S. J.: Fire: A Brief History. Seattle: University of Washington Press 2001

Ramenofsky, A.: Vectors of Death: The Archeology of European Contact. Albuquerque: University of New Mexico Press 1987

Reardon-Anderson, J.: Reluctant Pioneers: China's Northward Expansion, 1644–1937. Stanford: Stanford University Press 2005

Rocha Monroy, R.: Potosí 1600. La Paz: Alfaguara 2002

Russell, E. W. B.: People and the Land through Time: Linking Ecology and History. New Haven: Yale University Press 1998

Sandor, J. A.: Long-term effects of prehistoric agriculture on soils: examples from New Mexico and Peru. In: Holliday, V. T. (Ed.): Soils in Archeology: Landscape Evolution and Human Occupation; pp. 217–245. Washington: Smithsonian Institution Press 1992

Silver, T.: A New Face on the Countryside: Indians, Colonists, and Slaves. In: South Atlantic Forests. New York: Cambridge University Press 1990

John McNeill

STECKEL, R. H., and ROSE, J. (Eds.): The Backbone of History: Health and Nutrition in the New World. New York: Cambridge University Press 2002

STEINBERG, T.: Nature Incorporated: Industrialization and the Waters of New England. New York: Cambridge University Press 1991

SUNDERLAND, W.: Taming the Wild Field: Colonization and Empire on the Russian Steppe. Ithaca: Cornell University Press 2004

TUAN, Y.: Discrepancies between environmental attitude and behavior: Examples from Europe and China. Canadian Geographer *3*, 175–191 (1968)

WALTER, R. C., and MERRITS, D. J.: Natural streams and the legacy of water-powered mills. Science *319*/5861, 299–304 (2008)

WHITNEY, G. G.: From Coastal Wilderness to Fruited Plain: A History of Environmental Change in Temperate North America from 1500 to the Present. New York: Cambridge University Press 1996

WILLIAMS, M.: Americans and Their Forests: A Historical Geography. New York: Cambridge University Press 1992

        Prof. Dr. John R. MCNEILL
        Georgetown University
        History Department of History
        Washington DCBox 571305
        ICC 600
        Washington, DC 20057-1305
        USA
        Phone: +1 202 6875585
        Fax: +1 202 6877245
        E-Mail: mcneillj@georgetown.edu

**Research Training Program
„Interdisciplinary Environmental History"**

Preface

Bernd Herrmann ML and Christine Dahlke (Göttingen)

With 1 Figure

Since summer 2004, the German Science Foundation (DFG) has been promoting a Research Training Group in environmental history at the Georg August University in Göttingen. Eleven staff members, a postdoc and an average of thirteen PhD students are working together in projects focussing on Central European issues in environmental history, predominantly of the 18th and 19th century. Participants of the program are purposefully elected from a broad variety of disciplines instead of being solely history-centred.

In the Research Training Program 1024 "Interdisciplinary Environmental History: Natural Environment and Societal Behaviour in Central Europe" issues of social and economic history and cognitive scientific interest are combined in project studies and metatheoretical investigation in a way that creates a broader general standard concerning scientific problems and their solution for the qualification in the field of environmental history. It is also within the scopes of the program to improve the professional ability of the PhD students in interdisciplinary studies and professional skills for future academics.

The research program concentrates on suitable case or regional studies. Apart from the reconstruction of environmental preconditions and historical patterns of exploitation in particular areas, the awareness of consequences and "side effects" of exploitation since the Middle Ages are also important features. Epistemologically two pillars of Environmental History are defined in the program as "reconstruction" and "reception". Four fields of research – linked up with each other – have been set up for the first 6 years:

- Section A: Exploitation and experience of space and environment in the Middle Ages;
- Section B: Diking of nature, live stock diseases, pest control and river control(s) from Early Enlightenment to mid-twentieth century;
- Section C: Conflicts about natural resources, 18th–20th century;
- Section D: Constructions and reifications of the environment.

The structure of the research program is illustrated by Figure 1.

However, the program was recently reconsidered, since issues of Section D needed stronger integration into the other sections as thought before. Thus the program currently is slowly switching its focus to merely three major fields:

- Section A: Exploitation and experience of space;
- Section B: Crop failures and famines in Central Europe, genesis, dimensions, consequences;
- Section C: Concepts of environment and nature, patterns of valuation.

Bernd Herrmann and Christine Dahlke

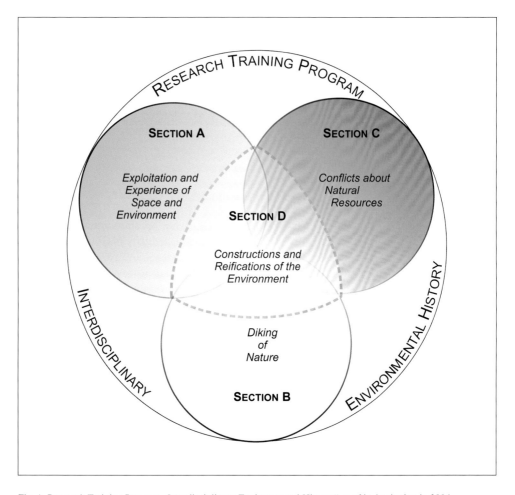

Fig. 1 Research Training Program „Interdisciplinary Environmental History" as of its beginning in 2004

In general, the study program aims at the exchange between natural and social sciences and humanities about their methods, approaches and cognitive interests concerning environmental history. It is designed to impart the "T-competence" necessary for interdisciplinary exchange.

Further development of the proven concepts and local experience in research training programs of interdisciplinary research are mainly practised in projects: working target-orientated, the postgraduates are engaged in a dialogue, explain problems from the perspective of their respective subjects and overcome terminological and epistemological differences. As a result, personal academic competence as well as group identity is increased. Through common topics, excursions, workshops, colloquia and special seminars the different approaches of the engaged sub-disciplines become visible towards reconstruction and reception of past environments. The members of the program are confident that this approach helps to form a disciplinary understanding of environmental history and to shape the contours of future academic professionals in environmental history.

The following contributions of the PhD-students offer insight into the broad scope of the program and the subjects dealt with. Length of contributions may differ due to the time period of membership and the current status of work.

>Prof. Dr. Bernd HERRMANN
>Georg-August University Göttingen
>Johann-Friedrich-Blumenbach-Institut für Zoologie und Anthropologie
>Abteilung Historische Anthropologie und Humanökologie
>Bürgerstraße 50
>37073 Göttingen
>Germany
>Phone: +49 551 393642
>Fax: +49 551 393645
>E-Mail: bherrma@gwdg.de

>Dr. Christine DAHLKE
>Georg-August-Universität Göttingen
>DFG-Graduiertenkolleg Interdisziplinäre Umweltgeschichte
>Bürgerstraße 50
>37073 Göttingen
>Germany
>Phone: +49 551 393890
>Fax: +49 551 393645
>E-Mail: cdahlke@gwdg.de

"Slavs, Waters and GIS" – Methods and Base Data to Search for Watercourses and Floodplains in a Meso Scale Study

Anne KLAMMT (Göttingen) and Martin STEINERT (Leipzig)

With 7 Figures and 2 Tables

Abstract

This paper focuses on acquisition and procurement of geodata in a meso-scale study in landscape archaeology. The research encompasses the early medieval land use in northern Central Europe. In this article special emphasis lies upon the evaluation of free available source 3arcsec SRTM and the calculation of Local Drainage Directions and a Topographical Position Index. The evaluation had been carried out by contrasting the results with models derived from calculations on basis of a digital landscape model and a digital elevation model, both at the scale of 1:250,000.

Zusammenfassung

Der vorliegende Artikel befasst sich mit Fragen der Geodatengrundlage im Rahmen einer großräumigen landschaftsarchäologischen Studie. Das Forschungsvorhaben befasst sich mit Aspekten der Raumnutzung im nördlichen Mitteleuropa während des frühen Mittelalters. In dem Aufsatz wird insbesondere die Verwendbarkeit der frei verfügbaren SRTM90-Daten, die Ableitung von Tiefenlinien sowie der Einsatz eines topographischen Positionsindex diskutiert. Die Bewertung wurde auf Grundlage eines Vergleiches mit Ableitungen aus einem digitalen Höhenmodel und eines digitalen Geländemodells (beide M 1:250 000) durchgeführt.

This article focuses on problems of procurement of datasets and suitable software solutions. The procedures and their results derive from research for a current Ph.D. dissertation at the Department of Pre- and Protohistory at the Georg-August University Goettingen in cooperation with the Geographical Institute at the University of Leipzig. The project aims to compare the early medieval settlement history in three adjacent regions in northern Germany. Special emphasis lies upon the early medieval period when three ethnic groups, differing in social structure and cultural behavior, were living in this region. Due to historical and political circumstances the areas developed differently, at least since the turn of the millennium. Despite a long and fruitful history of research, the knowledge about growth and decline of settlement networks is still sketchy. Likewise, the relation between economic structure, environmental conditions and changes of social organization is largely unknown up to the High Medieval colonization. Especially difficult seems the overall evaluation of the Slavonic culture and its attachment to settlement areas close to water. Rarely, these waterbodies were systematically categorized according to there type (lake, river or beck) or connection to other lines of communication nor did their former courses play a major role. A second commonly mentioned feature of Slavonic settlement is the preference of sandy, easily tillable soils (GRABOWSKI 2007). As pointed out recently, these two features have to be regarded inextricably linked –

i.e. the location near a border of various habitats (SAILE 2007). In a broader sense, we want to explore this approach by including the terrain into our study. Within this paper we want to explore as first steps the impact of modern land use on the archaeological record and ways to reconstruct the former landscape on a meso-scale size (SAILE and LORZ 2006). Our main focus lies upon the experiences with procurement and use of cheap (Corine Land Cover 2000) or even freely available (3 arcsec SRTM) data bases and application programs (Local Drainage Directions, Topographical Position Index). Of special interest were the potentials of the 3 arcsec SRTM elevation model. Although there are several tests concerning its accuracy, only few were published on the effects of the well-known systematic errors in modeling. As the data presented by soil-maps comprises very special problems of the more theoretical kind, we excluded these from this article.

1. Framework: Archaeological Record and Test Areas

Since our research encompasses at a vast geographical area of some 31,000 km² we are dependent on the use of "off the shelf" data. An edition of the data would by far exceed the scope of our study. This is a situation comparable to experiences of compliance investigations (DORE and WANDSNIDER 2006, p. 78). The archaeological data was obtained from local records and several publications, as far as an exact geographical location of the site is known. At the same time we are searching for easily accessible digital maps of recent land use, soils, digital terrain models and methods of procurement and analysis.

In order to test them we choose two small areas of 690 km² (Fig. 1). Test Area I is situated in the south western fringe of the chain of glacial lakes in Schleswig-Holstein, reaching into the adjacent Mecklenburg-Vorpommern. The landscape is flat to undulating and structured by the river Stecknitz and several small watercourses. The artificial *Elbe-Lübeck-Kanal* runs from north to south by using several former riverbeds and depressions. Test Area II is characterized by a flat depression of the partially heavily modified river Elbe in the west and a slightly ascending terrain to the east. The areas were chosen by random with the precondition of almost the same quantity of sites of the Roman Iron Age and Slavonic period (201 in Test Area I and 199 in Test Area II) and a different proportion of forest and pasture/grassland.

2. Source Criticism by the Use of Corine Land Cover 2000

Preceding all further investigations, the evaluation of the impact of different search strategies and find situations on the spatial distribution of the archaeological sites seems to be of initial importance. Therefore all sites of the Roman Iron Age in the area have also been recorded in order to contrast (compare) the distribution. The overwhelming majority of the sites of both periods is known from surface finds. They were discovered by accident or more frequently during archaeological surveys, mainly carried out by field walking. The western part of Test Area I in Schleswig-Holstein was systematically surveyed by archaeologists from 1936 to 1939 (KERSTEN 1951). Later on, several new sites were reported by non professionals (SCHMID-HECKLAU 2002). Very few sites were excavated: urngraves, earthworks (almost exclusively restricted to small trenches) and few sites by rescue excavations preceding building activities. Approximately the same conditions determine our knowledge of

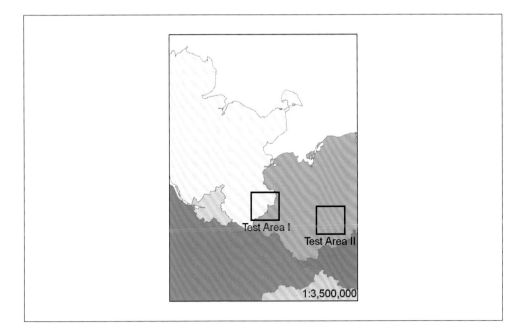

Fig. 1 Test Areas I and II

Test Area II, with one important addendum. Since 2004, excavations are carried out in the Slavonic stronghold at Friedrichsruhe, district Parchim (http://www.elbslawen.de) (Fig. 6 and 7 marked as black triangle).

Due to this situation, the assumption was that ploughed up sites on arable land will be statistically overestimated. Analogically, we supposed that there will be too few known sites on moorland and grassland as well as forests. This was checked by a rather simple comparison between the ratios of the sites on these different kinds of landscape in relation to the total area each of the types occupied in the Test Areas. In our first tests we used the dataset Corine Land Cover 2000. Corine Land Cover is the result of a Europe-wide remote-sensing on a scale of 1:100,000. Its easy availability made it a promising source. A structural weakness of remote-sensing is the identification of sparsely forested areas like gardens, pastures and moorland (*European Environment Agency* 2006, p. 24). Thus, we supposed the occurrence of mistakes, especially in the mapping of fens and natural grasslands. Later we had the opportunity to contrast the results with a query using a digital landscape model (DLM) based on a topographical map in the scale 1:25,000.[1] Naturally, this is a more detailed and very reliable, but truly more expensive record. The comparison by overlaying the Corine Layer and the DLM Layers with the sites marked as points resulted in an astonishing small differences (Tab. 1 and 2). Observing the minor differences, two facts have to be considered. First, we did not respect the slightly different aggregation of the various categories of fallow land,

---

[1] For realizing the opportunities of exchange with the Roman-Germanic Commission (RGK) within the scope of the research-project "Slawen an der Mittelelbe" – http://www.elbslawen.de/ – we would like to thank Dr. F. LÜTH and Dr. A. POSLUSCHNY.

grassland, moorland, heathland etc. Secondly, agricultural economy is dynamic and the use of areas may change. We expected a large quantum of alteration in Test Area II following the political changes during the last decade in eastern Germany. Again Corine Land Cover proved to be very useful, as it provides not only the land use in the state of 2000 but also of 1990. A query of changes between grassland and arable land adds up to a moderate area of 21,8 km² in total (3,2 % of the land area) and affected only five archaeological sites of 199.[2] Thus, the Corine Land Cover turned out to be quite reliable and useful cheap base data that is accessible for the entire European Union.

In order to explore whether there is a significant relation between tilled land and occurrence of sites, we compared the supposed with the observed quantity of sites (Tab. 1 and 2). To calculate the supposed ratio (this would mean a non-specific distribution) of sites for each type of land use, the number of archaeological sites were multiplied with the percentage of area. After that the supposed quantity was divided by the number of observed sites. A quotient lower than 1 signifies avoidance of the type of terrain. A value close to 1 implies no differentiated allocation. In turn a quotient larger than 1 is a hint for preference (SAILE 2007).

The results falsified our expectation, as there are more sites on moor- and grassland as we supposed. Quotients of 0,9 and 1,1 indicate that – in relation to the entire area – archaeological sites can be found on moor- and grasslands in an average distribution. A distinction according to the archaeological period threw light on the very interesting fact that sites of the Slavonic period are equally frequently found on arable land as those of Roman Iron Age. This seems even more surprising when taking in account the more fragile and coarse pottery of the Roman Iron Age in comparison to the high fired Slavonic ceramics. Hence the chances to discover sites of the latter period should be as good if not even better.

In conclusion, we can rule out that the spatial distribution of sites on arable land and grass- as well as moorland and gardens is only a correlative of the procedure of survey. Nevertheless, the ratio between forest and quantity of sites approximately corresponds to our suggestions. Concerning our research, the afforested areas have to be treated as voids. Since the spatial distribution of sites is neither random nor solely determined by the activity of the plough, further efforts can be made due to the reconstruction and operability to the impact of terrain and related features.

3. SRTM Elevation Data and Computing Local Drainage Directions

The general problem is to decide whether the sheer proximity to a water course or the position between two different ecological habitats was the critical point for the medieval people to choose a place to settle. Until now, most analysis referring to the distance of early historical settlements to watercourses used more or less the modern courses. Only in small areas one was able to reconstruct the earlier states of water level. This was almost entirely restricted to the study of lakes and applied detailed geomorphological and sedimentological measurements (KIEFMANN and MÜLLER 1975, BLEILE 2002). A completely different approach is to reconstruct former waterbodies as a result of an elevation model. By the computing of the potential local drainage direction (LDD) one can get an idea of a landscape before the heavy

---

[2] For test area I – reaching from Schleswig-Holstein to Mecklenburg-Vorpommern (former GDR) – the ratio was even less with 1,6 % of the land area (10,6 km²) and two sites affected.

Tab.: 1 Sites quantity area I

Sites quantity area I			
Roman Iron Age	slavonic	all sites	
49	156	201	

Corine Land Cover

Land use in area I			
	land area	arable land	grassland
km²	657,8	347,3	76,4
%	100	52,7	11,6

Sites on arable land				
	Roman Iron Age	slavonic	all sites	calculation
observed	29	93	120	
expected	25,8	82,2	105	52,7 % of total
quotient	1,1	1,1	1,1	observed/expected

Sites on grassland, pastures etc.				
	Roman Iron Age	slavonic	all sites	calculation
observed	5	19	24	
expected	5,7	18,1	23,3	11,6 % of total
quotient	0,9	1	1	observed/expected

DLM

Land use in area I				
	land area	arable land	grassland	forested area
km²	653,4	310,4	91	231,1
%	100	47,5	13,9	35,4

Sites on arable land				
	Roman Iron Age	slavonic	all sites	calculation
observed	27	87	114	
expected	23,3	74,1	95,5	47,5 % of total
quotient	1,2	1,2	1,2	observed/expected

Sites on grassland, pastures etc.				
	Roman Iron Age	slavonic	all sites	calculation
observed	5	17	22	
expected	6,8	21,7	27,9	13,9 % of total
quotient	0,7	0,8	0,8	observed/expected

Sites in forested area				
	Roman Iron Age	slavonic	all sites	calculation
observed	8	25	32	
expected	17,3	55,2	71,2	35,4 % of total
quotient	0,5	0,5	0,5	observed/expected

Tab.: 2 Sites quantity area II

Sites quantity area II			
Roman Iron Age	slavonic	all sites	
57	156	199	

Corine Land Cover

Land use in area II			
	land area	arable land	grassland
km²	673,63	327,9	116,6
%	100	48,6	17,3

Sites on arable land				
	Roman Iron Age	slavonic	all sites	calculation
observed	35	71	101	
expected	27,7	75,8	97	48,6 % of total
quotient	0,8	1,1	1	observed/expected

Sites on grassland, pastures etc.				
	Roman Iron Age	slavonic	all sites	calculation
observed	10	41	46	
expected	9,8	26,9	34,4	17,3 % of total
quotient	1	1,5	1,3	observed/expected

DLM

Land use in area II				
	land area	arable land	grassland	forested area
km²	673,63	280,68	149,22	199,4
%	100	41,7	22,2	29,6

Sites on arable land				
	Roman Iron Age	slavonic	all sites	calculation
observed	26	60	81	
expected	23,7	65	82,9	41,7 % of total
quotient	1,1	0,9	1	observed/expected

Sites on grassland, pastures etc.				
	Roman Iron Age	slavonic	all sites	calculation
observed	16	51	60	
expected	12,7	34,6	44,2	22,2 % of total
quotient	1,3	1,5	1,4	observed/expected

Sites in forested area				
	Roman Iron Age	slavonic	all sites	calculation
observed	8	29	36	
expected	16,9	46,2	58,9	29,6 % of total
quotient	0,9	0,6	0,6	observed/expected

impact of hydraulic engineering (CONOLLY and LAKE 2006, pp. 257–260). In order to get an impression of the ability of a map of water courses based on LDDs we created them for both Test Areas. They were calculated on the basis of a digital terrain model (DTM) using the open source program SAGA-GIS. The DTM itself derives from the free available 3 arcsec SRTM elevation data. To close the voids in the dataset, again the SAGA-GIS module was used. After that the results were transformed into an ESRI Grid and the slope was derived from using ArcView 3.2a (Fig. 2 and 3).

Due to their free accessibility, the quality of the SRTM dates were broadly discussed. As our Test Areas are situated in a flat to slightly undulated landscape, we were particularly interested in the problem of height errors. The existence of such errors is known, but they seem to be in open areas of systematic nature. Minor problems occurred in woodlands. Here, the deciduous forests show fewer errors, because the satellite-based measurements took place during February (KOCH et al. 2002, CZEGA et al. 2005). As systematic errors they should not interfere the relative calculations of the LDDs. A problem is to control the LDDs without appropriate historical maps. If one compares the LDDs to a map of recent watercourses, it is impossible to decide whether the discrepancies originate from an inapt calculation or are

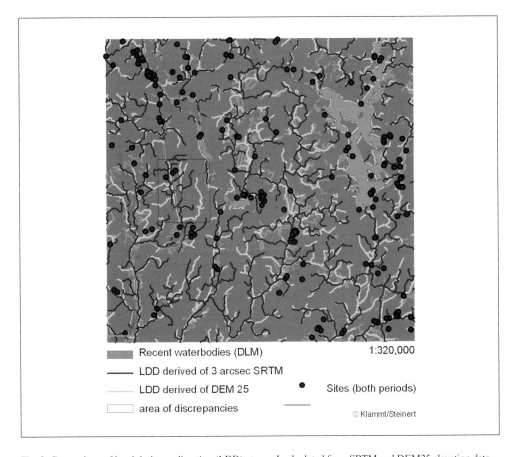

Fig. 2 Comparison of local drainage direction (LDD) at area I calculated from SRTM and DEM25 elevation data

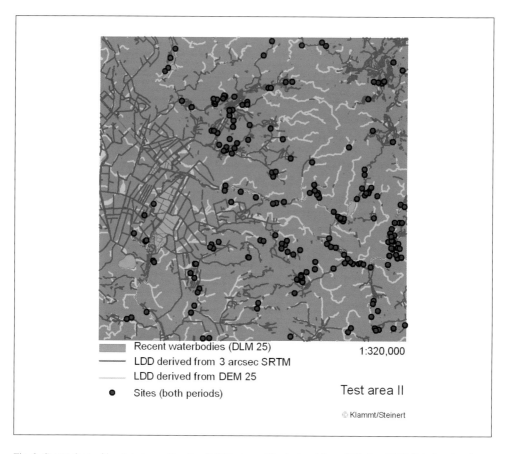

Fig. 3 Comparison of local drainage direction (LDD) at area II calculated from SRTM and DEM25 elevation data

caused by hydraulic engineering. Things changed as we received the opportunity to recalculate the LDD on the base of a digital elevation model in the scale 1:25,000 (DEM 25). Now the ability of the 3 arcsec SRTM data basis could be cross-checked. Presently we could only do this by optical control (Fig. 2 and 3). This, nevertheless, gives further insight, as major discrepancies occur in forested regions (Fig. 2, marked with a black frame). This puts us back to the problems from height errors of the SRTM elevation data in arboreous areas. Finally, LDDs derived from 3 arcsec SRTM data seem to be in open areas – which are exactly those ones, there the known archaeological sites are located – an alternative to the purchasable and often more cursory digital stream networks. An estimation of the distance of sites to the nearest water courses can now calculated according to a far differentiated basis. Already at first sight on to the map, there appears a more or less common preference to short distances between sites and water. Nevertheless, it is necessary to differentiate between the kind of sites (graveyards, settlements, fortifications and further more). In the realm of this article this will be too detailed.

4. The Use of Topographical Position Index for Calculating Floodplains

As floodplains have been repeatedly recognized as favorite places for Slavonic settlement areas, we are interested in methods of reconstructing or modeling valleys and designating possible floodplains. We tested the calculation of a Topographic Position Index (TPI). The TPI is the basis for a classification of the landscape according to Slope Position and Landform Categories (WEISS 2001). Once again we carried out the first trials based on 3 arcsec SRTM elevation data.[3] A so called TPI value grid was calculated using the freely available ArcView Extension TPI (JENNESS 2006). Within the scope of the project we were first of all interested in the slope position classification. The TPI is the difference between a cell elevation value and the average elevation of the neighborhood around that cell. A positive TPI value means that the cell is higher than the surrounding ones while negative values indicate a depression. Near zero TPI values could mean either a flat area or a mid-slope area. This may cause classification problems if only using TPI. In completion for a clear accuracy of discrimination

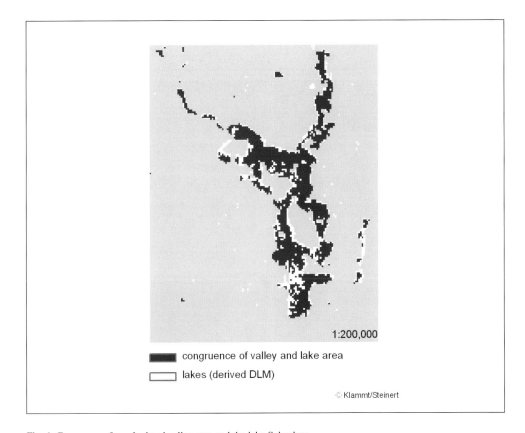

Fig. 4 Congruence for calculated valley area and the lake Schaalsee

---

3 Certainly the calculation of floodplains on such a basis is not appropriate to specify the local conditions that actually determined the decision to settle at this very place.

the slope is used. The TPI combined with the slope enables the user to define slope position classes like Ridge Top, Upper, Middle, Flat, Lower Slope and Valley Bottom. The TPI approach is very scale-dependent. Determined by the scale used to analyze the landscape the same point could be a ridge or a valley. In the TPI algorithm "scale" is determined by the parameter "Neighborhood". The threshold of the neighborhood defines which cells will be included in the TPI calculation. In the same way the neighborhood size affects the Slope Position classification. Small neighborhoods detect small and local mounds and depressions while large ones can show comprehensive features like mountains and valleys.

On first sight the resultant grid is quite convincing. The identification of e. g. river valleys was rather satisfying by optical investigation. To examine the quality, we checked if the TPI converted lakes to valleys. This test eventually showed some limits of the method, which were most probably caused by the noise all major lakes and waterbodies produce in the SRTM data (Fig. 4). It is no work around to clip these areas, as they are involved as neighbored cells in the calculation of the TPI. This is a minor problem as lakes can easily made visible by adding further shapefiles. More trouble causes the false calculation of forested areas. To demonstrate this, the area of a small bog, Kehrsener Moor, is singled out and compared to a satellite image (Fig. 5). Nevertheless, as stated above, the forested areas are also in the archaeological record more or less voids. Within a small part of Test Area II we confronted the results of calculating

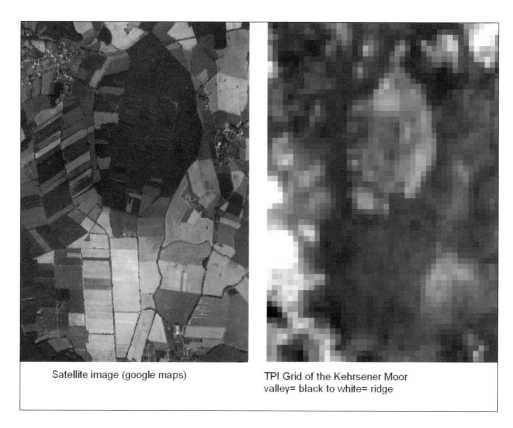

Fig. 5 Comparison of satellite image and topographic position at Kehrsener Moor

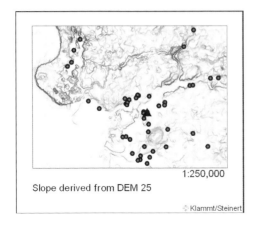 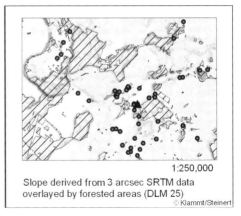

Fig. 6 Slope derived from DEM 25 an 3 arcsec SRTM elevation data of area II

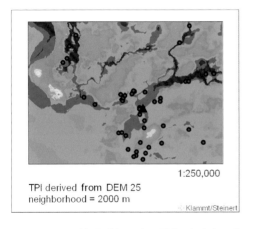 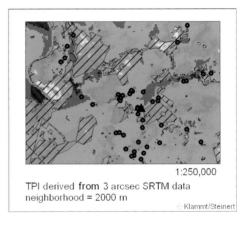

Fig. 7 Topographic Position Index (TPI) calculations from DEM 25 and 3 arcsec SRTM elevation data overlayed by forested areas (DLM25)

the TPI on basis of the 3 arcsec SRTM and on the DEM 25. For each calculation we defined the threshold with 2,000 m (Fig. 7). Although, as one can see looking on the derivation of slope for both datas, the DEM offers a far more detailed relief, the results are on this scale (1:250,000) still quite similar (Fig. 6 and 7).

5. Resume

This paper discussed two data bases and two applications. We tested them on two areas, both situated within the research area of a major study. The engagement with these issues arose in need for easy and quick solutions when working in a meso-scale region. Further data acquisition gave us the opportunity to evaluate some of the issues based on detailed and highly reliable data sources. Our first observations concerned the reliability and skills of the

Anne Klammt and Martin Steinert

Europe-wide obtainable remote sensing data Corine Land Cover 2000 for the estimation of the dependence of the discovery of sites and the land use. The results in comparison with the ATKIS Digital Landscape Model 25 were generally convincing. After that we tried to model potential watercourses by calculating local drainage directions. This was done using 3 arcsec SRTM elevation data by applying the open access SAGA-GIS. The derived LDDs were compared to LDDs calculated with the Digital Elevation Model 1:25,000. Discrepancies occur in forested areas, while open areas are showing astonishing similar results, despite the different resolutions of the elevation models. Finally we presented our results using a topographical position index for designate floodplains. We calculated the TPI with a free ArcView Extension. Using the 3 arcsec SRTM elevation model again, we were hampered once more by the height errors of forested area. Unlike the modeling of LDDs, the errors in this application are worse, but still limited to the forests. Until further investigations, we can support those who propose the 3 arcsec SRTM elevation model as a recommendable resource for geographical (and even archaeological) research. Nevertheless, all further applications like the modeling of potential watercourses and valleys should be carried out in the awareness of the clear limitation of the data.

6. Datasets and Programs

DLM ATKIS – Arbeitsgemeinschaft der Vermessungsverwaltungen der Länder der Bundesrepublik Deutschland (Ed.), Amtliches Topographisch-Kartographisches Informationssystem. http://www.atkis.de/
Corine Land Cover 2000 – Deutsches Zentrum für Luft- und Raumfahrt e. V. Deutsches Fernerkundungszentrum – http://www.corine.dfd.dlr.de
SAGA-GIS – http://www.saga-gis.uni-goettingen.de/html/index.php
SRTM version 2 – ftp://e0srp01u.ecs.nasa.gov/srtm/version2/SRTM3/Eurasia
ArcView Extension TPI – http://www.jennessent.com/downloads/TPI_jen.zip

References

Bleile, R.: Interdisziplinäre Forschungen zu Wasserstandsschwankungen der „Oberen Seen" (Mecklenburg) in spätslawischer Zeit (Ende 10.–12. Jh.). Greifswalder Geographische Arbeiten *26*, 179–182 (2002)
Conolly, J., and Lake, M.: Geographical Information Systems in Archaeology. Cambridge: University Press 2006
Czega, W., et al.: Validierung der freien C-Band SRTM Höhendaten in Hinblick auf Anwendungsmöglichkeiten in den Geo- und Umweltwissenschaften. In: Strobl, J., Blaschke, T., und Griesebner, G. (Eds.): Angewandte Geoinformatik 2005 – Beiträge zum AGIT Symposium Salzburg 2005. S. 106–111. http://www.saw-leipzig.de/sawakade/10internet/umwelt/Agit_2005_Handout.pdf (2005)
Dore, C. D., and Wandsnider, L.: Modeling for management in a compliance world. In: Mehrer, M. W., and Wescott, K. L. (Eds.): GIS and Archaeological Site Location Modeling; pp. 75–96. Boca Raton: Taylor and Francis 2006
European Environment Agency (Eds.): The thematic accuracy of Corine land cover 2000 – Assessment using LUCAS (land use/cover area frame statistical survey). A Technical Report No 7. Copenhagen.
http://reports.eea.europa.eu/technical_report_2006_7/en/technical_report_7_2006.pdf (2006)
Grabowski, M.: Zur slawischen Besiedlung in Wagrien und Polabien. In: Biermann, F., und Kersting, T. (Eds.): Siedlung, Kommunikation und Wirtschaft im westslawischen Raum. Langenweißbach: Beier und Beran 2007
Jenness, J.: Topographic Position Index (tpi_jen.avx) extension for ArcView 3.x, v. 1.3a. Jenness Enterprises. http://www.jennessent.com/arcview/tpi.htm. (2006)
Kersten, K.: Vorgeschichte des Kreises Herzogtum Lauenburg. Die vor- und frühgeschichtlichen Denkmäler und Funde in Schleswig-Holstein. Neumünster: Wachholtz 1951
Kiefmann, H.-M., und Müller, H. E.: Die Tiefenkarte des Großen Plöner Sees als Beitrag zur Erforschung einer frühen Kulturlandschaft und der regionalen Morphogenese. Offa *32*, 16–29 (1975)

Koch, A., Heipke, C., and Lohmann, P.: Bewertung von SRTM Digitalen Geländemodellen – Methodik und Ergebnisse. Photogrammetrie Fernerkundung Geoinformation 6, 389–398 (2002)

Saile, T.: Slawen in Niedersachsen – Zur westlichen Peripherie der slawischen Ökumene vom 6.–12. Jahrhundert. Göttinger Schriften zur Vor- und Frühgeschichte 30. Neumünster: Wachholtz 2007

Saile, T., and Lorz, C.: An essay on spatial resolution in predictive modelling: Pitfalls and challenges. In: Van Leusen, M., and Kamermans, H. (Eds.): Predictive Modelling for Archaeological Heritage Management: A Research Agenda; pp. 139–148. Nederlandse Archologische Rapporten 29. Amersfoort: ROB 2005

Schmid-Hecklau, A.: Slawenzeitliche Funde im Kreis Herzogtum Lauenburg. Studien zur Siedlungsgeschichte und Archäologie der Ostseegebiete 3. Neumünster: Wachholtz 2002

Weiss, A.: Topographic Position and Landforms Analysis. Poster Presentation, ESRI User Conference, 2001, San Diego, CA (2001)

Anne Klammt
Seminar für Ur- und Frühgeschichte
Nikolausberger Weg 15
37073 Göttingen
Germany
Phone: +49 551 393868
E-Mail: aklammt@gwdg.de

Martin Steinert
Helmholtz-Zentrum für Umweltforschung GmbH – UFZ
Department Landschaftsökologie
Permoserstraße 15
04318 Leipzig
Germany
E-Mail: martinsteinert@web.de

Imaging and Integrating Heterogeneity of Plant Functions: Functional Biodiversity from Cells to the Biosphere

Internationales Leopoldina-Meeting

vom 29. Juli bis 31. Juli 2007 im Forschungszentrum Jülich

Nova Acta Leopoldina N. F., Bd. 96, Nr. 357
Herausgegeben von Ulrich SCHURR (Jülich), Barry OSMOND (Canberra City), Ulrich LÜTTGE (Darmstadt), Uwe RASCHER (Jülich), Susanne VON CAEMMERER (Canberra City) und Achim WALTER (Jülich)
(2009, 192 Seiten, 60 Abbildungen, 5 Tabellen, 23,95 Euro, ISBN: 978-3-8047-2603-1)

Bildgebende Verfahren und Computertechniken spielen auch in der Botanik eine immer größere Rolle. Räumliche Heterogenitäten und zeitliche Dynamik von Strukturen und Funktionen sind essentiell, um das Verhalten von Pflanzen und ihre Wechselwirkung mit Böden und der Atmosphäre verstehen zu können. Ihrer quantitativen Analyse kommt deshalb eine Schlüsselrolle zu – sowohl für die Erforschung der Grundlagen pflanzlichen Verhaltens als auch für die Entwicklung innovativer Anwendungen in der Pflanzenproduktion. Neue Methoden der Bildaufnahme und die quantitative Bild(sequenz)analyse schaffen derzeit die Grundlage für ein völlig neues Verständnis der Bedeutung von Heterogenität und Dynamik in Pflanzen und Umwelt auf allen Skalen von der Zelle bis zum Ökosystem. Der Band verdeutlicht, wie dynamisch Wachstum, Photosynthese und Transport wirklich sind und welche Bedeutung räumliche und zeitliche Muster für Pflanzen haben. Die Beiträge belegen ein hohes Maß an Interdisziplinarität und Kommunikation zwischen Entwicklern von Verfahren, Anwendern aus den Pflanzen- und Umweltwissenschaften sowie Modellierern und Theoretikern. Es erweist sich, dass die Integration von modernen Methoden, innovativen experimentellen Ansätzen und Theoriebildung unser Verständnis von Pflanzen und von ihrem Verhalten in ihrer sich ständig verändernden Umwelt grundlegend wandeln wird. Alle Beiträge sind in englischer Sprache verfasst.

Wissenschaftliche Verlagsgesellschaft mbH Stuttgart

Water and Waters in the Latin Medieval Sources – an Evaluation of the Settlement Area of the Slavs by a Semantic Analysis

Jens POTSCHKA (Göttingen)

Abstract

The thesis analyzes medieval written sources with regard to the interactions between human beings and natural environment. What can the written sources tell us about perceptions, adaptations and experiences, about the subjective and collective imagination of the authors? The work is focused on the relations between the authors and water as a factor of the environment. The evaluation concerns the settlement area of the slaves from the early Middle Ages to 1250. Especially since the middle of the 12th century the settlers from the west came in droves to the whole area and transferred their technologies of cultivation of grain, of hydraulic power and of canalization to the new homeland (*aedificatio terrae*).

The evaluation was accomplished by a semantic method. In difference to comparable studies it is not possible for this dissertation to resort to a digital collection of source material, so that a new text corpus had to be collected. The narrative texts and the charters of the European Middle Ages published by the *Monumenta Germaniae Historica* (MGH) serve as the main source material. The analysis follows mainly two semantic patterns of classification: The structure of word fields and the distributions of collocations. At the same time the different ways of interpretation by the medieval authors will be analyzed with regard to the meaning of water as an acting object, the role of god as part of nature and, finally, the interpretive patterns for natural phenomena.

Zusammenfassung

In der Dissertation werden mittelalterliche Schriftquellen nach den Wechselwirkungen zwischen natürlicher Umwelt und dem Menschen analysiert. Was verraten die Schriftquellen über die Wahrnehmung, Erfahrung, Deutung und Bewertung der Autoren sowie über ihre subjektive und kollektive Vorstellungswelt? Die Arbeit fokussiert dabei thematisch das Verhältnis der Autoren zum Umweltfaktor „Wasser". Untersucht wird das Siedlungsgebiet der „Elb- und Ostseeslawen" vom Frühmittelalter bis 1250. Westliche Kolonisten besiedelten dieses Gebiet verstärkt seit der Mitte des 12. Jahrhunderts und transferierten dabei auch ihre kulturlandschaftlichen Techniken im Getreideanbau, bei der Nutzung von Wasserkraft oder des Kanalbaues in die neue Heimat (Landesausbau).

Die Auswertung erfolgte quantitativ anhand einer semantischen Methode. Im Unterschied zu vergleichbaren Studien kann in diesem Projekt für die Auswertung nicht auf eine digital vorhandene Quellensammlung zurückgegriffen werden, so dass ein neues Textkorpus zu generieren war. Dieses Textkorpus wurde vorrangig aus den in der *Monumenta Germaniae Historica* (MGH) publizierten Schriftquellen zusammengestellt. Für die Auswertung soll zwei Klassifizierungsmustern nachgegangen werden, einerseits der Analyse der Wortfeldstruktur und andererseits dem Auffinden von Kollokationen als semantischen Feldern. Auf interpretatorischer Ebene sollen die Bedeutung des Wassers als handelndes Objekt, die Rolle Gottes in der Natur und die autorenspezifischen Interpretationsmuster natürlicher Phänomene einer grundlegenden Betrachtung unterzogen werden.

Jens Potschka

1. Questions and Aims

This thesis analyzes medieval written sources with regard to interactions between human beings and natural environment. Applying the concept of mentalité the focus of the dissertation lies on the perception of nature by medieval men: Which natural and cultural conditions and processes of the environment found the attention of human beings? What can written sources tell us about perceptions, adaptations and experiences, about the subjective and collective imagination of the authors (GOETZ 2003)? The thesis concentrates exclusively on the complex subject area "water". This subject area contains on the one hand the natural water reservoirs (the sea, bodies of standing and flowing waters) and on the other hand men made constructions like ponds, springs, aqueducts and their economic uses; including fishing and aquaculture, the construction of water mills and fords, floating and navigation.

2. Emphases and Demarcations

The area of investigation is approximately the region of today's East Germany, which was settled since the end of the 6th century by an ethnic group that came from the southeast. This ethnic group (called the slavs) consisted of many different and heterogeneous sub-groups, who were enemies in part. In the following centuries manifold economical and social contacts arose between the slavs and their western neighbors. From the 10th century onwards the Slavs were christianized by and by and had to pay tribute. Especially since the middle of the 12th century the colonists from the west came in droves to the whole area. In the process the settlers transferred their technologies of cultivating landscapes to the new homeland. Thereby they improved primarily the cultivation of grain, the hydraulic power and the canalization. The term *aedificatio terrae* appears in the sources during the whole process. Apart from the settlers the number of newly founded monasteries rose instantaneously. The abbots and convents carried out the Christian "assignment" to evangelize the natives mostly by communicating with the local authorities. Especially the religious order of the Cistercians and of the Premonstratensians had clear expectations and conceptions how to cultivate the surrounding areas of the monasteries (SCHENK 2002). The volatile history of this area (the so called *Germania Slavica*) offers the medievalist an auspicious field, to highlight the trends of perception of the environment by the medieval men.

The dissertation is focused on the High Middle Ages, especially the 11th – 13th century. The period of investigation starts with the 7th century and ends at 1250. By that time the main acts and changes of the *aedificatio terrae* were finished and the number of written sources is manageable to produce a substantial and sensitive evaluation.

3. Method

The emphasis lies on the medieval terminology. Therefore the evaluation was accomplished by a semantic method.[1] In contrast to comparable studies it is not possible for this dissertation to resort to a digital collection of source material. So a new text corpus had to be collected.

---

1 The method is inspired by JUSSEN 2002, 2006.

The narrative texts and the charters of the European Middle Ages published by the *Monumenta Germaniae Historica* (MGH) serve as the main source material. The emphasis lies on the medieval terms for all things linked in a wider sense to the subject area "water" including the analysis of their qualities. One problem was finding the numerous references to the topic "water" within the *Monumenta Germaniae Historica*. For that purpose a list was created consisting of about one hundred central semantic fields, which were determined through spot checks. The whole sentence, which contains the word, was taken out of the Corpora and put in a database. All in all 47 charter-books, 23 chronicles, 19 annals, 3 vitae and 2 other texts were evaluated.

An innovation of the thesis is the handling of different types of sources. The historic-semantic research works published so far analyze very frequent words and word fields,[2] so that a complete capture of the written sources is barely realizable. Those works evaluate large databases, which are digital available. This dissertation claims to collect the entire body of written historical sources of the period and to analyze narrative as well as normative sources in an extensive and balanced way.

The analysis follows mainly two semantic patterns of classification: The structure of word fields and the distribution of collocations. Within the area of corpus linguistics, collocation is defined as a sequence of words or terms which co-occur more often than would be expected. Following these method the thesis filters out the semantic connections between the particular semantic field and the other word units of the sentence. How do word fields change diachronically? Which terms appear and which terms disappear? The first results show, that there are some breaks in the linguistic patterns and the use of terms during the Middle Ages. Those breaks could happen at the same time as other social, economic, political or cultural procedures. Perhaps cultural processes like the *aedificium terrae* in East Middle Europe can be confirmed linguistically. Another question is, whether the analysis can show different linguistic patterns for different geographical regions within area under investigation. Which of the word fields and collocations can be expected, and how can variations be explained? The different ways of interpretation by the medieval authors will be analyzed with regard to the meaning of water as an acting object, the role of god as part of nature and, finally the interpretive patterns for natural phenomena.

4. Example

What kind of results can be expected of the analysis on the basis of the collocations? The terms *lacus/us* and *stagnum/i* represent the semantic field "standing waters". The medieval authors seemed to use both terms equally and without differences. The result of the analysis:

In the contexts of *lacus* terms of the subject matter "economy" were more commonly used: molendinum, taberna, pons, navis, teloneum, census, decima, marca, forum/forensis, denarius. These terms existed about nine times more frequently in the contexts of lacus than in the contexts of stagnum.

In the contexts of *stagnum* terms of the subject matter "orientation" were commonly more used: occidens/occidentalis, oriens/orientalis, meridies/meridionalis, aquilo/aquilonalis.

---

2 Other terms are lexical field, semantic field.

Jens Potschka

These terms appeared comparatively about five times more frequent in the contexts of *stagnum* than in the contexts of *lacus*.

The conclusion of this result: The contexts of *lacus* and *stagnum* differ fundamentally in various parts. *Stagnum* was used by the medieval authors as a typical term to describe paths and areas on the basis of orientations and directions. In contrast they employed *lacus* in the contexts of economic incidents, especially for contributions.

References

GOETZ, H.-W.: Vorstellungen und Wahrnehmungen mittelalterlicher Zeitzeugen. Neue Fragen an die mittelalterliche Historiographie. In: HASBERG, W., und SEIDENFUSS, M. (Eds.): Zwischen Politik und Kultur. Kulturwissenschaftliche Erweiterung der Mittelalter-Didaktik. Bayerische Studien zur Geschichtsdidaktik Bd. 6, S. 45–58. Neuried: Ars una 2003

JUSSEN, B.: Ordo zwischen Ideengeschichte und Lexikometrie. Vorarbeiten an einem Hilfsmittel mediävistischer Begriffsgeschichte. In: SCHNEIDMÜLLER, B., und WEINFURTER, S. (Eds.): Ordnungskonfigurationen im hohen Mittelalter. Vorträge und Forschungen Bd. 64, S. 227–256. Ostfildern: Thorbecke 2006

JUSSEN, B.: Confessio. Semantische Beobachtungen in der lateinischen christlichen Traktatliteratur der Patristik und des 12. Jahrhunderts. Zeitschrift für Literaturwissenschaft und Linguistik *126*, 27–47 (2002)

SCHENK, W.: Zisterzienser als Gestalter von Kulturlandschaft. Forschungsstand, offene Fragen und Konzept zum Umgang mit dem landschaftlichen Erbe. In: OTTEN, T. (Eds.): Ora et labora. Quellen und Elemente der Nachhaltigkeit zisterziensischen Lebens. Festschrift für Georg Kalckert. S. 71–88. Köln: Rheinischer Verein für Denkmalpflege und Landschaftsschutz 2002

> Jens POTSCHKA
> Georg-August-Universität Göttingen
> DFG-Graduiertenkolleg Interdisziplinäre Umweltgeschichte
> Bürgerstraße 50
> 37073 Göttingen
> Germany
> Phone: +49 551 393640
> E-Mail: jpotsch@gwdg.de

City – Forest – Man
Environmental-Historical Studies of a Fundamental Urban Relation: Goslar and Hildesheim between Medieval Desertification Period and Thirty Years' War

Cai-Olaf Wilgeroth (Göttingen)

With 3 Figures

Abstract

Medieval and early modern life and working were under the pre-industrial living conditions not possible without the forests and their resource repertoire. This is especially true for the urban settlements of late medieval times (1350–1600 AD) – for urban cities as much as for smaller towns. Besides the trees as sources of energy (fuel, charcoal) and of raw materials (wood, timber, bark, resin, tar) the aspects of nutritional physiology should not be neglected (fruits, pasture, hunting, fertilizer). Additional possibilities of using the forests are related rather to their legally determined spatiality than to their biomass (mining, water, clearing, settlement). Only the forest as a multifarious landscape element were principally able to meet such a spectrum of natural demands. Here the civic needs for forestal products exceeded in quantity and quality by far the peasant-agrarian ones. Indeed the anthropological basic needs were quite the same in town and country and had to be met likewise (cooking, heating, building); here the urban settlements as demographic agglomerations only consumed greater amounts of resources. But in the civilizing model 'city' was involved a promise of a locally concentrated range of services and conveniences (shelter, administration, trade, luxury). From this was resulting a comparatively large number of qualitatively different forestal demands while possessing a relative small settlement area. Thus the urban communities were anxious for the legal integration and controlled utilization of sufficient forests. This consistently provoked conflicts with the surrounding agrarian hinterland and the upcoming territorial state given inevitably clashing spheres of interest in a more and more shrinking forest.

Zusammenfassung

Mittelalterlich-frühneuzeitliches Leben und Wirtschaften war unter den Existenzbedingungen der vor-industriellen Epoche ohne die Wälder und ihr Ressourcenangebot nicht denkbar. Dies galt insbesondere für die städtischen Siedlungen der spätmittelalterlichen Jahrhunderte (14.–16. Jahrhundert) – gleichermaßen für große urbane Zentren wie für kleinere Stadtgemeinden. Neben dem Baumbestand als Lieferant von Energie (Holz, Holzkohle) und Rohmaterial (Holz, Bast, Borke, Harz, Teer) muss hier auch die ernährungsphysiologische Seite des Waldes für Mensch und Tier beachtet werden (Waldfrüchte, Mast, Jagd, Dünger). Hinzu gesellten sich noch spezielle Nutzungsmöglichkeiten, bei denen der Wald mehr durch seine rechtlich umgrenzte Flächenhaftigkeit als durch seine Biomasse relevant wurde (Bodenschätze, Gewässernutzung, Rodung, Siedlung). Nur der Wald als vielfältigstes Landschaftsgebilde konnte prinzipiell ein derartiges Spektrum an naturalen Ansprüchen bedienen. Die stadtbürgerliche Angewiesenheit auf Waldprodukte überstieg dabei in Quantität wie Qualität die bäuerlich-ländliche um ein Vielfaches. Zwar fielen die anthropologisch determinierten Grundbedürfnisse in Stadt und Land gleichermaßen aus und mussten quasi identisch befriedigt werden (Kochen, Heizen, Bauen); wobei die Städte als demographische Ballungsräume zunächst einmal lediglich größere Mengen an Rohstoffen verbrauchten. Allerdings ging mit dem zivilisatorischen Modell „Stadt" eben auch das Versprechen eines räumlich verdichteten Angebots an Dienstleistungen und Annehmlichkeiten einher (Schutz, Verwaltung, Handel, Luxus). Daraus resultierte bei relativ geringer Siedlungsfläche eine vergleichsweise große Zahl qualitativ unterschiedlicher Ansprüche an den Wald. Die Stadtgemeinden bemühten sich deshalb um rechtliche Integration und nutzungstechnische Kontrolle ausreichender Waldungen. Dies produzierte immer wieder auch Konflikte mit dem agrarisch geprägten Umland und dem erwachsenden Territorialstaat angesichts zwangsläufig kollidierender Interessenssphären in einem immer mehr schrumpfenden Wald.

1. The Forest of Hildesheim in the 15th Century

„XXXVIII. [...] unde wente sek denne dat volk vorsamm(l)et, miteinander to wonende, dar in dat erste dorpschup mede begrepen werden, van den dorpschuppen wikbelde werden gemaket unde van den wikbelden in dem, alse sek dat volk vormeret mit uthbredinge der buwe, stede werden begreppen unde gestichtet, de stede also gelecht unde begreppen werden, dat de innewonere der stede mogen hebben to orer brukinge water unde weyde, acker unde holt. Alse de stad to Hildensem ok so begreppen unde gemaket is, unde [...] de vorscreven wolt van der stad den namen heft, also dat yd de Hildensemsch wolt is gheheten [...]"(DOEBNER 1890)

What the Magistrate of Hildesheim here in the thirty-eighth article of his counter-claim against a previous episcopal complaint in 1440 declared, in more than one respect is unique in Hildesheim's late medieval material of written sources. On the one hand, in a description settled between the literary genres of 'city's laud' (*laudes urbium*) and 'narration of origin' (aetiology) we find the civil authorities' ideal conception of the primary genesis and setting of a medieval town. Regarding their written documents such literary touches are rather untypical for the normally more businesslike aldermen (JOHANNEK 1999).

At the other hand, under an environmental-historical perspective it is especially interesting, how much a resource-economic perspective of the essential natural conditionality of medieval urbanity is reflected here: water to drink, pasture for the cattle, agricultural land for nutrition, and last but not least wood (forest) as the much cited 'central resource' of pre-modern times (GLEITSMANN 1981). Of course it could be argued, that the article under discussion is obviously focused on the Forest of Hildesheim and his between the bishop as the nominal ruler of the town and 'his' civic community disputed usage: „[...] wann unse medeborgere eder de unsen holten in dem Hildensemschen wolde, drouwet he [sc. the bishop] one, se mit den sinen dar dael to slande [...]". (DOEBNER 1890.) Maybe, a reference towards natural circumstances in this context should not be seen as so much unexpected.

But in the other points of dispute at that time also aspects less connected with natural conditions has been under discussion. Aspects for example, which can be understood as constitutive for the urban community and its functioning in more constitutional, political, or social respects. Regarding these points, however, the magistrate did not argue as much fundamentally and literarily, did not use a topological repertoire derivable from the genre of 'town descriptions', which in this abstract non-material direction had emerged in the 15th century at least rudimentarily (JOHANNEK 1999). Because, that an alderman in Hildesheim the phenomena 'city' and 'urbanity' mostly conceived more from a habitual than from a material perspective, can be recognized from most of the other contemporary sources (WILGEROTH 2004).

Insofar it seems to be a justified conclusion that the magistrate members of Hildesheim something eminently important and central wanted to express here: for them the struggle was about the indispensable conditions of living in their town. That they conceptualized their arguments in a literary vestment as an etiological commonplace, was to strengthen the implicitness of the argumentation for their contemporaries all the more. The message was clear: to deprive a city of wood and forest, could (should) only be understood as an unnatural act.

And, as to describe later on, all this was not only about wood, the main raw material and energy source of that time. A fact, which once prompted Werner SOMBART (1917) to label the centuries before the beginning industrialization as marked by a „notedly wooden character".

2. No Urbanity without Forests

Taking up first thoughts of Gerhard PFEIFFER (1972) and Jürgen SYDOW (1981) Ernst SCHUBERT (1986) had pointed out fundamentally, in what a widespread sense the forests were the 'economic basis of the late medieval city' and its inhabitants. Generally spoken, pre-modern cities and towns represented cultural aggregations of all the multiple nature-based demands of a completely agro-silva-culturally oriented society. To the same degree, to which different people and activities in the towns were concentrated, there could be found many common, but also a lot of dissimilar natural-based needs. Under pre-industrial resource conditions a considerable part of these needs were linked to the forests (SCHUBERT 2002). Besides many other things only this type of landscape could provide wood, the indispensable central resource, in sufficient quantity and quality. Without the forests – as SCHUBERT (1992) puts it – no urban life at all! This diagnosis could be widened nearly until the mid-19th-century, because the pre-modern society had not been subjected to principle changes concerning their dependence on the forestal raw materials and by-products. Here environmental history leaves the classical division of history, because the Middle Ages and the early modern times lose their conventional borderline around 1500 under an resource-functional perspective (GLEITSMANN 1981).

Though, in an environmental-historical urban context, transformations of the basic conditions of urban wood and forest usage are definitely observable – transformations, which in particular took place between the late medieval heyday of urbanity and its increasing early modern 'mediatization' by the upcoming territorial sovereignties. In these nearly three hundred years between 14th century desertification period and pre-modern 16th century formation of the territorial state the towns and cities were confronted with fundamental changes: from originally inexhaustible appearing forests towards signs of wood shortage and forestal resource deficiencies. Changes, which were more and more generated on economic-political grounds from outside, though to some extent they were caused by the cities themselves and were as the consequence of an unsustainable utilization factual existent in the natural landscape.

Quite disparately reacting to such fluctuating conditions the urban communities in these particular three hundred years gave themselves a distinct urban, non-rural 'wooden character'. Concerning this, it should be considered, if the urban agglomerations with regard to the forest utilization were different from their village counterparts only in quantitative perspective or under qualitative aspects too. Certainly, both types of settlement showed forms of the general spectrum of cultural-determined, anthropological basic needs and therefore common demands, which in the cities simply had to be accommodated in a much higher degree. But owing to urbanity the palette of handcrafts, small trades, special businesses and other conveniences necessarily were much broader and more 'extravagant' than in any imaginable village of that time. After all, from this broad offering the settlement model 'city' obtained its civilizing advance and drew legitimatory profit: everything is available at one place. Not least by this promise, the cities had to keep consistently, they were inevitably bound to the forest and its 'products'.

In theoretical terms of environmental history the forests supplied people for most of their businesses with the necessary energy, the required organic and inorganic raw materials, as well as with a good deal of the essential reproduction energy and elements (physiological). Expressed less abstract, this meant first of all fuel: aside from the small-scale cases mineral coal and peat the only (renewable) energy source for all endothermic processes (cooking, heating, working). Besides this, wood – in all its great variety – was needed to fabricate almost every equipment and construction from bridges over houses to the point of fortifications and recreational facilities, which included e. g. furniture as much as theatre stages. But also the diverse metallic objects as nails, anvil, or weapons first of all consumed enormous amounts of wood to roast and melt out the ore. Bricks and adobes for urban representative buildings and churches had to be baked, as much as pottery and glass for their inhabitants required fuel for hardening and liquefying; stone, clay and sand pits for this purpose were often located in the forests too. For each of these and many other businesses special demands for wood qualities were to be found – as many as wood working experts to carry them out. Only in the cities such a great variety of businesses, craftsmen and a demanding consumership were assembled in such a direct way – surrounded by stone walls with wooden armourings or doors, and integrated in a civil community, whose original constituting motivation was the improvement of civilizing achievements including reliable and secure resource-economic supply conditions.

The day-to-day aspects of the forest usage concerning terms of nutritional physiology – even if hard to find in our sources – should not be neglected: From its origins hunting means first and foremost the provision of food (protein/fat), albeit already in the Middle Ages it had been depraved to a societal event. Uncultivated wild fruits and berries, roots and leaves, (beech) nuts and acorns served as sources of nutrition and energy not only for the pasturing cattle. The people enhanced their agro-culturally dominated 'menu' by these 'delicacies' too. Agricultural crops? Therefore the people had used the forest yet before – indirectly while fertilizing their fields by the excrements of their cattle, excrements which resulted to a large degree from forest-grown forage, and fields, which had been won by clearing the woodland. The skin of these animals was transformed into leather by urban tanners using the bark of oaks; whereas their meat was cured by the butcher with the aroma of smouldering beech wood. Then it dangled as gammon on a turned stake above the other with boiled salt conserved victuals. These again were stored in cooper-made oak barrels, in which also the omnipresent beer could be transported – brewed with hops, which had been raised up in long ranges of wooden hop poles standing on formerly densely wooded cleared land. By the way, the necessary caulking of these barrels, the wooden 'Containers' of the Middle Ages (SCHUBERT), was made with tar or pitch from smouldered pine wood. But based on this condensate also different types of pastes and elixirs could be mixed by the pharmacist. And after a revelry with opulent meals and excessive drinking, he also knew about digestive and analgetic effects of forestal herbs; for pastilles and candies he stirred them with the only available sweetening of that time – honey, possibly collected from nearby lime trees through apicultural domestication of wild bees.

Such a catalogue of direct and indirect forest utilizations by and relations of the urban communities could be extended almost ad infinitum, just as the number of concerned businesses, crafts, consumers, and users. We won't do this here. Only to be mentioned decidedly are the necessarily extramural, wood-related proto-industries mining, smelting, glassworks, and salterns. In late medieval times they were often directed from civic elites – eager for profit, but also anxious about the securing of theirs cities' self-sustaining conditions. The according wood and fuel demand for our research period has to be added on the intramural 'tally stick', although – particularly in the course of the beginning territorialization – the urban communities then no longer were the only large-scale demanders of such products; instead, they were loosing their production supervision more and more for the sovereigns' benefit.

Nevertheless, it should have become clear that 'from the cradle to the grave' (wooded coffin) the forests as multifunctional and therefore ever since (and increasingly) cultivated areas in a natural landscape had to provide mostly everything, which made urban life comparatively comfortable.

Who asks for complete numerical series or exact quantities of wood consumption – for instance, to draw really exact conclusions concerning the forestal conditions and their *longue duree* development – will only find limited answers in the contemporary written sources. For these early times paleo-botanical or natural scientific approaches must be consulted. Because, adequate recordings are rare until the 16th century's second half and their areal and temporal covering is rather small. Thus, solid and broad calculations based on serial sources such as those of Peter-Michael STEINSIEK (1999) for the Harz Mountains or of Winfried SCHENK (1996) for Mid-Germany are possible not until the 17th century. Also a silvicultural or botanically exact nomenclature, which can be found later on in corresponding documents and might be interpreted environmental-historically, is to be missed sorely. Beyond mere borderline circumventions roughly the same must be said about all kinds of detailed forest descriptions.

But: No lack of knowledge or interest is to be assumed from this. Even in the Middle Ages the people knew, what was meant by the term 'wood' or the landscape formation 'grove' in this context or that location. A rational pragmatism, inherent in late medieval daily life, had not yet been displaced by the scientific rationalism of 18th century 'forestal enlightenment'. Initially, the people of post-desertification times again have had more then enough forests with sufficient wood stocks almost freely at hand, so that a scrupulous surveying and counting probably had not been deemed necessary. Insofar, until the later 16th century, corresponding statistical information came only incidentally and sparsely.

Anyway, general dimensions and magnitudes can be estimated to that effect, that the urban demand for wood and forestal by-products had been great. So great, that the civic communities to satisfy their 'hunger' for wooden resources had to find ways and means, which dissolved the immanent disproportion between their city's space (area) and size (demography) on the one hand and the necessarily small, directly available nearby forest reservoir. Only a few towns and cities could access enough own and adequately structured forests to provide themselves completely. And, to aggravate the situation, the proceeding urbanization (and territorialization) demanded more and more from the landscape. The geographical sciences here have developed the striking concept of 'ecological footprint', which roughly spoken indicates, how much landscape – taken as area of production, growth, and possible exploitation – a given community altogether engrosses while meeting their natural resource demands. Because every city – pre-modern as well as modern – concentrates comparatively many consumers on comparatively small space, the resulting areal requirements must go far beyond the city's walls (REES et al. 1996). The 'urban footprint' was absolutely and relatively by far greater than that of a contemporary village. Hence, no city could exist without more or less integrated surrounding and back-country areas (hinterland), out of which it must be provided as a 'central place' with essential trading goods and raw materials, but which it has to offer – quasi as service in return – a lot of 'civilized' services and advantages (KIESSLING 1989, HILL 2004). For long time this was a kind of tacit agreement, which was challenged fundamentally for the first time in the 16th century.

Both, the official (magistrate) and private (citizens) strategies to overcome the described area-consumptive disproportion had to be adapted to the city's natural, political, and traffic-topographical situation, as much as to its commercial and socio-economical imprint. Therefore the implemented actions varied between additional forest acquisitions, the expansion of wood trading relationships or their politico-economical manipulation, a power-political conditioning of the wood supplying hinterland, and an intensified as well as increasingly rationalized forest management. Typically combinations were on the agenda; the chronologically different setting of priorities too. Generally a decreasing possibility of acquiring new forests came along with increasing management efforts related to existing ones. This may have been due to a more and more rigid resource occupation and limitation by the territorial states (Wilgeroth 2008), but – concerning the invention of laws – could also be seen as a mere characteristic of early modern societal or legislative developments.

3. State of Research

Urban historical(-geographical) research has already proposed 'wooden typologies'. But there are still not enough case studies to come to reliable categories related to the mechanisms of wood supply, forest dependency, and the resulting spectrum of chosen strategies. And: According to historical experience the latter could be modified – caused by political shiftings for example – which then blows up the chosen categorization. So, each urban community should be examined as an individual environmental-historical case, its 'wooden character' must be explicated without any classifying prejudice. Furthermore systematic comparisons are needed to reach adequate phenomenological selectivity.

But not only aiming for comparative typing, also interested in everyday life and mentalities a vast field of historical research opens up here for an anthropocentric environmental history, – a field, whose 'tilling' Schubert (1986) has called for, and which today Joachim Radkau (1997) must regard as only scarcely 'tilled', so that he speaks from 'the riddle of urban fuel supply in the wooden age'. A riddle, which for Radkau often arose, because the so advantageously, since – abstractly spoken – as a sustainable resource usable forest was subjected to a notorious overexploitation. Not only by the cities – but particularly there the resulting wood shortage must have provoked early questions of sustainability (Schubert 1986, Popplow 2002).

Initial steps towards studying the urban forest context are until now – apart from a few late-medievally oriented exceptions (Timm 1960, Heimann 1992, Sander-Berke 1995) – only done for the later early modern times (Siemann et al. 2002). Studies focusing on the transition period between medieval and early modern times from an environmental- or forest-historical perspective are rare at all; such publications then analyze urban aspects only to a different degree (Schenk 1996, Sonnlechner 2002, Borgemeister 2005, Warde 2006). General studies on either the medieval wood(s), forests, and hunting bans or the early modern structures are of course available without difficulty. However, they are concentrated predominantly on peasant-agrarian or forest-political matters (Kiess 1958, Allmann 1989, Epperlein 1993, Günther 1994, Dasler 2001, Rösener 2007).

4. Research Outline

Consequently, the forthcoming dissertation thesis is a mixture of historiographical approaches originally pursued in the histories of everyday life, mentalities, economy, techniques, and society. From a Lower Saxon historical regional point of view its research objects are the imperial city of Goslar with its pre-dominant mining-historical imprint and the episcopal town of Hildesheim with its more civil-agricultural character. The environmental-historical perspective is derived from analyzing the societal connections, expressions, and activities connected with the culturally formed landscape and ecosystem 'forest'. The corresponding social grouping is constituted by the citizens and inhabitants of each city, their inner-relationship, such as their 'foreign affairs' with the neighboring peasant, noble and territorial-governmental (hinter)land(s). Concerning their utilization concepts and property ideas citizens and non-citizens were inevitably 'banished' to meet (and confront with) each other in the relative few woods and forests. These conflicts could be negotiated by consent or conflict, but somehow or other they finally had to be integrated – either in a cooperative or coexistent way. Thereby the resource-centered dealing both with each other and with nature can throw light upon the prevailing conceptions of nature and the existing environmental awareness.

Besides this the concrete urban utilization structures within and related to the forests can become clear. Daily arising questions of the supply and consumption of wood, the acquisition and cultivation of forests, as well as the organization of hunting and pasturing have to be answered with regard to their inner- and outer-urban interdependencies.

The underlaying working hypothesis is: Under a typologizing perspective Hildesheim, as a community concerned with pasturing and handcrafts, situated in the loosely wood covered, hilly Börde region, forms the forest-functional normal case; Goslar, the city of mining and smelting at the edge of the Harz Mountains, represent an extreme case. The different individual settings of the urban communities must have provoked in each case special ideas, motivations, and actions in the wooden resource context. Or have there not been such fundamental differences between Harz and Börde as it is assumed by this typology?

By a primarily text-historical approach the dissertation project asks not so much for the real conditions of the natural environment, but rather for its perceived conditions and the consequent culture-nature-relationship, as it was concretely established by men's everyday activities. What was regulated, what not? And how? And when? Which arguments were brought forward? Which scope for action was thought available for oneself and how much was granted others? And what – by the way – was a forest for the contemporaries? All this, from a decided urban perspective.

Respecting the two pillars 'perception' (humanities) and 'reconstruction' (natural sciences) of the research training group 'Interdisciplinary Environmental History' in Göttingen this project tries to reconstruct the societally constituted relations to a natural environment, in which, however, societally imprinted perceptions of this environment were automatically included. This started at the mostly undifferentiated wood terminology of municipal accounts and it ended with brawl and murder during forestal outrage conflicts. It is all about interpretations of nature, the ability to determine them also for others concerning the utilization of resources, and the resulting daily practices. Radkau (2000) described this from the very elevated perspective of his 'world environmental history' as a matter of 'nature and power'. Whether the concept of 'power' on the micro- and meso-level of urban environmental-historical research shows a sufficient capability to explain the variable negotiation of environmental relations or not, must be a point of further discussion.

5. Methodical Examples

To give an impression, which types of written sources are used in the project, and what kind of information can be derived from them, three simple examples with their hermeneutical approaches should be presented. Heuristically they can be categorized as "standardizing", "registering", or "conflictive" sources.

First of all it must be emphasized, that the notorious *de normalibus non in actis* applies to such a historically 'quotidian' field of research as the forest and its wood more than ever. The information derived from medieval charters, manuscripts, and registers in most cases only provides 'tessarea', which firstly must be combined laboriously and then cautiously complemented by scientific findings. Already for the medieval charters can be said, what for the files and records of the 16th century is increasingly right: The conflicts about the forest, not its everyday occurrence, provoked documentation, which nowadays can give us insights.

But there is an additional burden too: Who wants to discover something about Goslar's or Hildesheim's late medieval forest relationships, will have to concentrate on the beginning early modern times. Then firstly, under the more and more increasing pressures of territorialization or those of natural dynamics, in many places becomes virulent, what before was functioning without problems, unworthy to be mentioned for the urban communities (and what therefore was not written down). Now, one argued outstandingly about the ostensibly poor forests and the meager wood. Nearly uncountable documents were produced, in which the old, the traditional way of everyday forest life, clashes with the new, the constraints of change. Forced by individual early 16th century political turning points both cities had to find new ways into no longer easily accessible forests and for their wood supply. Especially urban centers (which they were), for hundreds of years dominating forces in a (supra)regional competition for resources, here must have had their problems. They did not want to give up old habits and suffered from such altered conditions. For the case of Goslar SCHUBERT (2004) spoke of the phenomenon of a 'synchronism of non-synchronisms'.

After the bishop's defeat in the so called 'feud of the bishopric' the aldermen of Hildesheim made hay while the sun shined: using the power vacuum of the sovereign in his own territory and capital, they hoisted themselves *de iure* in the saddle of the resource administration in the nearby Forest of Hildesheim, a territorial forest area. Where the urban population has already been participating in the common forest utilization there as a member of the 'community of heirs' (common), the still powerful magistrate of Hildesheim now made clever use of the decision-making mechanism of the wood court ('holting') belonging to the common's organization. As supreme judges since that time Hildesheim's mayors directed the fortunes of the forest completely in an urban spirit. Demonstrably since 1531 the old-town mayors were by

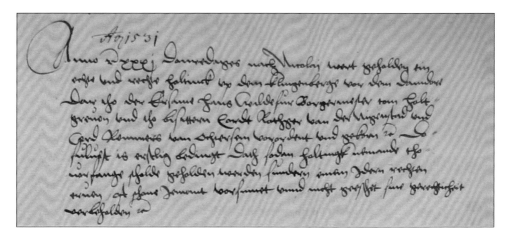

Fig. 1 Norm. Municipal Archives Hildesheim, Best. 50 Nr. 2674: 'book of the holting' (1526–1735), pag. 9

office also 'earls of the wood' ('holtgreven') and did everything in their influence to exclude unpleasant rivals by official judgment from the forest usage, such as to dictate the(ir) way of using the wooden resources to the remaining heirs. That the written documentation of the court sentences, originally witnesses of an oral decision-making culture, were initiated under urban aegis, shed bright light upon these city's interest (Fig. 1).

Compared with this, the responsible aldermen of Goslar have not been so successful trying to undo the vast losses of Harz forests, from which their city had to suffer since 1525 and which were finalized by the city's great adversary, the Duke of Brunswick, in the Riechenberg Treaty of 1552. Whether one liked it or not, the magistrate after that had to take compensatory measures, if he wanted to secure the urban wood supply at least on a rudimentary level. A new phenomenon since the mid-16th-century is therefore – not only in Goslar – the generally increased mobilization of wooden reserves outside the specific forest areals. As an example can be given a detail of a special register accounting the sale of fuel wood previously cut down on the ramparts. What had just been a kind of 'fortifying cosmetics' originally, now had to serve as a way of resource procurement. Backed up prosopographically such registers allow insights into the city's wood economy, its socio-economic baselines, and its administrative or personal structures. Obviously in 1599 even the Cloister Neuwerk ('Closter Nienwergh'), once Goslar's second largest forest owner, was forced to buy its fuel wood from the magistrate, after the Duke has *de facto* usurped the cloistral forests too (Fig. 2).

Already starting in the 1540s Goslar has tried to purchase a forest area in the neighborhood called the 'Four Mountains'. Originally an estate of the Cistercian convent of Walkenried it had become disposable since 1538, because the cloister had been heavily affected by the Peasants' War. Goslar and Walkenried came to an agreement, and the townspeople, after paying an annual tribute, were allowed to make use of the undergrowth for their fuel demands. Some of the Harz forest losses could be compensated regarding Radkau's riddle. Step-by-step, based upon further contracts, the magistrate established in this deciduous woodland a controlled 'mid-forest management' with a local forest administration. 1579 at the latest there is evidence for his intention to cultivate there for the city's official demands a new reserve of fuel, timber and lumber. But since 1593 the citizens had to suffer there from undisguised forest invasions by bailiffs, sub-servants and subjects of the Dukes of Brunswick: Wood devastations were on the daily agenda and had been carried out by sheer force. Goslar's Four-Mountains' forester often could only stand by helplessly and watch, while he was merely able to take the offences down on his record. These recordings were monthly

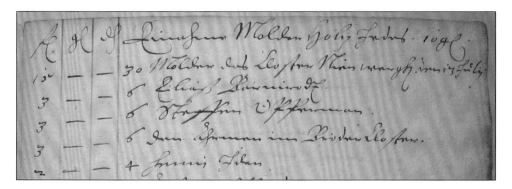

Fig. 2 Number. Municipal Archives Goslar, Sign. B 6078: 'fuel wood from the ramparts' (1599), fol. 2r

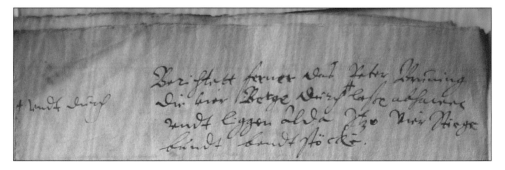

Fig. 3 Conflict. Municipal Archives Goslar, Sign. B 2452, 'Four Hills [...]' (1562-1662), report of the forester, 1608 May 28th.

handed over as formal reports to the aldermen. Besides implicit descriptions of the forest stand ('bendtstöcke') these documents reflect the Duke's way of treating (and acting in) the woods ('laßen abhawen durch und durch') as much as – read between the lines – they give an impression of completely different urban ideas of a better (sustainable?) forest utilization (controlled cutting; immediate wood removal) (Fig. 3).

The special attraction of sources like these for environmental-historical studies reveals itself not by applying elaborate theories to them, but rather through a solid source criticism, which is environmentally sensitized and supplemented with the knowledge of ecological and (!) societal connections. With each charter, manuscript, file or register at hand one should ask again for the intention, motivation and situation of its respective author. What was his individual perspective on the natural environment? An environment, which had not persisted unaffected by all this.

References

ALLMANN, J.: Der Wald in der Frühen Neuzeit. Eine mentalitäts- und sozialgeschichtliche Untersuchung am Beispiel des Pfälzer Raumes 1500–1800. Berlin: Duncker und Humblot 1989
BORGEMEISTER, B.: Die Stadt und ihr Wald. Eine Untersuchung zur Waldgeschichte der Städte Göttingen und Hannover vom 13. bis zum 18. Jahrhundert. Hannover: Hahnsche Buchhandlung 2005
DASLER, C.: Forst und Wildbann im frühen deutschen Reich. Die königlichen Privilegien für die Reichskirche vom 9. bis zum 12. Jahrhundert. Köln: Böhlau 2001
DOEBNER, R.: Urkundenbuch der Stadt Hildesheim. Teil 4: von 1428 bis 1450. Urkunde Nr. 358, S. 300. Hildesheim: Gerstenberg 1890
EPPERLEIN, S.: Waldnutzung, Waldstreitigkeiten und Waldschutz in Deutschland im hohen Mittelalter: 2. Hälfte 11. Jahrhundert bis ausgehendes 14. Jahrhundert. Stuttgart: Steiner 1993
GLEITSMANN, R.-J.: Aspekte der Ressourcenproblematik in historischer Sicht. Scripta Mercaturae *15*, 33–89 (1981)
GÜNTHER, R.: Der Arnsberger Wald im Mittelalter. Forstgeschichte als Verfassungsgeschichte. Münster: Aschendorff 1994
HEIMANN, H.-D.: Der Wald in der städtischen Kulturentfaltung und Landschaftswahrnehmung. In: ZIMMERMANN, A., and SPEER, A. (Eds.): Mensch und Natur im Mittelalter, 2. Halbbd. S. 866–881. Berlin 1992
HILL, T.: Die Stadt und ihr Markt. Bremens Umlands- und Außenbeziehungen im Mittelalter (12. – 15. Jahrhundert). Stuttgart: Steiner 2004
JOHANNEK, P.: Die Mauer und die Heiligen. Stadtvorstellungen im Mittelalter. In: BEHRINGER, W., und ROECK, B. (Eds.): Das Bild der Stadt in der Neuzeit 1400–1800. S. 26–38. München: C. H. Beck 1999
KIESS, R.: Die Rolle der Forsten im Aufbau des württembergischen Territoriums bis ins 16. Jahrhundert. Stuttgart: W. Kohlhammer 1958

KIESSLING, R.: Die Stadt und ihr Land. Umlandpolitik, Bürgerbesitz und Wirtschaftsgefüge in Oberschwaben vom 14. bis 16. Jahrhundert. Köln: Böhlau 1989
PFEIFFER, G.: Wasser und Wald als Faktoren der städtischen Entwicklung in Franken. Jahrbuch für Fränkische Landesforschung *32*, 151–170 (1972)
POPPLOW, M.: Meeting Review „Abholzung und Aufklaerung" – „Holzmangel im 18. Jahrhundert" (Potsdam, 07./08.06.2002). In: H-Soz-u-Kult, 08.07.2002, <http://hsozkult.geschichte.hu-berlin.de/tagungsberichte/id=58>
RADKAU, J.: Das Rätsel der städtischen Brennholzversorgung im „hölzernen Zeitalter". In: SCHOTT, D. (Ed.): Energie und Stadt in Europa. S. 43–75. Stuttgart: Steiner 1997
REES, W. E., WACKERNAGEL, M., and TESTEMALE, P.: Our ecological footprint: reducing human impact on earth. Gabriola Island, BC: News Society Pupl. 1996
RÖSENER, W.: Der Wald als Wirtschaftsfaktor und Konfliktfeld in der Gesellschaft des Hoch- und Spätmittelalters. Zeitschrift für Agrargeschichte und Agrarsoziologie *55*, 14–31 (2007)
SANDER-BERKE, A.: Baustoffversorgung spätmittelalterlicher Städte Norddeutschlands. Köln: Böhlau 1995
SCHENK, W.: Waldnutzung, Waldzustand und regionale Entwicklung in vorindustrieller Zeit im mittleren Deutschland. Stuttgart: Steiner 1996
SCHUBERT, E.: Der Wald: wirtschaftliche Grundlage der spätmittelalterlichen Stadt. In: HERRMANN, B. (Ed.): Mensch und Umwelt im Mittelalter. S. 257–274. Stuttgart: DVA 1986
SCHUBERT, E.: Städte im Aufbruch und Wandel. In: FLÜELER, M., Landesdenkmalamt Baden-Württemberg und der Stadt Zürich (Eds.): Stadtluft, Hirsebrei und Bettelmönch: die Stadt um 1300. S. 381–392. Stuttgart: Theiss 1992
SCHUBERT, E.: Alltag im Mittelalter. Natürliches Lebensumfeld und menschliches Miteinander. Darmstadt: WBG 2002
SCHUBERT, E.: Die geschwinden leuffte und die unmerklichen Wandlungen in der Welt um 1550. In: DETTMER, H.-G. (Ed.): Der Riechenberger Vertrag. S. 27–64. Goslar: Verlag Goslarsche Zeitung 2004
SIEMANN, W., FREYTAG, N., und PIERETH, W. (Eds.): Städtische Holzversorgung. Machtpolitik, Armenfürsorge und Umweltkonflikte in Bayern und Österreich (1750 bis 1850). München: C. H. Beck 2003
SOMBART, W.: Der moderne Kapitalismus. Historisch-systematische Darstellung des gesamteuropäischen Wirtschaftslebens von seinen Anfängen bis zur Gegenwart. Bd. II/2. Das europäische Wirtschaftsleben im Zeitalter des Frühkapitalismus, vornehmlich im 16., 17. und 18. Jahrhundert. München, Leipzig: Duncker und Humblot 1917
SONNLECHNER, C.: Verwaltung von Natur. Ressourcenmanagement und das geschriebene Wort in spätmittelalterlichen und frühneuzeitlichen Grundherrschaften. In: POHL, W., und HEROLD, P. (Eds.): Vom Nutzen des Schreibens. Soziales Gedächtnis, Herrschaft und Besitz im Mittelalter. S. 375–394. Wien: Verlag der Österreichischen Akademie der Wissenschaften 2002
STEINSIEK, P.-M.: Nachhaltigkeit auf Zeit: Waldschutz im Westharz vor 1800. Münster: Waxmann 1999
STEINSIEK, P.-M.: Review: BORGEMEISTER 2005. Niedersächsisches Jahrbuch für Landesgeschichte *78*, 521–522 (2006)
SYDOW, J.: Bemerkungen zur Holzversorgung in Städten. In: SYDOW, J. (Ed.): Städtische Versorgung und Entsorgung im Wandel der Geschichte. S. 77–98. Sigmaringen: Thorbecke 1981
TIMM, A.: Die Waldnutzung in Nordwestdeutschland im Spiegel der Weistümer: einleitende Untersuchungen über die Umgestaltung des Stadt-Land-Verhältnisses im Spätmittelalter. Köln: Böhlau 1960
WARDE, P.: Ecology, Economy and State Formation in Early Modern Germany. Cambridge: Cambridge University Press 2006
WILGEROTH, C.-O.: Stadtgestaltung im mittelalterlichen Hildesheim. Unpublished Master Thesis. University of Göttingen 2004
WILGEROTH, C.-O.: Bonam sylvarum partem in vicinia ... Politisch-generierte Ressourcenknappheit und städtische Kompensation: Goslar, die Landesherren und Walkenried im 16. Jahrhundert. Niedersächsisches Jahrbuch für Landesgeschichte *80*, 51–116 (2008)

Cai-Olaf WILGEROTH
Georg-August-Universität Göttingen
(formerly) DFG-Graduiertenkolleg Interdisziplinäre Umweltgeschichte
Bürgerstraße 50
37073 Göttingen
Germany
Phone: +49 551 399977
E-Mail: cai-olaf.wilgeroth@biologie.uni-goettingen.de

Life Strategies of Microorganisms in the Environment and in Host Organisms

Leopoldina-Symposium
Deutsche Akademie der Naturforscher Leopoldina in Zusammenarbeit mit dem Max-Planck-Institut für Marine Mikrobiologie Bremen und der Universität Bremen

vom 5. bis 8. April 2006 in Bremen

Nova Acta Leopoldina N. F., Bd. *96*, Nr. 356
Herausgegeben von Rudolf AMANN (Bremen), Werner GOEBEL (Würzburg), Barbara REINHOLD-HUREK (Bremen), Bernhard SCHINK (Konstanz) und Friedrich WIDDEL (Bremen)
(2008, 112 Seiten, 22 Abbildungen, 1 Tabelle, 21,50 Euro, ISBN: 978-3-8047-2499-0; Supplementum: 8 Seiten, 3 Abb.)

Viele Prokaryoten (Bakterien wie Archaeen) in Wasser, Boden oder Wirtsorganismus zeigen häufig spezifische, hoch entwickelte Anpassungen zur Nutzung des umgebenden Milieus als Energie- und Nährstoffquelle sowie als Überlebensraum. Bei prokaryotischen Organismen spielen sich die Anpassungen an die abiotische oder biotische Umgebung fast nur auf der Ebene des Stoffwechsels und der molekularen bzw. makromolekularen Strukturen und Wechselwirkungen ab. Auf dieser Ebene zeigen sich tatsächlich typisch prokaryotische Leistungen: eine Nutzung „ungewöhnlicher" Energiequellen, grundlegende Syntheseleistungen, enzymatische Katalysen, Anpassungen an extreme Verhältnisse, wie hohe Temperaturen, Kolonisations- und Invasionsmechanismen, und Regulationsvorgänge. Der Band behandelt die Rolle von Mikroorganismen in den globalen Elementkreisläufen, wie z. B. dem Schwefelkreislauf, oder aber auch dem Methankreislauf, der Kohlendioxidfixierung oder der Ammoniakoxidation, Symbiosen mit Bakterien-Eukaryoten-Interaktionen auf der Grundlage von Schwefel- oder Stickstoffstoffwechsel, Modelle zum prokaryotischen Ursprung von Organellen in Eukaryoten sowie die Beteiligung von Mikroorganismen an Infektions- und Pathogenitätsvorgängen.

Wissenschaftliche Verlagsgesellschaft mbH Stuttgart

Cattle Plague in Early Modern Germany: Environment and Economy / Knowledge and Power in a Time of Crisis

Dominik Hünniger (Göttingen)

Abstract

Despite the immense impact of epizootics on agrarian societies in the past, historical studies on animal diseases in Early Modern Germany are rare. The dissertation, which is summarized here, focuses on major aspects of disease control and disease concepts in 18th century Germany with special reference to the duchies Schleswig and Holstein. Environment, economy, knowledge and power are the four central categories in the analysis of the contemporary bureaucratic correspondence as well as the medical and religious advice literature of the time. The dissertation follows insights from the social history of medicine in its focus on the "framing of diseases" and the interests and behavior of different actors and their crisis communication.

Reflecting upon recent research on early modern state formation processes, which stresses the contested nature of early modern legislation and administration, I ask how different groups of actors shaped disease containment policies in a communicative process. Animal disease control in early modern Europe can only be understood as a complicated interactive process of continuous crisis management, in which legislation was altered in response to the practicalities of implementing it on the ground. Epizootics happened in a social, economic and political environment where all parties involved and affected had not only to tackle the immediate impact of the diseases but were eventually forced to contextualize disease within a wider variety of circumstances.

Zusammenfassung

Gemessen an den bedeutenden Auswirkungen von Viehseuchen auf die ländlichen Lebenswelten der Frühen Neuzeit ist die bisherige Vernachlässigung dieser Problematik in der deutschsprachigen historischen Forschung sehr erstaunlich. Gegenstand der hier zusammengefassten Dissertation sind die Bekämpfungsmaßnahmen und Bewältigungsstrategien der großen Viehseuchen der ersten Hälfte des 18. Jahrhunderts. In der Dissertation werden Viehseuchen und ihre Bekämpfung im Spannungsfeld von Umwelt, Ökonomie, Herrschaft und Wissen untersucht. Die Studie besitzt mit der Auswertung der zeitgenössischen deutschsprachigen Veröffentlichungen einen überregionalen Fokus, konzentriert sich aber schwerpunktmäßig auf die obrigkeitlichen Maßnahmen sowie die lokalen Reaktionen und Konflikte in den Herzogtümern Schleswig und Holstein. Parallel zu neueren Forschungen zur Sozial- und Kulturgeschichte menschlicher Seuchen stehen die Interessen und Handlungen der verschiedenen Akteure und ihre Kommunikation sowie die sozialen und kulturellen Konstruktionsprozesse rund um das Seuchengeschehen im Mittelpunkt.

Die administrativen Abläufe bei der Seuchenbekämpfung bestätigen die Ergebnisse neuerer Forschungen zu Herrschaftsausübung und -wahrnehmung in der Frühen Neuzeit. Es wurde deutlich, dass alle Beteiligten nicht nur die Seuchenabwehr zu beachten hatten, sondern dass zugleich die alltäglichen wirtschaftlichen und sozialen Handlungen möglichst aufrecht zu erhalten waren. Beide Ziele mussten gegeneinander abgewogen werden. Die Seuchenbekämpfung in der Frühen Neuzeit stellte sich somit generell als kontinuierlich nachregulierendes Krisen-„Management" dar.

Dominik Hünniger

1. Introduction

Cattle plague raged through Europe during the eighteenth century with three peaks in incidence. The first occurred from 1711–1717, the second from 1745–1757, and the third from 1769–1786 (SPINAGE 2003, years of outbreaks in the different territories of Europe varied of course). Their devastating impact on the economy and society can hardly be overrated in the light of estimated mortality rates of between 70 and 90%. However, until recently the historiography on veterinary medicine in Germany has concentrated on the development of veterinary services and has been written by practising vets. In general, studies on early modern Germany are rare, and hardly any attempt has been made to reveal the effect various diseases had on the everyday life of rural populations and the enormous challenge epizootics posed to early modern administrations (HÖLZL and HÜNNIGER 2008, ENGELKEN et. al. 2007). My dissertation (HÜNNIGER 2008) sheds new light on important aspects of disease control by focusing on changes and continuities in official legislation, as well as the role of communication between the actors involved. The events and developments in the Duchies Schleswig and Holstein are analyzed in a micro-historical case study which looks at environmental and economic circumstances, social and power relations and the perception of the epizootic crisis in the light of existing world-views. Hence this dissertation complements and occasionally questions the strand of environmental history that reconstructs material flow and changes in the longue durée (SIEFERLE 2006) by taking the dialectics of structures and practices seriously (LANGTHALER 2006).

Additionally, the extensive contemporary medical advice literature is analyzed from the point of view of recent research on questions of the "framing of diseases" in history. This focus on the interests and behavior of different actors and their communication is a central concern of the social history of epidemics today (DINGES 1995).

2. Perceptions and Diagnoses

Early Modern agricultural advice literature, household manuals and medical treatises referred to animal diseases only occasionally. By contrast the 18th century witnessed a major rise in publications, concerning number and the degree of detail. In monographs or articles in journals and newspapers the nature and meaning of epizootics, especially cattle plague, was debated extensively. This exchange between scholars, livestock owners and authorities can be found in the bureaucratic correspondence as well. In this communicative process, a body of knowledge was created that was both novel and related to older traditions and concepts.

First of all, when people tried to explain catastrophes and crises in Early Modern Times, divine wrath was almost always seen as the first source of disasters. This "economy of sins"/"Sündenökonomie" (BEHRINGER 2003, p. 114) features also very prominently in the cattle plague advice literature. Of course theologians stressed this aspect more than physicians; still the latter almost always included at least some reference to God's punishment into their publications well until the end of the century. Not only did all actors concerned view epizootics in one way or another as a divine punishment, at the same time the Christian God provided the most secure source of help in containing the disease. Thus physicians, authorities and livestock owners hoped and prayed for divine help to their containment policies and medical remedies. It is very important to note the contemporary compatibility of accepting the divine source of plagues and other disasters *and* applying a wide range of containment measures and medication. Governments, scholars and people in general, almost never retreated to fatalism and passive acceptance. On the contrary, disasters could be and were seen as ordeals that had to be overcome by mobilizing a wide range of resources and skills. Thus the medical and religious advice literature stresses the enlightenment's pedagogical zeal very prominently.

Secondly, although publications flourished, the university trained physicians had to defend their work on animal diseases as it was still considered dishonoring for their profession to deal with the "brute creation". Therefore these learned men invoked either famous predecessors (sometimes even dating back to classical antiquity) or stressed the immense economical losses these diseases created and the importance of finding cures and remedies. For example, Johann Christoph RIEDEL, Professor of Medicine at Erfurt University, invoked his general interest in the well-being of humanity and his special duty as a member of the Leopoldina in his work on the cattle plague (RIEDEL 1749). Although these justifications became less frequent and vigorous in the course of the 18th century, the establishment of the first European veterinary schools and university programs in the 1760s can only be seen as the first steps in a much longer and more complicated history of the veterinary profession, yet to be written (KOOLMEES 2002).

Finally, contempt for any veterinary professionalization was not the main source of difficulties. The medical profession in general was at a loss when it came to determining the nature and causes of "the true horned cattle plague". The diagnoses seem to have been almost as varied as the number of publications. This was partly due to the most widely accepted early modern disease concept of Galenism that viewed diseases not necessarily as specific entities but rather as "a matter of degree or a process" (DELACY 1999, p. 267f.). Still two main disease "identities" were popular in the 18th century. The first was the identification of the disease with plague, the other with fever, although the former was sometimes conceptualized merely as a very severe form of fever. Even more interesting than this debate is the authors' usage of concepts of "contagion" and "miasma" as it affirms recent studies on early modern disease concepts that argue for a re-conceptualization of the notions of these terms in pre-modern times. Local studies showed the coexistence and interchangeability of both concepts and their application according to context and preferences of measures (KINZELBACH 2006). In general much of the medical advice literature applied both disease "theories" widely and only appears fuzzy from a modern point of view. Within the framework of contemporary medical thinking and general world-views, this literature just related illness "to ideas about how the world works" (DUFFIN 2005, p. 14). Nevertheless, contemporary medical and metaphysical thought shaped not only the perception of epizootics but also a wide range of control measures, like state prayers and fasts, the application of drugs and herbs, as well as any other containment measures.

3. Power Relations and Containment Policies

Animal disease control in early modern Europe can be understood as a complicated interactive process of continuous crisis management, in which legislation was altered in response to the practicalities of implementing it on the ground. Epizootics happened in a social, economic, and political environment where all parties involved and affected had not only to tackle the immediate impact of the diseases but were eventually forced to contextualize disease within a wider variety of circumstances.

Reflecting upon recent research on early modern state formation processes, which stresses the contested nature of early modern legislation and administration (Asch and Freist 2005), I ask how different groups of actors shaped disease containment policies in a communicative process. State formation in the early modern period was not a process of simple straightforward modernization but rather a continuous one with many ruptures and idiosyncrasies. Subjects expected authorities to cater for their well being and administrators sought popular approval whenever possible. Therefore certain key aspects of epizootics control policies were deeply contested, namely, trade regulations, quarantine measures and the use of animal products. For instance, the quarantine regulations attracted considerable resistance from subjects who petitioned the authorities to moderate the rules. Although the authorities considered a complete quarantine of farms and villages to be the most effective means of containing disease, they were unable to shut off whole villages for a long time because it produced serious problems as can be seen in letters and supplications of villagers from many parishes. These letters provide a detailed insight into the organization of everyday live during times of crisis. An example of these tensions is apparent in the petition of a cobbler from Tetenbüll in Eiderstedt, Dirk Asmus. Asmus' work depended on the use of cattle hides, and he complained of the economic difficulties that arose because he could no longer earn money from trade, and he asked for a concession to allow him to send already manufactured boots and shoes out of the quarantined village. Another major problem was that an infected village was almost under a state of siege, so there were severe food shortages (especially flour). Food was not distributed equally, and many peasants complained that the guards took more than their fair share of provisions, calling them "impertinent eaters and drinkers" because they demanded bacon and meat instead of porridge. Psychological hardship affected members of small religious denominations, such as the Mennonites and Remonstrants, who were unable to attend the usual services at Friedrichstadt about 15 miles away. Instead they had to take a special oath to enable them to attend the Lutheran church in the village. These are just some of the effects quarantine had on those who had to endure the quarantining of their villages. Similar complaints, as well as petitions demanding changes to the ordinances reached the authorities from every corner of the Duchies and these increased in number and frequency, the longer an epizootic prevailed (Hünniger 2010).

Without oversimplifying I would argue that at the beginning of an epidemic the initial laws prohibited activities that the authorities considered to be possible agents for the spread of the disease, entirely. However, in the course of time, once an epidemic had prevailed for several months, these strict measures had to be adapted to accommodate other interests as well. When a serious epizootic continued for several years, as in the case of the 1745–1757 outbreak, the authorities eventually surrendered to the fact that efforts to contain the cattle plague had failed, as it had almost spread throughout the whole of one territory or even many countries.

Regulations for epizootics clearly showed that the process of early modern state formation was a complicated combination of bottom-up and top-down developments.

4. Knowledge and Communication

Recent studies on the perception and containment of human diseases in Early Modern Europe focused on the role of communication, especially the influence of rumors during times of epidemics (Werkstetter 2005). Not unlike human diseases, epizootics triggered many colorful accounts of origin and spread of diseases that gained popularity throughout many European countries. One story that can be found in many publications and that is still a prominent feature of epidemic narratives until today is that of "Patient Zero", the initial carrier of the disease (GILMAN 1988). In the Early Modern Italian literature e. g. a "Dalmatian ox" was the culprit on which this narrative of origin was laid. In the Duchies Schleswig and Holstein an unnamed Dutch cattle dealer was evoked as the carrier of the disease time and again throughout the 18th century.

In addition authorities handled their information according to specific circumstances. Decrees for cattle plague control sometimes began with statements about the origin and course of the disease in neighboring countries and information about these outbreaks was communicated very early on. By contrast many governments kept silent for sometimes very long periods of time when outbreaks in their own territories occurred. As trade with animals and their products was economically so important and governments dreaded trade bans, the authorities had to be very careful about the time and nature of their information policies. Thus it was not uncommon that sovereigns urged their local administration to keep silent or vague about outbreaks *vis-à-vis* neighboring territories. On the other hand subjects were advised to report outbreaks immediately and denunciation within villages was common.

Additionally, local authorities in Schleswig and Holstein collected information and statistics on mortality systematically and from a very early stage. Nevertheless, as most territories were governed by more than one political body and central administration was distant, many letters and decrees reached their addressees much later than intended, when either the designated measures or the information was already defective.

Still, documented experiences from and knowledge of earlier outbreaks was a very valuable source of containment policies. Despite certain reservations many different authorities exchanged decrees, reports or treatises, and local physicians and livestock owners reported their observations and experiences accordingly.

5. Conclusion

As can be seen from this short introduction on Early Modern epizootic containment policies and their perception, the history of animal plagues is clearly a central issue in human-environment relations. This analysis of disease perception and the containment policies during cattle plague provides a multifaceted picture of contemporary economic practices and power relations, as well as perceptions of the environment and the generation of knowledge in times of crisis.

Additionally, the specific environmental conditions of the Duchies (most prominently the distribution of marshes and geestland) and their impact on economic and social realities proved to be a major factor in shaping the spread of disease as well as the circumstances of containment policies.

Epizootics happened in a social, economic and political environment where all parties involved and affected had not only to tackle the immediate impact of the diseases but were eventually forced to contextualize disease within a wider variety of circumstances. The evidence illustrates that a dreadful crisis like an outbreak of cattle plague shook the foundations of 18th century society and people addressed this by mobilizing their social, economic, cultural and legal resources. Only by looking at all four aspects, is one able to understand the impact of epizootics on agrarian societies in Early Modern Europe.

Dominik Hünniger

Sources

Kreisarchiv für den Kreis Nordfriesland in Husum
Landesbibliothek Kiel, Manuscript Collection: Ordinances
Landesarchiv Schleswig-Holstein, Schleswig

References

ASCH, R. G., and FREIST, D. (Eds.): Staatsbildung als kultureller Prozess. Strukturwandel und Legitimation von Herrschaft in der Frühen Neuzeit. Köln: Böhlau 2005
BEHRINGER, W.: Die Krise von 1570. Ein Beitrag zur Krisengeschichte der Neuzeit. In: JAKUBOWSKI-TIESSEN, M., and LEHMANN, M. (Eds.): Um Himmels Willen. Religion in Katastrophenzeiten. S. 51–156. Göttingen: Vandenhoek & Ruprecht 2003
DELACY, M.: Nosology, mortality, and disease theory in the eighteenth century. Journal of the History of Medicine and Allied Sciences 54, 261–284 (1999)
DINGES, M.: Neue Wege in der Seuchengeschichte. In: DINGES, M., and SCHLICH, T. (Eds.): Neue Wege in der Seuchengeschichte. S. 7–24. Stuttgart: Steiner 1995
DUFFIN, J.: Lovers and Livers. Disease Concepts in History. (The 2002 Joanne Goodman Lectures) Toronto et al.: University of Toronto Press 2005
ENGELKEN, K., HÜNNIGER, D., und WINDELEN, S. (Eds.): Beten, Impfen, Sammeln. Zur Schädlings- und Viehseuchenbekämpfung in der Frühen Neuzeit. Göttingen: Universitätsverlag 2007
GILMAN, S. L.: Disease and Representation. Images of Illness from Madness to AIDS. Ithaca: Cornell University Press 1988
HÖLZL, R., und HÜNNIGER, D.: Global denken – lokal forschen. Auf der Suche nach dem ‚kulturellen Dreh' in der Umweltgeschichte. Ein Literaturbericht. WerkstattGeschichte 48, 83–98 (2008)
HÜNNIGER, D.: Wahrnehmung, Deutung und Bekämpfung von Rinderseuchen in der ersten Hälfte des 18. Jahrhunderts am Beispiel der „Hornvieh-Seuche" in den Herzogtümern Schleswig und Holstein. Diss. phil. Georg-August Universität Göttingen 2008
HÜNNIGER, D.: Policing epizootics. Legislation and administration during outbreaks of cattle plague in eighteenth century Northern Germany as continuous crisis management. In: BROWN, K., and GILFOYLE, D. (Eds.): Healing the Herds. Disease, Livestock Economies, and the Globalization of Veterinary Medicine; pp. 76–91. Athens: Ohio University Press 2010 (in press)
KINZELBACH, A.: Infection, contagion, and public health in late medieval and early modern German imperial towns. Journal of the History of Medicine and Allied Sciences 61/3, 369–389 (2006)
KOOLMEES, P.: Trends in veterinary historiography. In: SCHÄFFER, J., and KOOLMEES, P. (Eds.): History of Veterinary Medicine and Agriculture. Proceedings; pp. 235–243. Giessen: DVG 2002
LANGTHALER, E.: Agrarsysteme ohne Akteure? Sozialökonomische und sozial-ökologische Modelle in der Agrargeschichte. In: DIX, A., and LANGTHALER, E. (Eds.): Grüne Revolutionen. Agrarsysteme und Umwelt im 19. und 20. Jahrhundert. S. 216–238. Innsbruck et. al.: Studienverlag 2006
RIEDEL, J. C.: Untersuchung der jetzt grassierenden Vieh-Seuche, nebst kurtzen Unterricht, von dem Verhalten und Gebrauch, derer bey derselben nöthigen Artzeney-Mittel. Erfurt: Ritschel 1749
SIEFERLE, R. P., KRAUSMANN, F., SCHANDL, H., und WINIWARTER, V.: Das Ende der Fläche. Zum gesellschaftlichen Stoffwechsel der Industrialisierung. Köln et al.: Böhlau 2006
SPINAGE, C. A.: Cattle Plague. A History. New York: Kluwer 2003
WERKSTETTER, C.: Die Pest in der Stadt des Reichstags. Die Regensburger Contagion von 1713/14 in kommunikationsgeschichtlicher Perspektive. In: BURKHARDT, J., and WERKSTETTER, C. (Eds.): Kommunikation und Medien in der Frühen Neuzeit. (Historische Zeitschrift Beihefte N. F. 41). S. 267–292. München: Oldenbourg 2005

Dominik HÜNNIGER M. A.
Academic Coordinator
Georg-August-Universität Göttingen
Lichtenberg-Kolleg
Geismar Landstraße 11
37083 Göttingen
Germany

Phone: +49 551 3910626
Fax: +49 551 391810626
E-Mail: dominik.huenniger@zvw.uni-goettingen.de

Cameralistic and Utilitarian Conceptions of Happiness and their Implications in Respect of Today's Environmental Crisis

Jörg Cortekar and Rainer Marggraf (Göttingen)

Abstract

Modern neoclassical standard theory is constitutionally impeded to deal with certain subjects by some basic suppositions to economic theory. Doubtlessly, one of these subjects was, and still is the field concerning any aspect of natural environment. When it comes to the environment, many recent studies in experimental economics have shown that individuals often behave in a different way than predicted by the *homo economicus* model and its assumptions. As long as only a few individuals behave in this "defective" manner there is supposed to be no problem. But if divergent behavior is no longer the exception to the model the aggregate's behavior is being influenced with the consequence that statements based on the *homo economicus* are no longer valid. Hence, the question is which modifications have to be made to the model to achieve better predictions of how human beings behave. The history of economic thought may provide some hints to potential supplements in order to extend the model of *homo economicus*. Determining where to find them and ways of making them fruitful is the objective of this paper.

Zusammenfassung

Der ökonomischen Standardtheorie liegen Annahmen zugrunde, die es ihr konstitutionell erschweren, bestimmte Themenfelder zu bearbeiten. Eines dieser Felder ist zweifelsfrei das der natürlichen Umwelt. So haben empirische Studien belegt, dass sich Individuen speziell im Umgang mit der natürlichen Umwelt oft anders verhalten, als dies im Rahmen des *Homo-oeconomicus*-Modells vorhergesagt werden würde. Dieses Phänomen ist wenig problematisch, solange ein solches Verhalten die Ausnahme darstellt. Weicht allerdings eine hinreichend große Menge an Individuen systematisch vom *Homo-oeconomicus*-Verhalten ab, sind auf dem *Homo-oeconomicus*-Modell basierende Aussagen nicht mehr valide. Es kann also die Frage aufgeworfen werden, wie der *homo oeconomicus* „verändert" werden müsste, um bessere Vorhersagen menschlichen Verhaltens zu erhalten. Ein Blick in die Vergangenheit kann sich bei der Suche potenzieller Erweiterungen dabei durchaus als fruchtbar erweisen. Wo sich Ansätze finden lassen und wie diese aufgegriffen werden könnten, ist Ziel dieses Beitrags.

1. Why Should We Look Back?

Today's environmental crisis is omnipresent and one will probably fail in trying to find a scientist who denies any anthropogenic influence in what is commonly known as global warming. The question to be asked is, what can we do? It is obvious and without controversy that a single scientific discipline cannot solve a wide-ranging problem such as global warming alone. This paper presents possibilities to extend the approach to that problem from an economic viewpoint.

It is often said that a discipline, which allows within its framework an economically 'efficient ecological catastrophe' should better not deal with environmental aspects at all. But it also has to be acknowledged that today some policy instruments such as emission trading schemes are implemented (e.g. in Europe and RGGI in the US) in order to reduce CO_2 emissions and that the successful implementation is based *in pars pro toto* on environmental economic research. But: environmental economics stands on the fundament of mainstream neoclassical economics with the *homo economicus* at its basis. It can be asked, if the model of rational choice, assuming that all relevant information is available, is sufficient for analyzing topics in which decisions are made under uncertainty (which is often inherent in environmental problems). In this model direct incentives are set for every relevant agent, which is often not possible due to special problems of access, i.e. a wide range of affected persons, large external effects, and ecological interdependencies. Finally, the basic assumption of a strictly self-interested economic agent seems to be inappropriate from an empirical as well as a normative point of view, which derives from special dimensions of (social) responsibility. Given this background one should ask, which aspects are really relevant to explain human actions as well as attempting to systematically influence them (which is important to know for a political planer) with special respect to environmental issues.

To answer this last question, a look back to the starting point of the rise of economic theory in the mid 18th century might be helpful. At that time, the unity of the social sciences was still predominant, and concepts of the spring of human action were evaluated by their ability to explain or predict human behavior rather than by their applicability to exact mathematical modeling as it is today. It will be scrutinized in this paper, what lessons can be learned from the history of economic thought with respect to human behavior. In doing so, this paper contrasts two historic schools of thought, the German Cameralism and classical Utilitarianism to look what can be made fruitful to expand economic standard theory.

2. Concepts of Happiness and What They Stand For

When reading the works of cameralistic (e.g. von Justi or Jung) or utilitarian (esp. Smith and J. St. Mill) authors it can hardly be denied that happiness reflects the specific understanding of man, society as a whole and the state. But how do these concepts differ? And what are the consequences of that difference? We begin to contrast these two theories by looking at the German Cameralism first.

2.1 The Cameralistic Concept

Almost all works of the *Staatswissenschaften* of that time are shot through with passages stating that happiness is the sole end of society. Individual and collective elements were inseparably mingled within the conceptual framework of happiness; the one is not imaginable without the other. Hence, the terms "happiness of the people" and "happiness of the state" were mostly used synonymously (cf. von Justi 1759, 1761, 1762, von der Lith 1766, Jung 1792, Engelhardt 1981). Either way, an essential part of happiness is material welfare, which consists of goods rather than bullion.[1] But human beings were looked at with great

---

[1] In Mercantilism the measure of welfare was the amount of bullion. In Cameralism the quantity of goods is of far more interest. It was said, that bullion is just relative wealth whereas goods are the real one (cf. von Justi 1760). This is one difference between these two theories and marks a paradigm shift in economic thinking.

skepticism throughout the *Staatswissenschaften* (cf. VON JUSTI 1760, 1764/1970, or JUNG 1799). This argument becomes clear with the statement that when economic decision-making is solely given to private hands, every individual strives for its own best, which is argued to be in conflict with societies' best or the "happiness of the state" mentioned above. Hence, it is the states' duty and responsibility to ensure the basic requirements (i.e. security, freedom, and so forth) for the necessary material welfare and linked with that economic growth.[2] But how can the state to be enabled to realize these preconditions for economic prosperity? The answer was science. Considerable knowledge of conditions and interrelations of economic production was considered necessary for the state to control, direct, and raise the output of national economies[3] (cf. further explanations VON JUSTI 1761/1970, 1764/1970). Since the state is in charge of economic development it is its task to allocate the different (natural) resources available to create economic output as high as possible (which implies, that the current allocation at that time was not the best one possible). What to take into consideration to do so was elaborated in the *Policeywissenschaften*, which is likewise a governor's *vade mecum*[4] to run the state and national economy successfully. In these manuals it was argued that all elements influencing economic output necessarily had to have the right proportion to one another.[5] The probably most important of these elements is the shape of the landscape; e.g. enough farmland of high quality is needed to feed the population and woodland to secure supply with energy and building material. Since population supply depends on both and given the same acreage the number of people supplied by each is different there is a *trade off* between farmland and woodland so that the 'right' proportion between these two had to be found[6] (cf. DARJES 1756, VON JUSTI 1760, 1782, VON PFEIFFER 1783, BERGIUS 1767). Realizing the best alternative proportion possible, the landscape needed to be rebuilt for the necessities of economic production and consequently there are several precise instructions in the works of the *Policeywissenschaften* on what to do. Even population was considered as a resource under quantitative as well as qualitative aspects, which had to be directed at societies' best. Besides *Policeywissenschaften*, *Naturgeschichte* (so-called *historia naturalis*) is another scientific discipline, which exclusively dealt with natural resources and gave hints on how to use different natural resources most efficiently (cf. BECKMANN 1767, BAYERL 2001, MEYER and POPPLOW 2004). As was shown very briefly, environmental aspects were discussed in different but mostly very explicit ways, and negative repercussions of anthropogenic action were ignored in cameralistic literature conspicuously even though they existed (e.g. the large-scaled deforestation throughout Europe, which lead to forestry at the beginning of the 18th century; cf. VON CARLOWITZ 1713).

---

2   This might be one explanation of the fact that even though "happiness of the people" and "happiness of the state" were used synonymously, it is the latter form, which is being found more often. Another possible reason will be mentioned later.
3   The first chairs at universities to create this knowledge were established in 1727 in Frankfurt an der Oder and Halle. At the end of the 18th century almost every university has had a cameralistic chair or faculty (cf. HENNINGS 1988).
4   The *vade mecum* is being completed by financial sciences, i.e. tax system, tax rates etc. (cf. VON JUSTI 1762, VON DER LITH 1766, and JUNG 1789), which are of no further importance here.
5   A fact that is impossible to know for single men and reflects the great skepticism concerning human beings and their ability to act for the sake of society.
6   This is just one simplified example and even though it may sound odd, whatever is considered the 'right' proportion depends on different variables such as geographic site, topographic condition, economic structure etc.

2.2 The Utilitarian Understanding

The utilitarian idea of man and society[7] is the clear contrast to that of Cameralism. The question to be answered in historical as well as in modern Utilitarianism is 'what people want'. The handling of this question, however, is not as simple nor yet as intuitively obvious as that elementary thought implies (cf. SEN and WILLIAMS 1982). This statement pushes the individual implicitly into the center of Utilitarianism, which becomes more obvious when reading the utilitarian textbooks.

Authors of classical Utilitarianism such as BENTHAM, SMITH, or MILL answered the question with men pursuit of happiness or a happy line of conduct respectively, without giving detailed information of what happiness is or why it is desirable. To realize this pretentious goal people act or behave in certain manner. Even though BENTHAM, SMITH, and MILL were giving slightly different explanations of human behavior,[8] they all argue that every individual knows best what is necessary for him or her to be happy (cf. SMITH 1759/2002, BENTHAM 1789/1970, and MILL 1861/1969). Hence, men should be free to choose within a range of alternative actions. With respect to economic decision-making this system of 'natural freedom' is being considered to maximize material well-being because every individual takes action in accordance with external incentives set by markets (prices, loans, rents etc.), which coordinate complementary particular interests via SMITH's renowned 'invisible hand' (cf. the invisible hand SMITH 1776/1976). Knowledge of any interrelation between determinants influencing economic growth is not necessary. With individual freedom of choice being an important precondition for economic prosperity, it should only be restrained by governmental interventions, if the choice of one person could possibly harm another person's freedom to pursue his or her well-being. With every individual striving for its own best, they act to other people best automatically. In Utilitarianism and classical Political Economy 'responsibility' for economic growth is in the hands of the people rather than the state, and since every individual knows best, how to allocate given resources optimally, no guidance is necessary – and for that reason, no instructions are given in Political Economy of how to use or handle certain (natural) resources.

Since people are connected to each other via complementary interests, people are embedded in societal conditions and can realize their own happiness only in a larger context with others and society, respectively. For that reason, people feel a certain moral responsibility for others. This argument is very prominent in MILL's *Utilitarianism* and SMITH's *Theory of Moral Sentiments*, whereas BENTHAM argued that people always act or behave on behalf of their own happiness as an exclusively self-interested agent[9] (cf. BENTHAM 1954). The obligation of man to their fellow creatures is outlined in SMITH, but was more clearly elaborated in MILL. As already mentioned, people depend on each other, this forces them to cooperate, and as long as they do so, there would at least be a temporary feeling of obligation (MILL 1861/1969). Consequently, MILL argued (in opposition to BENTHAM) that the utilitarian standard is not the agent's own happiness, but that of all people concerned. Hence, people not

---

[7] The idea of man with its focus of autonomic individual action and individuals being the basis of society is the foundation of today's social sciences called methodological individualism, a principle, which still was unknown in Cameralism.
[8] This should not be discovered in detail; for a more comprehensive analysis see e.g. HOTTINGER 1998, ROSEN 2003, or CORTEKAR 2007.
[9] BENTHAM's idea of man is considered the foundation of modern *homo oeconomicus*.

only take happiness of others into consideration, but they act as disinterested and benevolent spectators, which means to lay all particular interests aside, sacrificing their own greatest good for that of others, if necessary (cf. MILL 1861/1969). Creating this idea of man it becomes obvious, that especially SMITH and MILL do not only deviate from radical versions of *homo economicus*; they even go beyond current results in empirical behavioral research (e.g. their emphasis of humans being able to feel sympathy for mankind as a whole).

3. What to Learn from the Past?

Like all social sciences modern economic theory is conceptually based on methodological individualism, which makes it difficult to learn anything from Cameralism. One thing coming to mind is the assumption that societies organized undemocratically with a strong central executive body (whichever form they may have) to direct economic processes do not tend to environmental friendly policies.[10]

Utilitarianism as a foundation of today's social sciences, however, can be made fruitful for economic theory, especially when it comes to the environment. There are still chasms in recent research on human nature, which might be filled if inspirations given by utilitarian writers are taken seriously by present-day research because many recent studies in experimental economics have shown, that individuals often behave in a different way than predicted by the *homo economicus* model and its assumptions (cf. ROTH 1995, CAMERER and THALER 1995, or LEDYARD 1995). As long as only a few individuals behave in this "defective" manner there is supposed to be no problem. But if divergent behavior is no longer the exception to the model, the aggregate's behavior is being influenced with the consequence that statements according to *homo economicus* – at least in its orthodox form – are no longer valid. With respect to environmental issues a sufficient amount of people act in accordance with social standards or a certain social order to provide a public good such as "clean air" even if financial or material incentives go in the opposite direction (cf. e.g. GILLROY and SHAPIRO 1986, DUNLAP 1987, LEPPELSACK 1985, and KESSEL and TISCHLER 1984). Obviously, people voluntarily sacrifice for the "good cause" because of moral standards. Well-defined laws of property and freedom of market exchange minimize the necessary scope and extension of such standards, but they by no means eliminate them. As individual property rights become confused, and as markets are replaced or subverted by governmental interventions, the dependence of order on some extended range of moral responsibility increases (cf. BUCHANAN 1979). Aspects of morality and how it influences human behavior is part of New Institutional Economics (where moral standards are modeled as informal institutions), and the little considered research program of Constitutional Economics.

If (*i*) direct financial incentives set by government do not necessarily affect human behavior in the intended manner, and (*ii*) people often automatically (even without incentives) behave in a way they are supposed to because they feel a certain moral obligation to society, it seems that governmental intervention should be reconsidered in environmental policy terms. When societies face new challenges such as today's environmental problems people need to

---

10 Cf. WICKE 1993. WICKE exemplifies his arguments by analyzing states of the former Eastern Bloc. But this does not necessarily imply, that societies organized like that could under no circumstances have environmental friendly policies.

be guided by governments. This guidance, however, should not suppress beneficial activities of individuals, e.g. by discouraging them by means of improper incentives. People are able to (and even want to) act to the benefit of society but they need to know how. One possible way is to provide information, e.g. in terms of labeling (cf. KRARUP 2005) or environmental education (cf. AZEITEIRO et al. 2004 or FABER and MANSTETTEN 2003), and to let people decide which would affect individual freedoms.

References

AZEITEIRO, U., GONCALVES, F., FILHO, W. L., MORGADO, F., and PEREIRA, M.: World Trends in Environmental Education. Frankfurt (Main): Lang 2004

BAYERL, G.: Die Natur als Warenhaus – Der technisch-ökonomische Blick auf die Natur in der frühen Neuzeit. In: HAHN, S., und REITH, R. (Eds.): Umwelt-Geschichte: Arbeitsfelder – Forschungsansätze – Perspektiven. S. 33–52. Wien: Verlag für Geschichte und Politik 2001

BECKMANN, J.: Anfangsgründe der Naturhistorie. Göttingen 1767

BENTHAM, J.: An Introduction to the Principles of Morals and Legislation. Ed. by J. H. BURNS and H. L. A. HART. London: Athlone Press 1789/1970

BENTHAM, J.: The psychology of economic man. In: STARK, W. (Ed.): Jeremy Bentham's Economic Writings. Vol. 3, pp. 419–450. London: Allen and Unwin 1954

BERGIUS, J. H. L.: Policey und Cameral Magazin. Vol. 1, Frankfurt (Main): 1767

BUCHANAN, J. M.: What should Economists do? Indianapolis: Liberty Press 1979

CAMERER, C., and THALER, R.: Ultimatums, dictators, and manners. Journal of Economic Perspectives 9, 209–219 (1995)

CARLOWITZ, H. C. VON: Sylvicultura Oeconomica: Anweisung zur Wilden Baumzucht. Leipzig: 1713

CORTEKAR, J.: Glückskonzepte des Kameralismus und Utilitarismus – Implikationen für die moderne Umweltökonomik und Umweltpolitik. Marburg: Metropolis Verlag 2007

DARJES, J. G.: Erste Gründe der Cameral=Wissenschaften. Jena 1756

DUNLAP, R. E.: Polls, pollution, and politics revisited: Public opinion on the environment in the Reagan era. Environment 29/6, 6–11 (1987)

ENGELHARDT, U.: Zum Begriff der Glückseligkeit in der kameralistischen Staatslehre des 18. Jahrhunderts (J. H. G. v. Justi). Zeitschrift für Historische Forschung 8, 37–79 (1981)

FABER, M., and MANSTETTEN, R.: Mensch – Natur – Wissen: Grundlagen der Umweltbildung. Göttingen: Vandenhoeck & Ruprecht 2003

GILLROY, J. M., and SHAPIRO, R. Y.: The polls: Environmental protection. Public Opinion Quarterly 50, 270–279 (1986)

HENNINGS, K. H.: Die Institutionalisierung der Nationalökonomie an deutschen Universitäten. In: WASZEK, N. (Ed.): Die Institutionalisierung der Nationalökonomie an deutschen Universitäten. S. 43–54. St. Katharinen: Scripta Mercaturae-Verlag 1988

HOTTINGER, O.: Eigeninteresse und individuelles Nutzenkalkül in der Theorie der Gesellschaft und Ökonomie von Adam Smith, Jeremy Bentham und John Stuart Mill. Marburg: Metropolis Verlag 1998

JUNG, J. H.: Lehrbuch der Finanz-Wissenschaft. Leipzig 1789

JUNG, J. H.: Die Grundlehre der Staatswirtschaft. Marburg 1792

JUNG, J. H.: Gemeinnütziges Lehrbuch der Handlungswissenschaft. 2nd Ed. Leipzig 1799

JUSTI, J. H. G. VON: Systematischer Grundriß aller Oeconomischen und Cameral-Wissenschaften. Frankfurt (Main), Leipzig 1759

JUSTI, J. H. G. VON: Die Natur und das Wesen der Staaten. Berlin, Stettin, Leipzig 1760

JUSTI, J. H. G. VON: Die Grundfeste zu der Macht und Glückseeligkeit der Staaten. Königsberg, Leipzig 1761

JUSTI, J. H. G. VON: Gesammelte politische und Finanz-Schriften. Vol. 2. Reprint. Aalen: Scientia Verlag 1761/1970

JUSTI, J. H. G. VON: Ausführliche Abhandlung von denen Steuern und Abgaben. Königsberg, Leipzig 1762

JUSTI, J. H. G. VON: Gesammelte politische und Finanz-Schriften. Vol. 3. Reprint. Aalen: Scientia Verlag 1764/1970

JUSTI, J. H. G. VON: Grundsätze der Policeywissenschaft. 3rd Ed. Göttingen 1782

KESSEL, H., and TISCHLER, W.: Umweltbewusstsein – Ökologische Wertvorstellungen in westlichen Industrienationen. Berlin: Ed. Sigma 1984

Krarup, S.: Environment, Information, and Consumer Behaviour. E. Elger 2005
Ledyard, J.: Public goods: A survey of experimental research. 1995. In: Kagel, J., and Roth, A. (Eds.): Handbook of Experimental Economics; pp.111–194. Princeton: Princeton University Press 2004
Leppelsack, W.: Ergebnisse aus der sozialwissenschaftlichen Umweltforschung. Berlin: Texte Umweltbundesamt 1985
Lith, J. W. von der: Neue vollständig erwiesene Abhandlung von denen Steuern und deren vortheilhaftester Einrichtung. Ulm, Stettin 1766
Meyer, T., and Popplow, M.: "To employ each of Nature's products in the most favorable way possible" – Nature as a commodity in eighteenth-century German economic discourse. Historical Social Research 29/4, 4–40 (2004)
Mill, J. S.: Utilitarianism. In: Priestley, F. E. L., and Robson, J. M. (Eds.): Collected Works of John Stuart Mill. Vol. *10*, pp. 203–259. London, Toronto: University of Toronto Press 1861/1969
Pfeiffer, J. F. von: Grundsätze der Universal-Cameral-Wissenschaft. 2nd Part. Frankfurt (Main) 1783
Rosen, F.: Classical Utilitarianism from Hume to Mill. London: Routledge 2003
Roth, A.: Bargaining experiments. In: Kagel, J., and Roth, A. (Eds.): Handbook of Experimental Economics; pp. 253–348. Princeton: Princeton University Press 1995
Sen, A. K., and Williams, B.: Utilitarianism and Beyond. Cambridge (Mass.), London: Cambridge University Press 1982
Smith, A.: The theory of moral sentiments. In: Haakonssen, K. (Ed.): Cambridge Texts in the History of Philosophy. Cambridge: Cambridge University Press 1759/2002
Smith, A.: An inquiry into the nature and causes of the wealth of nations. In: Campbell, R. H., and Skinner, A. S. (Eds.): The Glasgow Edition of the Works and Correspondence of Adam Smith. Vol. *1*. Oxford 1776/1976
Wicke, L.: Umweltökonomie. 4th rev. Ed. München: Vahlen 1993

Dr. Jörg Cortekar
Department of Agricultural Economics
and Rural Development
Georg-August-Universität Göttingen
Platz der Göttinger Sieben 5
37073 Göttingen
Germany
Phone: +49 551 394829
Fax: +49 551 394812
E-Mail: jcortek@uni-goettingen.de

Prof. Dr. Rainer Marggraf
Department of Agricultural Economics
and Rural Development
Georg-August-Universität Göttingen
Platz der Göttinger Sieben 5
37073 Göttingen
Germany
Phone: +49 551 394829
Fax: +49 551 394812
E-Mail: rmarggr@gwgd.de

Escherichia coli – Facets of a Versatile Pathogen
On the Occasion of the 150th Birthday of Theodor Escherich (1857–1911)

Leopoldina-Symposium
Deutsche Akademie der Naturforscher Leopoldina
in Zusammenarbeit mit der *European Molecular Biology Organization* (EMBO) und der *Federation of European Microbiological Societies* (FEMS)
vom 9. bis 12. Oktober 2007 im Bildungszentrum Kloster Banz, Bad Staffelstein

Nova Acta Leopoldina N. F., Bd. *98*, Nr. 359
Herausgegeben von Gabriele BLUM-OEHLER (Würzburg), Ulrich DOBRINDT (Würzburg), Jörg HACKER (Würzburg – Berlin) und Volker TER MEULEN (Würzburg – Halle/Saale)
(2008, 180 Seiten, 22 Abbildungen, 7 Tabellen, 22,95 Euro,
ISBN: 978-3-8047-2519-5)

Aus Anlass des 150. Geburtstages von Theodor ESCHERICH, dem Entdecker des Bakteriums *Escherichia coli*, werden hier neue Forschungsergebnisse aus den Gebieten der Genomik, Pathogenese bakterieller Erkrankungen und Wirts-Bakterien-Interaktionen zusammengestellt.
Der Kinderarzt und Mikrobiologe ESCHERICH beschrieb 1885 erstmals das „*Bacterium coli commune*". Das später nach seinem Entdecker *Escherichia coli*, kurz *E. coli*, genannte Bakterium entwickelte sich zum beliebtesten „Haustier" der Molekularbiologen. *E. coli* stellt mittlerweile molekularbiologisch den am besten untersuchten Organismus dar und wird von Wissenschaftlern weltweit als Modellorganismus genutzt. Behandelt werden außer der Bedeutung von *E. coli* in der molekularbiologischen Forschung vor allem Fragen der Genregulation, Beziehungen zwischen Kommensalismus und Pathogenität und das Problem der Virulenzfaktoren.

Wissenschaftliche Verlagsgesellschaft mbH Stuttgart

Contested Forests – Environmental Crimes between Science and Rural Society: Bavaria 1780–1860

Richard HÖLZL (Göttingen/Regensburg)

With 3 Figures

Abstract

The article presents an overview of topic, design and examplary results of my dissertation "Der Wald im Konflikt". The project explores ways to conduct environmental history of forests beyond the limits of the dominant perspective of 'Holznot' (timber and fuel scarcity). Instead of assessing the question, whether a scarcity of forest resources actually existed in preindustrial Germany, it connects approaches from the history of science with cultural anthropology. The aim is an environmental history that concentrates on the production of scientific environmental knowledge in combination with everyday human practices of using the environment. Central to the project is the study of offences against forest regulations. This 19th century everyday crime generated archival sources that yield insights into the conflicts that occurred during the implementation of forest reform and the perspectives of local non-bourgeois populations. In official records und published accounts these conflicts have been covered up by layers of contemporary interpretation. To illustrate the potential of microhistoric approaches the text presents three court cases that were heard before the county court of Viechtach, Bavaria, in the years 1827, 1831, and 1819.

Zusammenfassung

Der Artikel gibt einen Überblick über das Thema, den Forschungsansatz und exemplarische Ergebnisse meiner Dissertation „Der Wald im Konflikt". Diese zielt auf eine Umweltgeschichte des Waldes im 19. Jahrhundert ab, die sich von der dominanten Forschungsperspektive, der ‚Holznotdebatte', löst. Anstatt nach der tatsächlichen Existenz eines fundamentalen Ressourcenmangels im vor- und frühindustriellen Deutschland zu fragen, verbindet sie Ansätze der neueren Wissenschaftsgeschichte mit historisch-anthropologischen Fragestellungen. Das Ziel ist eine Umweltgeschichte, die sich an der Schnittstelle von wissenschaftsbasierter Forstwirtschaft und alltäglichen Praktiken der Waldnutzung orientiert. Die Analyse des Alltagsdelikts Forstfrevel ist dabei zentral, da durch die hier produzierten historischen Quellen nicht-bürgerliche und alltagsbezogene Perspektiven rekonstruiert werden können, die von den zeitgenössischen bürgerlichen Interpreten verdeckt wurden. Am Ende des Textes werden drei Forstfrevelprozesse, die 1819, 1827 und 1831 vor dem Landgericht Viechtach im Bayerischen Wald verhandelt wurden, vorgestellt. Sie können das Potential des verfolgten Ansatzes aufzeigen.

1. Scientific Forestry Invented

The decades after 1750 saw the establishment of a new, centralized, and science-based approach to the natural resources of forests: scientific forestry (*rationelle Forstwirtschaft*). Located at the institutions of the ‚republique de lettre', e.g. the academies of sciences or learned societies, and passed on in journals, letters, and early scientific works the discourse on forestry reform developed the idea of a ‚modern forest'. At its heart lay a special environmental narrative: Germany's forests had been devasted by population growth and unregulated, communal practices of local populations, e.g. pasture. Scientific forestry would provide a solution for the anticipated resource calamity: ‚sustainable yield forestry' (STUBER 2008, HÖLZL 2010).

Centralized management in the hands of professional foresters and scientific concepts such as *Schlagwirtschaft*, artificial afforestation, or the introduction of foreign species of trees promised to maintain timber and fuel resources for the future (LOWOOD 1991). Abolishing agro-forestal practices was deemed indispensable for the success of sustainability. Based on results of my Dissertation, this article assesses the possible benefit of new approaches to the environmental history of forests. Inspired by the historiography of sciences and cultural anthropology, a culturally minded approach to environmental history may yield insights into the background of environmental reform agendas and the conflicts sidelining their implementation stage.

2. Village Ecologies

The logics of pre-industrial village economies integrated forests on an extensive basis into the daily quest for sustenance. Forests provided pasture and litter for cattle, pig, or sheep, timber for buildings, machines and tools as well as raw materials for local handicrafts. Forests added reserves of nutrients, fodder, fuel and timber to the local economy and subsidized central farmlands. However, particular forms of field-forest intersections depended on the respective networks of property, social stratification, cultural dispositions, traffic infrastructure, climate, soil, or elevation. The effects of these diverse ecologies were open forests interspersed with grassland – highly valued by modern restoration ecology due to a high diversity of species (SELTER 1995, BECK 1993, 2003 OHEIMB 2006).

3. Negotiating Forest Reform

The idea of the ‚modern forest', that focussed on fuel and timber production and restricted access of local forest users, entered the stage of implementation around 1800. It was underpinned by a state-driven move to professionalize forest personnel, to concentrate on material and financial revenue of wood and timber production and to introduce business planning for individual forest districts (GÖTZ 1996, GREWE 2004, BAUER 2002). Attempts of implementation resulted in a wave of civil law suits by holders of usufruct rights as well as a tremendous increase in forest offences (HÖLZL 2007a). The broad based challenge of state regulation of resource use erupted in violent resistance during the 1848/49 revolution. The state reacted by criminalizing large parts of the rural population and by modifying reform policies. In the course of the century civil law suits to retain forest access rights brought before court by the landholding peasantry were successful. Usufruct practices became accepted as traditional and moral rights. Economists integrated the idea of the state guaranteeing basic resource supply into political theory. And nationalist historians began to talk of the ‚freedom of German forests' (LEHMANN 2001). A fundamental change in local forest use, however, took place with the breakthrough of artificial fertilizer and fossil fuels by the end of the 19th century.

4. Design and Sources of the Dissertation

The dissertation explores ways to conduct environmental history of forests beyond the limits of the dominant perspective of ‚Holznot' (timber and fuel scarcity) (Radkau 2002, Grewe 2004). Instead of assessing the question, whether a scarcity of forest resources actually existed in pre-industrial Germany, it connects approaches from the history of science with cultural anthropology (Tanner 2004). The aim is an environmental history that analyzes the production of scientific environmental knowledge in combination with everyday human practices of using the environment (Hölzl and Hünniger 2008). It thereby achieves a new reading of environmental conflicts that have hitherto been regarded either as socio-political in content or as indicators of resource scarcity. The primary sources of the project are archival records preserved in the Bavarian State Archives in Würzburg, Landshut, and Munich concerning civil and criminal court proceedings as well as topographical descriptions, administrative reports and published works in the fields of 18th and 19th scientific forestry, economics and ethnology.

5. Exemplary Results

Central to the project is the study of offences against forest regulations. This 19th century everyday crime generated archival sources that yield insights into the conflicts occurring during the implementation of forest reform. It recovers perspectives of local non-bourgeois populations, which in official records und published accounts have been covered up by layers of contemporary interpretation.

Following exemplary results are located within the kingdom of Bavaria in the 19th century. The problem of forest offences reached its crises here in the late 1830's and 1840's with over a 180,000 cases per year on a population of 4 to 4.5 Million inhabitants (Fig. 1).

The social status of forest offenders has been debated in historical literature. Some historians, like Dirk Blasius (1976) or Bernd Stefan Grewe (2004), argued that mainly the poor members of the rural lower classes were committing forests offences the prime mover being material distress and poverty. Others have emphasized that in parts of Germany forest offences also concerned the rural middle classes, in particular the landed peasantry (Prass 1996). The social stratification of forest offenders cannot be derived from the official statistics of the 19th century. A close analysis of court records is necessary. The following records are taken from the Bavarian State Archive Landshut (Landgericht ä. O. 1116). In general court records give us the socio-economic background of the offender. They distinguish between different classes of farmers according the size auf the farms, as well as so-called inhabitants (*Inwohner*) denoting families living with a landholding farmer and managing small plots of arable land on their own. A final category includes day laborers and cottagers without land. The following analysis of the quarterly court sessions of the year 1830/31 shows offences from the forest beat Drachselsried. Almost two thirds of the accused belonged to the rural middle classes: peasant farmers, a baker, a miller and a smith. Only two of 74 offenders were sentenced to forest work, because of attested poverty. What we found here, are certainly not the down and out of the countryside. (Fig. 2)

However, this result needs some qualification: In the environs of early industrial regions the share of the rural underclass could be considerably higher.

Richard Hölzl

Fig. 1 Absolute number of offences against forest regulations in Bavaria (without the Palatinate) 1835 to 1860 (MAYR 1867)

German historiography has long equalled forest offences with wood theft. Recently, the scope was extended to the theft of forests resources taking into account that the use subsistence economy made of the forests was far more diverse than a mere interest in timber and fire wood. A close look on the records of the county court of Viechtach provides us with a different and diverse picture:

Surprisingly, only 12% of the offences were actual wood theft; if we include illegal pasture (16%) and the theft of litter (1%) we end up with hardly a third of the offences being the theft of forest resources. More than two thirds of all the offences were "bad practices" (71%), i.e. trespasses of the codes of conduct scientific forestry claimed as the only means to avoid a massive shortage of timber and for a sustainable use of forest resources. The following analysis shows the kinds of trespasses that occurred in this sample (Fig. 3)

Some offenders had used a wrong measure, erecting their wood piles two or three inches higher then decreed. Some had not yet felled the trees they had bought, when the forester came to asses the value of the timber. One offender had taken a different tree than the one he

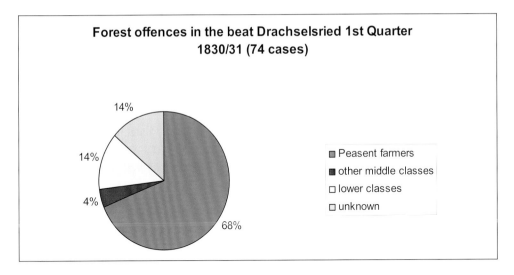

Fig. 2 Social stratification of defendants in a sample of forest offence charges

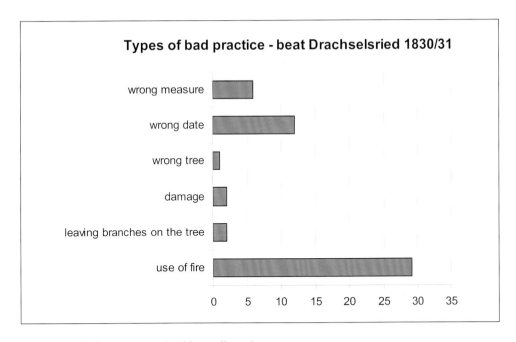

Fig. 3 Types of offences in a sample of forest offence charges

was authorized to. Some had damaged trees while felling their timber, or left the branches and tree tops attached to the tree, while it was resting to be taken home in winter. The majority was accused of using fire in the woods. In two cases the slash and burn customs the farmers commonly used in their steep plots got out of hand and encroached on the state forest. But the majority of cases were camp fires for warming meals during timber work.

Richard Hölzl

Both, the social stratification of the offenders and the closer qualification of the offences, call for a new and differentiated interpretation of forest conflicts in 19th century Germany. The close relation between material need respectively actual wood shortage and the increase in offences becomes at least dubitable. Forest offences are less an indicator for a lack of forest resources than for the contested nature of the new centralized and science-based resource exploitation regime.

As a final step I want to present three cases that were heard before the county court of Viechtach in the years 1827, 1831, and 1819. They illustrate the potential of microhistoric approaches for my argument. I have elaborated on the historical context of these cases in 19th Bavaria elsewhere (HÖLZL 2007a). The cases are taken from records of the Bavarian State Archive Landshut (Landgericht ä. O. 1116, A 1101):

On October 8, 1827 the members of the municipality Arnbruck were accused of not having removed the branches and tree tops from their timber trees to the scheduled date. The foresters claimed that they had not been able to measure the timber to establish its value. The new growth had also been hampered. Furthermore, the indictment of the local forest authorities claimed that leaving the branches attached to the tree in the forest would breed the dangerous bark beetle. The representatives of the villagers, the publican J. SCHAFFER and the farmer J. GRASSL, argued: "On several occasions we have been ordered to cut off tops and branches at a time when it is detrimental to our timber trees. We could not do this, in order to prevent damage. This month is the right time. We will now go to work immediately. It will not be much of a disadvantage to the forest, if the felled trees stay there with the branches and tops a few months longer. In similar cases we have been discharged several times. We hope for the same judgment today." The forester had the role of a prosecutor in forest court hearings. He replied that they, nevertheless, had trespassed the governmental orders. However, the villagers of Arnbruck were found not guilty.

What has happened here? Traditional ways of treating timber to be used for building collided with the growing fear of large scale destruction by forest pests. The villagers knew that the trees were sapped best by leaving branches and tops attached to the tree for a few months until the trunks were taken home. This was necessary for achieving the proper material for construction. On the other hand, the trained foresters argued that this practise was acting against best practice rules of forest police. Indeed, between 1797 and 1822, 7 decrees of the government had been released, which gave precise instructions what to do when an epidemic of forest pest hit a forest, but nothing was stated about preventive measures. Still, the lines of conflict are evident. J. M. BECHSTEIN, a well known forest instructor, wrote in his *Forest Insectology* of 1818 about timber that is kept in the forests during summer: "Those who fell the timber [...] must be ordered with the threat of severe penalty to immediately make the timber 'forest compatible', or remove the bark or burn it. This kind of timber, or the fire wood that is hauled away late, become the work places, out of which the forest pest spreads." (BECHSTEIN 1818, pp. 182–183; WINDELEN 2007, THOM 2007.)

Local forest users since the beginning of the 19th century were increasingly confronted with science-based concepts that were implemented by a modernized administration and well trained forest personnel. Local practices collided with thoroughly developed instructions of wise use of resources. The measures for the prevention of forest pests are one example; mathematically accurate surveying of forests is another.

On February 23rd, 1831 the municipality of Oberried had to stand trial being accused of illegal cattle and goat grazing in the Royal Forest. The representatives of the village, five farmers, did not deny this. They explained to the court that they had every intention to trespass the landmarks erected in 1828/1829 further on. They argued: This borderline was drawn without their knowledge and the survey had cut off an important part of their formerly open pastures. The transcript of the hearing goes on: "They facilitate their objection by claiming that they neither had seen a document about the survey that announced this reduction of their pasture rights, nor signed anything. Further they announced that they would go ahead using the pastures to the usual extent. They would not let themselves be prohibited to cross the newly drawn line." The Royal forester answered as follows: "The municipality had been granted the right to graze their cattle in the said forest on July, 6th 1665. In 1806 the forest border was accurately surveyed by the ducal Surveyors BAUMANN and REBER; following their maps the borders were geometrically detected and fixed […] and marked by 164 stones that carry the letters K and N in 1829. The minutes have been signed by all adjacent owners. There is no reason for the villagers of Oberried to claim a formerly extensive pasture right, because the described proceedings have regulated the borders as good as they have been since time out of mind." There was no verdict in this hearing and the judge invited the villagers to take civil action to retain their rights.

In the 17th century the village Oberried was granted the right to graze its cattle in the Royal forest. Most likely, they exercised pasture in this place long before. In 1806 these Royal forests where surveyed and twenty years later the pasture rights were limited to accurately delineated patches. The villagers regarded these lines as acts of arbitrariness whereas from the viewpoint of scientific forestry a geometrically accurate survey was the starting point for a rational use of forest resources. Furthermore, as G. L. HARTIG, one of the "forest classics" of the time, stated clearly: "Of all the evils forests are subjected to exaggerated forest pasture is one of the worst; […] the inevitable consequence is that the forests are ruined or become devoid of trees, and the future generations will suffer wood shortage. It is an important duty of the forest administration to abolish pasture altogether or at least limit it so far that it will be of no harm." (1840, p. 253.) According to scientific foresters local forest users were torpedoing the attempts of scientific forestry to create rational and sustainable forest operating plans. Their actions had to be measured, guided and thoroughly monitored.

The transcripts of the court hearings not only give us a clue about the background of the offences against the forest laws, but also provide an idea of how ordinary rural people dealt with the situation before court, how they defended themselves and what kind of tactics they used (CERTEAU 1988). To defend their claim on unrestricted pasture the villagers of Oberried evoked a composition of arguments including tradition and custom, official documents and semi-juridical concepts: unrestricted use 'since time out of mind', a granted privilege of the duke and the lack of a signature certifying their consent to the surveying, marking and zoning of the Royal forest adjacent to their village. Tradition and custom are ways of structuring reality that were important in the daily lives of the villagers. 'Doing the things the way they were done since time out of mind', was an important argument that could not simply be ignored by the authorities.

On the other hand, the forest laws were the rules that governed courts. In the vast majority of the cases the defendants were convicted. The testimony of the forester as an eye witness of the offence outweighed the affirmations of the defendants. All in all, the modern legal system combined with and containing new science-based concepts of resource use was alien to the local forest users. Before court they tried to put forward alternative modes of structuring reality (BERGER and LUCKMANN 2003). 'Custom' and 'tradition' are one example. 'Nature' is another. The patterns of behavior as well as the ways of arguing before court seem to draw on the 'habitus' of the rural population of the time (BOURDIEU 1993).

On September 30, 1819 35 villagers including the local school teacher and the catholic parish priest from Drachselried, Unterried and Arnbruck, were tried and convicted in Viechtach. They had gone into the Royal Forest and had felled fire wood for the winter without waiting for the forester's permission. The rationale the defendants brought forward in the court house was almost identical in all of the cases. J. JUNGBECK, a peasant farmer, was the first in the dock. He argued: "The permission of the forest office came particularly late this year and only at the end of June. And since harvest begins in July and later on there is the sowing of the fields, when it is impossible for a farmer to take the time for felling the required wood for wintertime, and since all cultivation work of a man of the country must follow the different seasons, I was forced, in order to avoid the danger of obtaining no wood at all this years, to go and cut the fire wood without permission." JUNGBECK argued that the cycle of the seasons, nature in fact, would not allow him to comply with forests laws.

The micro-historic examples from 19th century Bavaria show the potential of a culturally minded approach to environmental history. Conflicts that had hitherto been attributed to a scarcity of forest resources can be interpreted quite differently: The contest about forests was strongly connected to the introduction of a science-based and centralized system of resource exploitation that met with resistance among local forest users. The micro-historic approach also renders the specific reaction of local population towards a new resource exploitation regime visible and, therefore, makes an evaluation of social acceptance and costs of environmental reform possible.

References

BAUER D.: Von der ungeregelten zur nachhaltigen Forstwirtschaft. Eine Analyse der Prozesse in Bayern an der Schwelle zum 19. Jahrhundert. München: Frank 2002
BECHSTEIN, J. M.: Forstinsectologie oder Naturgeschichte der für den Waldschädlichen und nützlichen Insecten nebst Einleitung in die Insectenkunde überhaupt, für angehende und ausübende Forstmänner und Cameralisten. Gotha: Henning 1818
BECK, R.: Unterfinning. Ländliche Welt vor Anbruch der Moderne. München: Beck 1993
BECK, R.: Ebersberg oder das Ende der Wildnis: Eine Landschaftsgeschichte. München: Beck 2003
BERGER, P. L., und LUCKMANN, T.: Die gesellschaftliche Konstruktion der Wirklichkeit: eine Theorie der Wissenssoziologie. Frankfurt am Main: Fischer [19]2003
BLASIUS, D.: Bürgerliche Gesellschaft und Kriminalität: zur Sozialgeschichte Preußens im Vormärz. Göttingen: Vandenhoeck und Ruprecht 1976
BOURDIEU, P.: Sozialer Sinn. Kritik der theoretischen Vernunft. Frankfurt am Main: Suhrkamp 1993
CERTEAU, M. DE: Kunst des Handelns. Berlin: Merve 1988
GÖTZ, T.: Der Staat im Wald. Forstpersonal und Forstausbildung im ersten Drittel des 19. Jahrhunderts aus umweltgeschichtlicher Perspektive. Das Beispiel des Rheinkreises. In: ERNST, C., GREWE, B.-S., und KUNTZ, J. (Eds.): Beiträge zur Umweltgeschichte I. S. 36–77. Trier: 1996
GREWE, S.: Der versperrte Wald. Ressourcenmangel in der bayerischen Pfalz (1814–1870). Köln, Wien: Böhlau 2004
HARTIG, G. L.: Lehrbuch für Förster und für die, welche es werden wollen. Stuttgart, Tübingen: Cotta [8]1840
HÖLZL, R.: Der Wald als ökologisches, soziales und kulturelles Konfliktfeld. Alltagsgeschichtliche Beispiele aus Bayern im 19. Jahrhundert. In: ALLEMEYER, M.-L., JAKUBOWSKI-TIESSEN, M., und RUS RUFINO, S. (Eds.): Von der Gottesgabe zur Ressource. Konflikte um Wald, Wasser und Land in Spanien und Deutschland seit der frühen Neuzeit. S. 109–134. Essen: Klartext 2007a
HÖLZL, R.: „wie sothane Conservation und Anbau des Holtzes anzustellen" – Anfänge der Forstwissenschaft im 18. und 19. Jahrhundert. In: FANSA, M., und VORLAUF, D. (Eds.): Holzkultur. Von der Urzeit bis in die Zukunft. Band zur Ausstellung im Landesmuseum Natur und Mensch Oldenburg. S. 43–48. Oldenburg: Isensee 2007b

Hölzl, R.: Umkämpfte Wälder. Die Geschichte einer ökologischen Reform 1760–1860. Frankfurt (Main): Campus 2010 (forthcoming)

Hölzl, R., und Hünninger, D.: Braucht die Umweltgeschichte einen ‚kulturellen Dreh'? Auf der Suche nach produktiven Erweiterungen in der Geschichtsschreibung der Mensch-Natur-Beziehungen. Werkstatt Geschichte *48*, 83–98 (2008)

Lehmann, A.: Der deutsche Wald. In: François, E., und Schulze, H. (Eds.): Deutsche Erinnerungsorte. Bd. *III*. S. 187–200. München: Beck 2001

Lowood, H.: The Calculating Forester: Quantification, cameral science and the emergence of scientific forestry management in Germany. In: Frangsmyr, T., Heilbronn, J. L., and Rider, R. E. (Eds.): The Quantifying Spirit in the Eighteenth Century. Berkley: University of California Press 1991

Mayr, G.: Statistik der Gerichtlichen Polizei im Königreich Bayern und in einigen anderen Ländern. München: K. Bayerisches Statistisches Bureau 1867

Oheimb, G. von, Eischeid, I., Finck, P., Grell, H., Härdtle, W., Mierwald, U., Riecken, U., und Sandkühler, J.: Halboffene Weidelandschaft Höltigbaum: Perspektiven für den Erhalt und die naturverträgliche Nutzung von Offenlandlebensräumen. Münster: BfN-Schriften-Vertrieb im Landwirtschaftsverlag 2006

Prass, R.: Verbotenes Weiden und Holzdiebstahl: Ländliche Forstfrevel am südlichen Harzrand im späten 18. und frühen 19. Jahrhundert. Archiv für Sozialgeschichte *36*, 51–68 (1996)

Radkau, J.: Natur und Macht: eine Weltgeschichte der Umwelt. München: Beck 2002

Selter, B.: Waldnutzung und ländliche Gesellschaft: landwirtschaftlicher „Nährwald" und neue Holzökonomie im Sauerland des 18. und 19. Jahrhunderts. Paderborn: Schöningh 1995

Stuber, M.: Wälder für Generationen. Konzeptionen der Nachhaltigkeit im Kanton Bern (1750–1880). Köln: Böhlau 2008

Tanner, J.: Historische Anthropologie zur Einführung. Hamburg: Junius 2004

Thom, K.: Raupenleim und Fanglaterne – Forstliche Schädlingsbekämpfung im 19. Jahrhundert. In: Fansa, M., und Vorlauf, D. (Eds.): Holzkultur. Von der Urzeit bis in die Zukunft. Band zur Ausstellung im Landesmuseum Natur und Mensch Oldenburg. S. 54–58. Oldenburg: Isensee 2007

Windelen, S.: Die Entdeckung schädlicher Tiere und Insekten im Wald. In: Fansa, M., und Vorlauf, D. (Eds.): Holzkultur. Von der Urzeit bis in die Zukunft. Band zur Ausstellung im Landesmuseum Natur und Mensch Oldenburg. S. 49–53. Oldenburg: Isensee 2007

        Richard Hölzl, M. A.
        Georg-August-Universität Göttingen
        Seminar für Mittlere und Neuere Geschichte
        Platz der Göttinger Sieben 5
        37073 Göttingen
        Germany
        Phone: +49 551 3912471
        E-Mail: rhoelzl@gwdg.de

Ergebnisse des Leopoldina-Förderprogramms VI
Tagung und Berichte der Stipendiaten

am 14. November 2008 in Halle (Saale)

> Nova Acta Leopoldina N. F. Supplementum 21
> Herausgegeben von Gunter S. Fischer (Halle/Saale), Andreas Clausing (Halle/Saale) und Volker ter Meulen (Halle/Saale – Würzburg)
> (2008, 200 Seiten, 107 Abbildungen, 1 Tabelle, 21,80 Euro, ISBN: 978-3-8047-2518-8)

Deutschlands älteste Akademie, die Deutsche Akademie der Naturforscher Leopoldina, bemüht sich in besonderem Maße um die Förderung von Nachwuchswissenschaftlern. Seit 1992 vergibt sie zur Unterstützung der beruflichen Weiterentwicklung herausragender junger Wissenschaftlerinnen und Wissenschaftler ein Stipendium, ausgestattet durch Zuwendungen des Bundesministeriums für Bildung und Forschung, das es den Ausgezeichneten ermöglicht, innerhalb von zwei bis drei Jahren eigenständig ein außergewöhnlich innovatives Forschungsprojekt an ausländischen Wissenschaftseinrichtungen umzusetzen. Über 320 Forscherinnen und Forscher konnten seit Beginn des Programms gefördert werden. Der vorliegende Band zeigt die Vielfalt der Projekte und liefert Beispiele für die erreichten Ergebnisse seit 2006. Damit werden Chancen und Ansprüche des Förderprogramms für künftige Bewerber deutlich.

Wissenschaftliche Verlagsgesellschaft mbH Stuttgart

Nature's Product? An Environmental History of the German Pulp and Paper Industry

Mathias Mutz (Göttingen)

Abstract

The article highlights some aspects of the interdependency of business and the environment using the German pulp and paper industry in the 19th and early 20th century as an example. It identifies three dimensions of interaction: a spatial one relating to location and infrastructural networks, a material one as input and output of resources, and finally an institutional one concerning conflict strategies, organisational choices, and mentalities. To cover space for the transport of goods, to control material flow of raw materials like water and wood, or to invest in waste water technology tied up resources of the firm and affected its prospect of success. The co-evolution of environment and enterprise must be seen as a central part of business development because of this. Nature became the subject of standardization and regulation in the course of industrialization and a trend towards vertical integration of the environment into the firm becomes evident.

Zusammenfassung

Der Beitrag beleuchtet einige Aspekte der Wechselbeziehungen von Unternehmen und Umwelt am Beispiel der deutschen Papierindustrie im 19. und frühen 20. Jahrhundert. Dabei werden drei Dimensionen der Interaktion unterschieden: eine räumliche mit Bezug auf Standort und Infrastruktursysteme, eine materielle als Ressourcen-In- und -Output und schließlich eine institutionelle bezüglich Konfliktstrategien, organisatorischen Strukturen und Mentalitäten. Die Überwindung von Raum beim Transport von Waren, die Regulierung von Stoffströmen bei Rohstoffen wie Wasser und Holz oder Investitionen in Abwasserreinigungsanlagen banden wichtige Ressourcen des Unternehmens und beeinflussten seine Erfolgsaussichten. Die Ko-Evolution von Umwelt und Unternehmen muss deshalb als wichtiger Teil der Unternehmensentwicklung angesehen werden. Durch die Industrialisierung wurde die Natur zum Objekt von Standardisierung und Regulierung, was zu einer zunehmenden vertikalen Integration der Umwelt in das Unternehmen führte.

Economy and ecology are somehow a classical contradicting couple. Therefore, the gap between the two is rarely crossed by economic or environmental history. This separation of spheres is rather questionable as economic processes depend on the natural environment for several reasons: Nature must be treated as a supplier of raw materials and disposal opportunity. Locational and technical aspects determine every business enterprise's scope of action. And conflicts arising through the utilization of natural resources shape corporate communication and strategic decisions. In a society relying on division of labor it is usually the business enterprise that becomes the 'intervening variable' between utilization and consumption of natural resources. They stand out for functioning as an interface between materialistic utilization and cultural perception of nature. This is the starting-point of the concept of an "eco-cultural business history" connected with Christine Meisner Rosen's and Christopher

Mathias Mutz

Sellers's (1999, p. 591) request "to craft an approach to business history that is at once *ecological* and *economical* and *cultural*".

The pulp and paper industry is an ideal example to illustrate this interconnectedness of business and environment. First of all, the paper industry has long been (in)famous for its intensive demand for resources and energy as well as its overall environmental impact (*Council on Economic Priorities* 1972, *OECD* 1973). But at the same time paper has become an inevitable feature of today's consumer and information society that gives relevance to the industry even beyond its economical significance (Oligmüller and Schachtner 2001, Bayerl and Pichol 1986, Hunter 1978). As the most important regional cluster of pre-1945 German pulp and paper industry the region of Saxony provides several well-documented case studies for an in-depth analysis (Schultze 1912, Blechschmidt 2000). This is especially true for the period of high industrialization between the 1850s and the great depression of the 1930s. During these decades the pulp and paper industry did not only emerge as big business, it also had to deal with extraordinary socio-economic changes.

The technologic and economic development of the period is a significant starting-point to highlight the relevance of natural factors for the evolution of paper production. During the 19th century the luxury article paper became a product for mass consumption. In 1800 German per capita demand was half a kilo. This figure rose to 27 kg in 1913 and surpasses 230 kg today (Oligmüller and Schachtner 2001, p. 11). This change was not only a matter of quantity, it deeply relied on a qualitative change in the use of paper. New applications as packaging material (corrugated cardboard, folding boxes) and sanitary paper (tissue handkerchiefs, toilet paper) created new markets. To supply this want, radical technical changes were necessary which were connected both to new machinery and new raw materials. Until 1800 paper production completely relied on rags which were treated in paper mills with hand-operated vats. After the invention of the paper machine the production of 'endless paper', i.e. paper rolls, became generally established until the mid-19th century in all of Europe. Increasing production capacities made resource shortage a major threat for the industry's further development. Experiments to improve raw material supply can be tracked to the late 18th century. An important first step was the introduction of chlorine bleaching around 1800, introducing new chemical procedures and sewages. In 1843, Friedrich Gottlob Keller from the Saxon Erz Mountains invented the groundwood process to grind wood which became the starting-point of producing wooden paper on a large scale. This "wooden revolution" (Mutz 2007a) gained momentum in the last third of the century through new chemical processes to produce cellulose. In Germany, Alexander Mitcherlich's sulphite process (patented in 1874) became the nucleolus of a completely new industry which produced some 300,000 tons of chemical pulp in 1900 (Kirchner 1907, p. 30). At the advent of the First World War foundations for industrial mass production have long been led including a tremendous change in the utilization of natural resources.

For further research on this "business-environment connection" (Rosen 2005) modern business history offers various conceptional ties. This way, communicative processes and their preconditions have become a major research interest (Berghoff 2004, Wischermann and Nieberding 2004). Nonetheless, it has been rarely recognized that any kind of transaction has a spatial structure and is often connected with the transport of goods. Environmental factors are effective here, especially in industries with an intensive use of resources, where the establishment of transport systems is decisive. At the same time natural resources shape technical processes and make the organization and management of knowledge related to natural and material characteristics an important factor in an industry's development. Intensive use of resources also produces a higher probability of environmental conflicts which business enterprises have to solve and avoid. State bureaucracy, other companies, neighbors, and also environmentalists act as "stakeholders" (FREEMAN 1984) whose concerns have to be integrated into the firm's actions. Generally speaking, three dimensions of interaction between business and environment can be identified: a spatial one relating to location and infrastructural networks, a material one as input and output of resources, and finally an institutional one concerning conflict strategies, organizational choices, and mentalities. The rest of this text will focus on these aspects by introducing several examples.

Even though infrastructure seems to be the most natural thing today, establishing transport and supply facilities is an important feature of industrialization from a business enterprise's point of view (MUTZ 2007b). Every firm needs to choose a location and needs to overcome distances efficiently in order to buy and sell goods. Pulpwood demand strongly changed the paper industries spatial structure when more and more firms established themselves in densely-wooded regions like the Saxon Erz Mountains in late 19th century. Older firms like KÜBLER and NIETHAMMER (1956) in Kriebstein founded subsidiary factories in these areas. Simultaneously, large amounts of time as well as money were invested to improve transportation facilities within the company including railways and streets. At the beginning of the 20th century the possibility to transport wood via waterways became the outstanding location factor for new production sites.

This does not only illustrate how natural contexts shaped production processes, but also how industrial requirements changed the environment. Extensive hydraulic engineering ventures were everyday business for every paper mill. Their intrusion into natural water balance reached from regulating rivers to building water-power plants and dams (MUTZ 2007b). Another example is the way how wood as raw material for paper had a direct impact on scientific forestry and the cultivation of forests. The new industrial demand for fast growing, small dimensioned wood altered the utilization strategies of state and private forest administrations (SCHUSTER 1961, JOSEPHSON 2005). On the one hand, this fostered monocultural tree planting, especially of spruce, characterizes German low mountain ranges like the Saxon Erz Mountains until today. On the other hand, there is a close connection between the rise of pulpwood demand and the establishing of thinning as a central technique of modern rational forestry (REINHOLD 1932). The failure of alternative surrogates can also be attributed to environmental factors. While the early methods to produce cellulose were developed for straw the production of straw paper never took off. As straw was more difficult to transport (because of its density) and less steady in supply (because of vicissitude in annual harvests), it was soon overtaken by wood as the new resource for industrial papermaking. Increased difficulties with effluents and especially air pollution and smell nuisance contributed to the squeezing out of the straw pulp industry (ALTMANN 1914).

But even if groundwood and wood cellulose were used the waste waters of pulp and paper mills were the major concern on the output side. Although papermaking has always been connected with air and water pollution (BAYERL 1981), these negative effects increased during the 19th century through an intensified utilization of harmful chemicals like chlorine and sulphur compounds. With a lack of willingness and technical solutions for wastewater treatment a large proportion ended in the receiving stream. Organic wood components boosted growth of algae and endangered fish populations in rivers near production facilities (SIEMEN 1993, SCHACHTNER 1992). Within a few years the situation became intolerable. In 1892 a newspaper article described the rivers and creeks of the Erz Mountains: "The once so clear water, in which until recently trouts played joyfully, has become so dirty from the unfamiliar work in the factory that the whole river bed, on whose ground once the pebble sparkled, is covered by a muddy mass. All objects of the river bed seem to be wrapped in grey paper." (N. N. 1893.) In addition to water pollution, the rising demand for energy was responded by the building of steam power stations which caused air pollution and smoke damage (e.g. RUDLOFF 2004). At the same time the expansion of paper consumption attached more importance to the disposal and recycling of paper (FUCHS 1998).

Both spatial and material aspects of utilization of the environment demand for an effective management of knowledge, transport facilities, and resource supply. This makes organizational studies an integral part of every study in environmental history (UEKÖTTER 1998). In the case of industrial enterprises these organizational requirements usually grew due to mass production and mass consumption. Every single enterprise had to adjust to changing societal backgrounds and natural environments and develop strategies for how to communicate with forest owners, neighboring farmers, state bureaucracy, or competing water users. Organizing resource supply and sewage disposal and their environmental implications became an integral part of business activities. In the case of wood supply a whole new trade system had to be established in the 1860s and afterwards which led to a growing international trade in pulpwood. 43% of the German pulpwood supply came from Russia, Scandinavia, and Eastern Europe in 1913. This trend accelerated in the interwar years when over 60% were imported (REINHOLD 1927, p. 143). While German forest administrations tried to impede imports through the establishing of prohibitive tariffs for economic reasons (MANTEL 1973, pp. 403–409), the development must also be interpreted as exporting some of the environmental costs of paper consumption.

The problem of water pollution was already disputed in the 1870s and in some cases even became a threat to the existence of paper and especially pulp mills. Starting in the 1880s public ordinances forced entrepreneurs to install so-called pulp catchers to clean the effluents mechanically. Many firms did this reluctantly arguing with economic cost-effectiveness and the loss of jobs. But although the technical solutions of the time must be viewed as insufficient from today's perspective compliance to state regulation became inevitable for water pollutants in late 19th century (SIEMEN 1993). For example, in the case of the pulp mill in Gröditz near Riesa (Saxony) the operating company had to implement expensive measures following the instructions of the "Reichsgesundheitsrat" (Federal Health Council) (GÄRTNER 1913). Nonetheless, the conflicts with neighbors and public authorities remained unsolved for decades.

In the course of industrialization entrepreneurs displayed attitudes that moulded their further relationship towards nature. Increasing steadiness of natural processes was not only a question of technical solutions, nature was also subject to standardization and regulation on a discursive level. Despite their everyday relevance ecological dependencies like floods, water shortage or ice drift were pushed away and natural elements like water became calculable commodities. Discussing the project of a dam with a reservoir on the Zschopau River a Saxon entrepreneur stated his expectations as follows: "When the water is held back on a Sunday during a period of water shortage, so that it does not pass unused during the Sunday rest, it has to be let off from the reservoir right on time, so that it arrives here early at 6 o'clock the next working day after a Sunday or a holiday in a such quantity that work can begin. [...] For any damage that might emerge, if the water is late and I have to wait for hours before

I can begin working, the operating corporation will be made liable for." (Reese 1913.) But the view of mastery and marketization could not satisfy in all contexts, as some dependencies were too obvious. Floods remained a severe risk which could even be increased through hydraulic engineering if a dam was managed inappropriate (Mutz 2007b). As many pulp grinders relied on water power, water shortage and flood water continued to be an economic factor by affecting raw material prices. Finally, interpretations from different industrial users contradicted each other. Ideas of control, dependence, and conservation of the environment often were entangled in reality.

To cover distances, to control material flow, or to invest in waste water technology tied up resources of the firm and affected its prospect of success. Co-evolution of environment and enterprise, i.e. depending on the environment and reshaping it at the same time, must be seen as a central part of business development. Arrangements to secure supply and disposal of materials and especially the regulation of conflicts aimed at controllability and steadiness of natural sequences in order to integrate the environment into the 'network of the firm'. This development towards a vertical integration of the environment becomes eminently visible in the case of infrastructure networks which Dirk van Laak (2001, p. 371) defined as an "intermediary object system between human and natural metabolism". On organizational levels increasing activities of associations and organizations related to environment-related topics like the "Deutscher Wasserwirtschafts- und Wasserkraftverband" (German Association for Water Management and Water Power) and its predecessors show a similar trend (DWWV 1930). Institutes for paper technology at several technical universities and institutes for cellulose chemistry at forestry academies led to an intensified scientific penetration of production processes. Nonetheless, it is too simple to speak of mastering nature through industrialization. At least the process is interminable and complex, as it is illustrated by the example of the German pulp and paper industry. Therefore, a historical perspective on different societal uses of natural resources can help to understand the material, organizational and mental difficulties of a process that remains formative for the 'business-environment connection' until today.

References

Altmann, P. E.: Die Strohstoff-Fabrikation. Handbuch für Studium und Praxis. Berlin: Krayn 1914
Bayerl, G.: Vorindustrielles Gewerbe und Umweltbelastung. Das Beispiel der Handpapiermacherei. Technikgeschichte *48*, 206–238 (1981)
Bayerl, G., and Pichol, K.: Papier. Produkt aus Lumpen, Holz und Wasser. Reinbek: Rowohlt 1986
Berghoff, H.: Moderne Unternehmensgeschichte. Eine themen- und theorieorientierte Einführung. Paderborn u. a.: Schöningh 2004
Blechschmidt, J., and Strunz, A.-M.: Papierindustrie in Sachsen. In: *Bildungswerk der Sächsischen Wirtschaft* (Ed.): Wirtschaft – Innovation – Bildung. Beiträge zur Darstellung von 100 Jahren Industrie- und Wirtschaftsentwicklung in Sachsen. S. 156–163. Chemnitz: Bildungswerk der Sächsischen Wirtschaft 2000
Council on Economic Priorities (Ed.): Paper Profits. Pollution in the Pulp and Paper Industry. Cambridge, MA: MIT Press 1972
DWWV (*Deutscher Wasserwirtschafts- und Wasserkraft-Verband*) (Ed.): Die Wasserwirtschaft Deutschlands. Berlin: DWWV 1930
Freeman, R. E.: Strategic Management. A Stakeholder Approach. Boston: Pitman 1984
Fuchs, B.: Papier-Recycling und Umweltschutz aus historischer Sicht. In: Kroker, W. (Ed.): Der Weg zum modernen Papier. S. 75–84. Bochum: Georg-Agricola-Gesellschaft 1998
Gärtner, A., et al.: Gutachten des Reichs-Gesundheitsrats, betreffend die Verunreinigung der grossen Röder durch die Abwässer der Zellulosefabrik von Kübler und Niethammer in Gröditz in Sachsen. Berlin: Springer 1913

Hunter, D.: Papermaking. The History and Technique of an Ancient Craft. New York: Dover 1978
Josephson, P. R.: Resources under Regimes. Technology, Environment, and the State. Cambridge, MA: Harvard University Press 2005
Kirchner, E.: Das Papier. Teil 3: Die Halbstofflehre der Papierindustrie, Abschnitt B/C: Die Strohstoff- und Holzzellstoff-Fabrikation. Biberach: Güntter-Staib 1907
Kübler & Niethammer (Ed.): Papier aus Kriebstein. Darmstadt: Heppenstedts Wirtschafts Archiv 1956
Laak, D. van: Infra-Strukturgeschichte. Geschichte und Gesellschaft *27*, 367–393 (2001)
Mantel, K.: Holzmarktlehre. Ein Lehr- und Handbuch der Holzmarktökonomie und Holzwirtschaftspolitik. Melsungen: Neumann-Neudamm 1973
Mutz, M.: Die hölzerne Revolution. Produktion und Konsum von Papier im 19. und 20. Jahrhundert. In: *Landesmuseum Natur und Mensch* (Ed.): Holz-Kultur. Von der Urzeit bis in die Zukunft. S. 59–64. Oldenburg: Isensee 2007a
Mutz, M.: Naturale Infrastrukturen im Unternehmen. Die Papierfabrik Kübler und Niethammer zwischen Umweltabhängigkeit und Umweltgestaltung. Saeculum. Jahrbuch für Universalgeschichte *58*, 61–89 (2007b)
N. N.: Aus dem Erzgebirge. In: Dresdener Anzeiger Nr. 206 vom 25. Juli 1893. Hauptstaatsarchiv Dresden, Bestand 10736, Nr. 13537, Bl. 87 (1893)
Oligmüller, J. G., and Schachtner, S.: Papier. Vom Handwerk zur Massenproduktion. Köln: DuMont 2001
OECD (Organisation for Economic Cooperation and Development) (Ed.): Pollution by the Pulp and Paper Industry. Present Situation and Future Trends. Paris: OECD 1973
Reese, F.: Schreiben an die Amtshauptmannschaft Döbeln, Wöllsdorf-Limmritz 4. 1. 1913. Staatsarchiv Leipzig, Bestand 20026, Nr. 5519, Bl. 144 (1913)
Reinhold, G.: Die Papierholzversorgung. Berlin: Hofmann 1927
Reinhold, M.: Forstwirtschaft und Papierindustrie. Forstwirtschaftliches Centralblatt *54*, 113–131 (1932)
Rosen, C. M.: The Business-Environment Connection. Environmental History *10*, 77–79 (2005)
Rosen, C. M., and Sellers, C.: The nature of the firm. Towards an ecocultural business history. Business History Review *73*, 578–600 (1999)
Rudloff, M.: Die Fabrik im Dorf. Interessenkonflikte zwischen industriellen und agrarischen Eliten am Beispiel der Firma Kübler und Niethammer in Kriebstein. In: Hess, U., Listewnik, P., und Schäfer, M. (Eds.): Unternehmen im regionalen und lokalen Raum 1750–2000. S. 289–299. Leipzig: Universitätsverlag 2004
Schachtner, S.: Konflikte um sauberes Wasser. Die Auseinandersetzungen zwischen Bergisch Gladbacher Papierfabrikanten und anderen Wassernutzern. Rheinisch-Westfälische Zeitschrift für Volkskunde *37*, 83–94 (1992)
Schultze, J.: Die Papierfabrikation im Königreich Sachsen unter besonderer Berücksichtigung ihrer Beziehungen zu den Holzschleifereien. Tübingen: Tränkle 1912
Schuster, E.: Der Einfluß der wirtschaftlichen Entwicklung des aufstrebenden Kapitalismus auf Holznutzung und Baumartenwahl in der Forstwirtschaft, dargestellt vor allem am Beispiel Sachsens. Archiv für Forstwesen *10*, 1208–1227 (1961)
Siemen, B.: Ökologische Aspekte der Sulfitzellstoff-Herstellung in Deutschland um die Jahrhundertwende. Hamburg: Dissertation 1993
Uekötter, F.: Confronting the pitfalls of current environmental history. An argument for an organisational approach. Environment and History *4*, 31–52 (1998)
Wischermann, C., and Nieberding, A.: Die institutionelle Revolution. Eine Einführung in die deutsche Wirtschaftsgeschichte des 19. und frühen 20. Jahrhunderts. Stuttgart: Steiner 2004

Mathias Mutz, M. A.
Georg-August-Universität Göttingen
Institut für Wirtschafts- und Sozialgeschichte
Platz der Göttinger Sieben 5
37073 Göttingen
Germany
E-Mail: mmutz@uni-goettingen.de

200 Years of Flora Development in the Natural Landscape Unit "Göttinger Wald"

Jessica SPICALE and Renate BÜRGER-ARNDT (Göttingen)

With 4 Figures

Abstract

During the past 200 years, large scales of woodland were transformed into meadow and agricultural crop land due to exhaustive cultivation and reorganisation of agriculture. This caused the development of infertile grassland and other marginal sites in central Germany. By the end of the 19th century, the decline of pasture farming and an increasing demand on timber products led to reforestation of vast formerly deforested territories. Today the remaining marginal sites as well as ancient areas of woodland are classified as valuable ranges of high species diversity. These facts lead to one of the basic research questions of the project: How do changes of land use affect the development of different biotopes, and how are the effects to be assessed? To find an adequate answer to these questions, one research unit is concerned with acquisition and digital editing of floristic surveys and historic land use patterns derived from archival sources. As a suitable investigation area the "Göttinger Wald" was detected. Several parts of the "Göttinger Wald" arose from afforestation of former meadows. First results of the study show a strong correlation between changes in landscape structure due to a changing agriculture and silviculture and species composition. The development of rather multiple used woodlands to completely stocked high forests led to an increasing proportion of shade-tolerant species. The commutation of forest rights and the dissolving of the Commons causing deep changes in landscape structure are revealed in a decreasing proportion of specific ruderal and segetal communities and forest edge communities.

Zusammenfassung

Die einstmalige Zurückdrängung der Wälder durch Raubbau und Umwandlung großer Landschaftsteile Mitteldeutschlands in Weide und ackerbaulich genutzte Fläche schaffte aus heutiger Sicht wertvolle Grenzertragsstandorte und Magerflächen. Der Rückgang der Weidewirtschaft und der hohe Holzbedarf sorgten Ende des 19. Jahrhunderts für die erneute Aufforstung großer ehemals entwaldeter Gebiete. Heute gelten nicht nur die verbliebenen Grenzertragsstandorte, sondern im Besonderen alte Waldstandorte als Gebiete großer Artenvielfalt und somit als besonders schützenswert. Wie ist also Landnutzungswandel in Bezug auf die Entstehung bzw. das Verschwinden von bestimmten Biotoptypen zu bewerten? Zur Beantwortung dieser Frage befasst sich ein Arbeitsblock des Projektes mit der Erfassung und digitalen Aufbereitung von Vegetationsdaten aus archivarischen Quellen. Das Projektgebiet Göttinger Wald, ein zum Teil aus der Aufforstung großflächiger Weidegebiete hervorgegangenes und ab Mitte des 18. Jahrhunderts wiederholt von namhaften Vegetationskundlern floristisch erfasstes Gebiet, bietet hier eine viel versprechende Basis. Erste Ergebnisse zeigen die Auswirkungen einer sich ändernden Landnutzung auf die Zusammensetzung der Flora des Naturraumes Göttinger Wald sehr deutlich. Die Überführung der Nieder- und Mittelwaldwirtschaft in eine Hochwaldwirtschaft spiegelt sich in einem zunehmenden Anteil schattentoleranter Arten wider. Die sich im Laufe des Verkoppelungsprozesses und der Ablösung der Waldweiderechte verändernde Landschaftsstruktur führte zu einem Rückgang der Pflanzengesellschaften der Ackerraine und Waldsäume.

1. Introduction

Due to changes of the natural environment which accelerated during the past decades, a deep interest in their causes arose. Recently, climate changes are one of the most tensing objects examined with respect to their impact on the environment. It is almost agreed that climate change is anthropogenic and it seems to affect all continents in a similar way though at different levels of intensity. In regard to a scientific work of that dimension, it may appear almost dispensable to deal with land use changes and their effects on biodiversity on a regional scale. But according to one of the major conclusions almost every speaker of the Leopoldina Workshop in Göttingen made, changes encroaching all continents do start with changes of their several elements. Anthropogenic influence on the elements occurs in different amounts and directions and shows various effects. However, the part of local driving factors of environmental changes on a landscape level is not at all sufficiently examined. Therefore, the chief aim of the presented study is to illuminate the anthropogenic effects of land use on the diversity of species and habitats. The analysis is based on an intersection of local conditions, land use patterns and diversity of biotopes, plant communities and species at different time segments. As a basic investigation object of the study, changes in species composition and diversity of vascular plants serve as an example for the reaction of biological resources on human impact. The area of investigation, the "Göttinger Wald", is located in Lower Saxony to the east of Göttingen. It passed through many different types of agricultural and silvicultural land use practices during the past 200 years. At present ancient areas of woodland as well as remaining marginal sites are classified as valuable ranges of high species diversity. Those areas are considered to be prior aims of nature conservation programs and landscape management. Several ranges of the "Göttinger Wald" arose from afforestation of former pastures and meadow which now enclose open ranges with various anthropogenic affected biotopes. Retracing the development of the flora is expected to be one proper instrument to reconstruct the effects on the ecological mode of biotopes and habitats caused by different land use practises in the past.

2. Data Basis

The survey and analysis of land use pattern appearing in the "Göttinger Wald" and of its change during the past 200 years is based on historic maps. The availability of these maps simultaneously determined the choice of certain time segments. Information about different land use types was transferred into digital data using maps dated 1784, 1878, 1910, 1965 and 2002. The data processing using the GIS program ArcGis 9.2 (ESRI) derived information about area proportion, number and dimension of every land use type for each time segment. By intersecting the information of the different time segments, changes in the proportion of land use types became evident. The further analysis of the results will be conducted by the interpretation of additional archival and regional textual sources.

The floristic analysis is mainly based on the interpretation of different floristic inventories between 1745 (Rupp ed. v. Haller) and 1964 (Fuchs). Since mid of the 18th century, the "Göttinger Wald" was part of frequent floristic inventory conducted by reputed academics. This is why the research site makes a wide data basis available to draw conclusions from a long period of time. The registered data provides information about specific names, habitus, site requirements, and proven positions of occurrence of vascular plants. Each proof was assigned to one of the four historic time-segments (ZS I 1784, ZS II 1878, ZS III 1910, ZS IV 1965). The data of the recent time segment (ZS V 2002) is based on the habitat mapping of Lower Saxony. Data of 991 species were entered into five databases each revealing species appearing in one time-segment. The data were analysed using the software FRIDOLINO (cf. Ebrecht 2005) with courtesy of the Institute of Silviculture (University of Göttingen). It revealed for instance shifting spectra of indicator plants, the percentage of each phytosociological group (Ellenberg 1996) or typical forest species (Schmidt 2003) appearing in each time segment.

3. First Results – Increase of Shade-Tolerant Species due to Changing Silviculture

During the last 200 years the forest structure of the "Göttinger Wald" was affected by afforestation and a changing silviculture. Generally multiple used woodlands were developed to completely stocked high forests (Früchtenicht 1926, Hölz 1987, Wagenhoff 1987). This process was initiated in 1860 with the commutation of forest rights held by the local population. The main interest of the forest administration was focussed on timber products rather than former utilizations such as topping systems, removal of litter or grazing forests. The forest thus gradually changed from a coppice selection system to sprout seedling forests and completely high stocked forests at last. This change in forest structure is clearly revealed by the results of the analysis of the floristic data: A change in species composition can be retraced where the percentage of the phytosociological group containing broadleaved woodlands and related communities increases (Fig. 1) and the percentage of forest species predominant in dense forest steadily rises (Fig. 2). This result is also confirmed by the analysis of the Ellenberg L-index. An increasing suppression by shade caused by a dense canopy of overwood had a supporting effect on the spread of shade-tolerant species. Furthermore, large areas of farmland and meadow were reforested, especially in the time between time segment II and III. Therefore, spectra of indicator plants shift towards shade-tolerance.

4. Changing Species Composition due to Changing Landscape Structure

During the time segments I to III agricultural and silvicultural land use on small areas with a mosaic structure was prevailing. The historic maps of these time segments show smooth transitions between wood- and farmland. Alleys and paths within agricultural cropland were lined with hedges and uncultivated headland. Therefore, up to time segment III, the percentage of species appearing at forest edges and clearances mount as well as the percentage of herbaceous vegetation of frequently disturbed sites (ruderal and segetal communities). On the phytosociological level of orders blackthorn shrubbery and related communities (*Prunetalia spinosae*) hold a great share (Fig. 3).

By the year 1910, the results of the land use analysis add up to a decreasing percentage of marginal grazing areas, a higher amount of forest areas with realigned boundaries and a sharp separation between forests and agricultural lands. These changes in landscape structure refer to measures taken in the process of dissolving the Common Land. In the area of the "Göttinger Wald" this process took place between 1860 and 1900 (Bussemeier 1986, Wrase 1973).

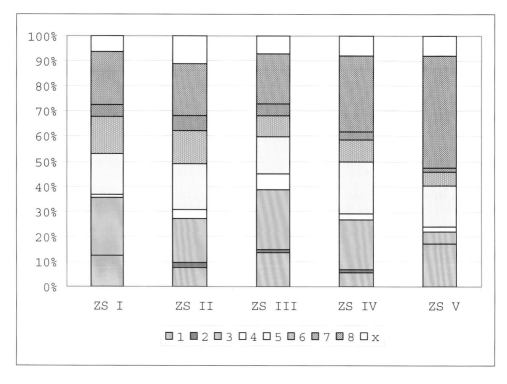

Fig. 1 Percentage of sociological groups (according to ELLENBERG 1996) on the basis of the overall species list compiled for each time segment. (1: Freshwater and mooreland vegetation, 2: halophytes and marine vegetation, 3: herbaceous vegetation of frequently disturbed sites, 4: rock pavements and alpine grassland, 5: anthropo-zoogenic heath and grassland, 6: woodland-related herbaceous perennial and shrub communities, 7: coniferous forests and related communities, 8: broadleaved woodland and related communities)

The effects of these changes are reflected by the floristic data as follows: The percentage of forest species appearing at forest edges and clearances heavily declines as well as the percentage of the phytosociological group containing herbaceous vegetation of frequently disturbed sites. Even though the dissolving of the Common Land certainly created more boundaries of possession, and therefore, potentially more space for the development of disturbance adapted and edge communities, it has to be considered, that the whole process was accompanied by an intensification of cultivation. Common pasture land and broad uncultivated strips were transformed into cropland, a new road net was constructed and a lot of hedges formally serving as living fences were removed. According to the analysis of the floristic data, this measure is revealed in a steeply decreasing percentage of the order of blackthorn shrubbery and related communities (*Prunetalia spinosae*). In the area of the "Göttinger Wald", specific forms of hedges, which were used for the purpose of hedging in cattle, mostly belonged to plant communities of this order. Furthermore, former common sheep meadows of lower nutrition quality enclosing shallow and oligotrophic areas often were reforested (BUSSEMEIER 1986). The vegetation of those marginal sites on calcareous soil is generally characterized by several plant communities of the sociological group of anthropo-zoogenic heath und grassland. Due to frequent afforestation of potential sites of these plant communities, the percentage of

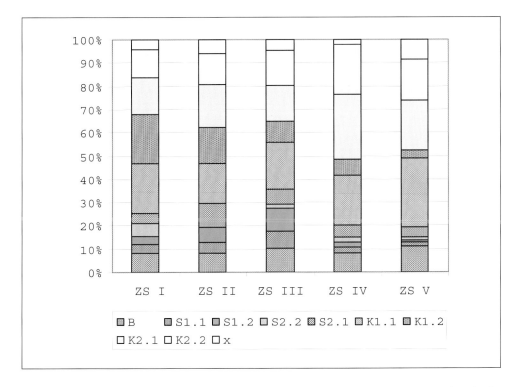

Fig. 2 Percentage of forest-plant classification (according to SCHMIDT 2003) on the basis of the overall species list compiled for each time segment. (B: Tree layer, S (Shrub layer) 1.1: predominant in dense forest, S 1.2: predominant at forest edges and clearances, S 2.1:associated with forests as well as with open landscape, S 2.2: incidental in forests, with emphasis on open landscape, K (Herb layer) 1.1: predominant in dense forest, K 1.2: predominant at forest edges and clearances, K 2.1:associated with forests as well as with open landscape, K 2.2: incidental in forests, with emphasis on open landscape, x: indifferent plants)

that certain sociological group was expected to decline as well. Nevertheless, the analysis of the floristic data reveals that the percentage of this group remains almost constant during all time segments. Considering the fact that every inventory is presumably influenced by personal preferences of the person raising the data (especially the ancient ones) and also by the mapping purpose (especially the recent one), this might be a systematic error. Habitats on marginal sites, generated by anthropogenic utilization are in the very centre of nature conservation interests today. In times of the earliest vegetation surveys analyzed in this project, the surveys may supposably underestimate the flora of marginal grasslands used as sheep meadow because of their former commonness. The results on the phytosociological level of orders (Fig. 4) show an increasing proportion of grassland species on calcareous marginal sites (*Festuco-Brometea* class, represented by the *Festucetalia valesiacae* and *Brometalia erecti* order) during time segments IV and V, which presumably is really a result of changing preferences in mapping. Calcareous marginal sites are one of the most attended sites in the habitat mapping of the "Göttinger Wald" and the increase of plant communities of this order could not be explained by changes in site conditions so far. Furthermore, Figure 4 reveals a development of the percentage of the *Nardo-Callunetea* class (represented by the *Nardetalia*

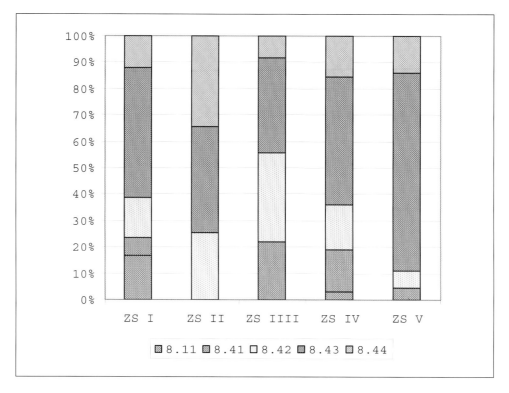

Fig. 3 Proportions of the sociological group (8) broadleaved woodlands and related communities on the level of phytosociological orders. (8.11: Saliceatalia purpureae, 8.41: Quercetalia robori-petraea, 8.42: Quercetalia pubescenti-petraea, 8.43: Fagetalia sylvaticae, 8.44: Prunetalia spinosae)

order) parallel to the percentage of the *Prunetalia spinosae* order. As aforementioned, the extent of former common sheep meadows declined between time segment II and III because of afforestation. The *Nardo-Callunetea* class is known to be semi-natural, for it is the typical flora of nutrient-poor and acid-soil (sheep-) pastures (MERTZ 2002). Similar to the development of the *Prunetalia spinosae* order the decrease of the *Nardo-Callunetea* class after time segment III is presumably due to afforestation in the process of dissolving the Commons.

In relation to changing and especially intensified agriculture since the third time segment, another interesting result referring to a reaction of the flora is given by the Ellenberg N- and F-index. The spectra of indicator plants shift towards a superior supply of nitrogen and humidity. Evidence suggests that on the one hand an increase of fertilization increased on intensively cultivated croplands and on the other hand humid areas of marginal productivity were abandoned, leaving space for natural succession. The intersection of land use patterns and flora development is not yet completed and it is intended to derive more detailed information about the respective correlations and their spacing. Especially on the level of single species characterizing specific site conditions the data of the historic inventories is expected to generate significant information about the development of diversity and species composition in relation to human impact on habitats.

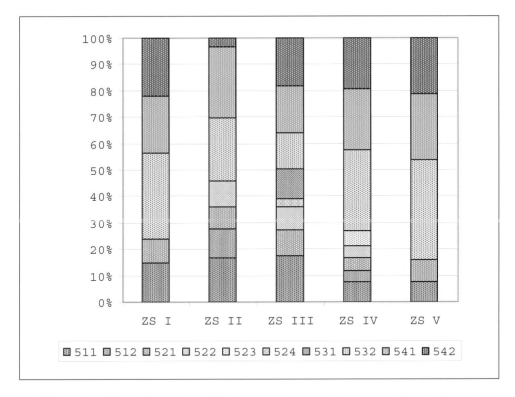

Fig. 4 Proportions of the sociological group (5) anthropo-zoogenic heath and grassland on the level of phytosociological orders. (5.11: Nardetalia, 5.12: Genisto-Callunetalia, 5.21: Sedo-Scleranthetlalia, 5.22: Corynephoretalia, 5.23: Festuco-Sedetalia, 5.24: Thero-Airetalia, 5.31: Festucetalia valesiacae, 5.32: Brometalia erecti, 5.41:Molinietalia, 5.42: Arrhenatheretalia)

References

Bussemeier, F.: Der Wandel der Landnutzung im Raum Göttingen vom Beginn der Verkoppelung bis zum 1. Weltkrieg. Examensarbeit. Betreut von Prof. Dr. Nitz. Göttingen: Georg-August-Universität, Geographisches Institut 1986

Ebrecht, L.: Vegetation, Standortverhältnisse und Ausbreitungsbiologie von Pflanzen auf Rückegassen und Waldwegen im Göttinger Wald und im Solling. S. 31, 1. Aufl. Göttingen: Cuvillier 2005

Ellenberg, H.: Vegetation Mitteleuropas mit den Alpen. 5. stark veränderte und verbesserte Auflage. Stuttgart (Hohenheim): Eugen Ulmer 1996

Früchtenicht, W.: Die Entwicklung der Göttinger Stadtforsten. Göttingen: Turm-Verlag W. H. Lange 1926

Fuchs, H.: Flora von Göttingen. Führer zu den wildwachsenden Pflanzen des Göttinger Muschelkalkgebietes. Göttingen: Vandenhoek & Ruprecht 1964

Hölz, A.: Historische Entwicklung des Waldes im Göttinger Raum – untersucht anhand der „Topogr. Karten" von 1784, 1876 und 1979. Diplomarbeit. Göttingen. FH Forst 1987

Mertz, P.: Pflanzengesellschaften Mitteleuropas und der Alpen. Erkennen, Bestimmen, Bewerten; ein Handbuch für die vegetationskundliche Praxis. Landsberg am Lech: ecomed 2002

Rupp, H. B.: Flora Jenensis 1718. Ed. III cur. A. v. Haller. Göttingen 1745

Schmidt, M., Ewald, J., Ellenberg, H., Fischer, A., Kreibitzsch, W.-U., Oheimb, G., und Schmidt, W.: Liste der typischen Waldgefäßpflanzen Deutschlands. (Mitteilungen der Bundesforschungsanstalt Forst-Holzwirtschaft 212, S. 1–34). Bonn: Eigenverlag der Bundesforschungsanstalt Forst-Holzwirtschaft 2003

Jessica Spicale und Renate Bürger-Arndt

WAGENHOFF, A.: Der Wald der ehemaligen Herrschaft Plesse – seine frühere Nutzung und seine heutige Bewirtschaftung. In: *Flecken Bovenden* (Ed.): Plesse-Archiv. S. 95–159. Göttingen: 1987
WRASE, S.: Die Anfänge der Verkoppelung im Gebiet des ehemaligen Königreichs Hannover-Hildesheim. 1973 (Nachdruck des Jahresheftes der Albrecht-Thaer-Gesellschaft *20*, S. 65–177 [1980])

    Jessica SPICALE, M. Sc.
    Georg-August-Universität Göttingen
    Fakultät für Forstwissenschaften und Waldökologie
    Abteilung Landschaftspflege und Naturschutz
    Büsgenweg 3
    37077 Göttingen
    Germany
    Phone: +49 551 3922326
    Fax: +49 551 393415
    E-Mail: jspical@gwdg.de

    Prof. Dr. Renate BÜRGER-ARNDT
    Georg-August-Universität Göttingen
    Fakultät für Forstwissenschaften und Waldökologie
    Professur für Naturschutz und Landschaftspflege
    Büsgenweg 3
    37077 Göttingen
    Germany
    Phone: +49 551 394486
    Fax: +49 551 393415
    E-Mail: rbuerge@gwdg.de

Urban Environments and their Perception Reflected in 18th and 19th Century Medical Topographies

Anna-Sarah Hennig (Göttingen)

Abstract

From the end of the 18th to the beginning of the 20th century so-called medical topographies played an important role within the German-language medical literature. Medical topographies are detailed descriptions of a particular region – usually a city –, composed by physicians with the aim of recording everything that might have an influence on the inhabitants' health.

This study presents an analysis, carried out from an environmental-historical perspective, of 19 medical topographies of German major cities, investigating the authors' perceptions and strategies with regard to a number of environmental factors. The particular focus of the study lies on the analysis of the three environmental media water, air and soil, while further taking into account descriptions of architectural constructions like streets, houses and public places as well as vegetation and animals. The study sheds light onto the role the topos of 'the detrimental city' plays within the medical topographies, and how it is constructed. Another question deals with the way the paradigm shift apparent in the medical field at the time such as the then triumphing field of bacteriology which was newly introduced by Robert Koch works is reflected in the physicians' perceptions.

Zusammenfassung

Vom Ende des 18. bis zum Anfang des 20. Jahrhunderts nahmen sogenannte medizinische Topographien einen wichtigen Platz in der deutschsprachigen Medizinalliteratur ein. Dabei handelt es sich um von Ärzten verfasste detaillierte Beschreibungen der Gebiete, in denen sie tätig waren – überwiegend von Städten –, mit dem Anspruch, alles zu erfassen und zu schildern, was als die Gesundheit der Einwohnerinnen und Einwohner beeinflussend betrachtet wurde.

In der hier vorgestellten Untersuchung wurden 19 medizinische Topographien deutscher Großstädte unter einer umweltgeschichtlichen Perspektive ausgewertet, um die Wahrnehmungs- und Umgehensweisen der Autoren bezüglich ihrer Umwelt herauszuarbeiten. Als konkrete Teile dieser Umwelt werden in der Arbeit die Darstellungen der drei Umweltmedien Wasser, Luft und Boden untersucht. Auch die baulichen Strukturen, wie Straßen, Häuser und Plätze, und die Umweltaspekte der Vegetation und der Fauna werden in den Blick genommen. Untersucht wird, ob der Topos der „schädlichen Stadt" bedient und wie er konstruiert wird. Weiter interessiert, wie sich Wandlungen medizinischer Paradigmen wie der Siegeszug der Bakteriologie, eingeleitet durch die Arbeiten Robert Kochs, auf Untersuchungsschwerpunkte und Schlussfolgerungen medizintopographischer Autoren auswirken.

1. Introduction

Right until the beginning of the 19th Century medical philosophy was deeply informed by ancient ideas: it was believed that disease was caused by geographic phenomena, such as the quality of water, wind direction or degree of air pressure. In his *De aere, acquis et locis*, Hippokrates (c. 460–377 BC), who is often described as the founder of a rational-empirical understanding of medicine, places attention onto the importance of taking an empirical investigation of the environment in the diagnosis and aetiology of existing diseases: "The one who wants to carry out the art of medicine correctly, must comply with the following rules: First, he [sic] must consider the seasons of the year and their impacts. [...] He must further take into account the winds [...]. He must also consider the impacts of the waters, lakes and rivers; when a water's taste is not the same as another and its weight is different, it will have an equally diverse effect. When somebody arrives in a city he does not know, he must carefully consider the town's position towards the winds and the sunrise. [...] The ground is of further concern; whether it is bare and low in water, or whether it is overgrown and lush, and whether the area lies in a dell and is humid or placed on a peak and cold. [...] Following these points, every single detail needs to be thought through. When a person has correctly identified such conditions, the local diseases will not remain concealed from him, nor will the conditions of [the inside of] the bodies of the inhabitants."[1]

Such Hippocratic tradition is reflected in so-called 'medical topographies', which were written by physicians and available in print in German speaking countries since the late 18th century. The here presented study, carried out as part of ongoing research, analyses 19 medical topographies from the 18th and 19th century in order to investigate the conceptions of environment, health issues and medicine at the time. It presents the first systematic attempt at evaluating medical topographies of a number of German cities (Berlin, Bremen, Dresden, Hamburg, Frankfurt, Cologne, Munich, Vienna) from an environmental-historical perspective. The analysis sheds light onto the contemporary local environmental conditions and, moreover, investigates the authors' perceptions and proposed strategies with regard to their environmental surroundings, taking into account both natural as well as anthropogenic elements.

2. Medical Topographies

Already at the turn of the 17th to the 18th century LEIBNIZ had pointed out the usefulness of detailed records of medical issues and is therefore often referred to as the originator of medical topographies. In the 18th and 19th century, medical topographies and geographically oriented medicine in general flourished in Germany and reached their climax in the middle of the 19th century. Medical topographies are near-comprehensive accounts reporting on local health issues and medical problems and relating these to the geographic environment. They include descriptions of the region, the climate, geological and hydrological conditions as well as social, cultural and ethnic factors. Induced by the 'medicinische Policey' (a governmental set of

---

[1] My own translation, based on the German translation by DILLER (1970): "Wer der ärztlichen Kunst in der richtigen Weise nachgehen will, der muß folgendes tun. Erstens muß er über die Jahreszeiten und über die Wirkungen nachdenken, die von jeder einzelnen ausgehen können. [...] Ferner muß er sich über die Winde Gedanken machen [...]. Er muß auch über die Wirkungen der Gewässer nachdenken; denn wo sie sich im Geschmack und Gewicht unterscheiden, so ist auch die Wirkung eines jeden sehr verschieden. Wenn also jemand in eine Stadt kommt, die er nicht kennt, so muß er sich genau überlegen, wie ihre Lage zu den Winden und zum Aufgang der Sonne ist. [...] Weiter die Beschaffenheit des Bodens, ob er kahl und wasserarm ist oder dichtbewachsen und bewässert und ob das Gelände in einer Mulde liegt und stickig ist oder hochgelegen und kalt. [...] Hiervon ausgehend, muß man sich jede Einzelheit überlegen. Wenn nämlich jemand das richtig erkannt hat, so werden ihm, wenn er in eine ihm unbekannte Stadt kommt, weder die einheimischen Krankheiten verborgen bleiben, noch wie das Leibesinnere der Bewohner beschaffen ist." – Many thanks to Svenja WURM, M. A. (Heriot-Watt University, Edinburgh), for providing assistance with the translations.

rules regarding health issues), academic competition and the physicians' personal scientific interests, practitioners composed detailed descriptions of the regions where they lived and worked – mainly cities – with the purpose to record and depict everything that they regarded as having an impact on people's health; meteorological data, analyses of water and air quality and demographic details can be equally found as discussions of architectural trends, nutrition, the way of life and 'morality' of the citizens.

BRÜGELMANN (1982) reports on the production of 51 medical topographies in a narrower sense in German-speaking countries between 1779 and 1850; medical regional descriptions written in German-speaking and other European countries as well as from overseas are estimated to reach several hundreds in total (JUSATZ 1967). These accounts nearly always concentrate on cities, and – if formally distributed at all – were predominantly published as monographs, which is also the case for the 19 topographies forming the basis of this study. Other medical topographies appeared in medical journals, such as the German journal of the 'state' pharmacology, the *Zeitschrift für Staatsarzneikunde*, or in festschrifts. Similar accounts were requested by the paramedic services of the Prussian, Russian and Austrian army in the form of so-called *Garnisonsbeschreibungen* ('garrison descriptions') containing a health-oriented focus.

Such medical topographies constitute detailed descriptions depicting a plethora of phenomena that are concerned with the natural as well as cultural and social environment, frequently stretching over a considerably extensive period – several authors spent many years collecting data for their descriptions.

An analysis of these assessments of the environmental features together with an understanding of the environmental conditions of the region in question reveal insights into the physicians' perceptions of the environment, which is at the heart of this study.

First we should briefly discuss the concept of *perception*. Within psychology the term 'perception' refers to those cognitive-mental processes, in which an individual processes external and internal stimuli and generates impressions, conceptions, interpretations and evaluations. Perception, therefore, describes the process as well as the outcome of the processing of stimuli. Using this definition as the basis for our work, perception is here regarded in a fairly wide-ranging sense, not only referring to the act of cognitive processing of sensory information, but also describing mental products such as interpretations and evaluations.

The analysis of the findings will go beyond a description of isolated phenomena; instead they are put into relation with each other and compared. In this sense, discrepancies in content and form with regard to place and time will be analyzed to reveal palpable shifts of perceptions. This work thus steps into the realms of environmental, urban and medical history, while further attempting to add to the field of environmental psychology by giving it a historical dimension. Within environmental psychology, which engages with "the impacts of the physical-empirical and cultural (outside) world as well as the spatial-social influences on the experiences and behaviors of people",[2] the material environment in general and thus the urban habitat in particular are attached with great importance. Amongst others, the city is regarded as a significant factor in containing 'disease-causing' as well as 'health-ensuring' functions, and further serves as the foundation for empirical investigations of perceptions of this specific

---

2 My own translation of HELLBRÜCK und FISCHER (1999, p. 29): "den Auswirkungen der physisch-materiellen und kulturellen Außenwelt sowie den räumlich-sozialen Einflussfaktoren auf das Erleben und Verhalten der Menschen".

habitat and its attributions. Until now their historical dimension remains unexplored, and historical patterns of perception and evaluation of the urban environment are largely neglected. An interdisciplinary environmental-historical approach with a focus on aspects of perceptions within the historical relationships between person and environment will tackle this lack of investigation by sharing knowledge of different fields.

3. The Study

The primary aim of this study is not to reconstruct a comprehensive account of specific physical characteristics of the urban environment portrayed in the medical topographies, but rather to understand the perceptions of a specific group of experts (viz. physicians). The following questions are addressed:

- What are the authors' empirical methods and tools when examining aspects of their environment?
- How are their results and observations interpreted? Which phenomena are seen as interconnected and how and why are connections between particular phenomena assumed?
- Which problems are noticed and what kind of problem solving strategies are suggested?
- Which environmental conditions are considered beneficial to people's physical and mental health?
- Which environmental factors are regarded as bearing an impact on human experience and behavior?

The fact that such medical topographies appeared in an age that gave rise to medical innovations and new insights into medical phenomena, apparent for example in the highly influential and ground-breaking developments in the area of bacteriology at the time, raises further questions:

- How is this paradigm shift apparent in the medical field at the time – from miasma-based interpretations to bacteriologically informed explanations of health issues – reflected in the physicians' perceptions?
- Do the authors follow this trend and adopt a bacteriological framework?
- How much weight is granted to socio-economic and social-hygienic perspectives compared to climate-deterministic concepts?

During the 19th century German cities were under the significant influence of the flourishing industrialization and the urbanization that accompanied it. Caused by migration, many of them grew enormously and rapidly, becoming densely populated spaces with their own specific problems. Results indicate that the attention of the authors, just like that of leading hygienists of the early 20th century, focused much more on the social factors, such as housing conditions, than on, for instance, climate data. Medical topographies of big cities like Vienna, Berlin, Frankfurt, Bremen, Hamburg, Cologne, Munich and Dresden are thus particularly suitable for the investigation of how increasing urbanization affects the authors' perceived environment. They shed light onto the role the topos of 'the detrimental city' plays within the medical topographies, and how it is constructed, answering questions such as whether the authors explicitly distinguish between natural and anthropogenic environments and how 'nature' is conceived within the medical topographies. Especially with regard to the urban space which can be viewed as highly anthropogenic and therefore often contradicting the common understanding of 'nature' (a perspective that also shines through in the medical topographies) such questions proof highly relevant and lead to the further following ones:

– Is the city perceived as an 'unnatural' place of living?
– What exactly makes the city an 'unhealthy' place?
– And which are the preconditions for urban "salubrity", i.e. the factors that ensure the healthy state of a city?

The findings of the study are based on an analysis with a particular focus on the environmental media water, air and soil, while further taking into account descriptions of architectural constructions like streets, houses and public places as well as vegetation and animals. These aspects are not only at the centre of attention within current research of environmental history, but also constitute a significant part of each medical topography. Therefore we are able to avoid a retrospective projection of current ideas onto our historical data, as the objectives and structure of the topographies demonstrate that our current conceptualization of the environment is reflected in the historical understanding of the relationship between these environmental media and human medical conditions; subsuming the above mentioned aspects under the concept "environment" seems appropriate and medical topographies reveal themselves as a valid source for the investigation of the perception of such environmental factors.

References

BRÜGELMANN, J.: Der Blick des Arztes auf die Krankheit im Alltag 1779–1850. Medizinische Topographien als Quelle für die Sozialgeschichte des Gesundheitswesens. Phil. Diss., FU Berlin 1982
DILLER, H.: Hippokrates: Über die Umwelt. Berlin: Akademie-Verlag 1970
HELLBRÜCK, J., and FISCHER, M. (Eds.): Umweltpsychologie: ein Lehrbuch. Göttingen: Hogrefe 1999
JUSATZ, H. J.: Die Bedeutung der medizinischen Ortsbeschreibungen des 19. Jahrhunderts für die Entwicklung der Hygiene. In: ARTELT, W., und RÜEGG, W. (Eds.): Der Arzt und der Kranke in der Gesellschaft des 19. Jahrhunderts. Stuttgart: Ferdinand Enke 1967

        Anna-Sarah HENNIG
        Georg-August-Universität Göttingen
        DFG-Graduiertenkolleg Interdisziplinäre Umweltgeschichte
        Bürgerstraße 50
        37073 Göttingen
        Germany

Der Briefwechsel von Johann Bartholomäus Trommsdorff (1770–1837)
Lieferung 11: Trott – Ziz und Nachträge

Acta Historica Leopoldina Nr. *18*/11
Bearbeitet und kommentiert von Hartmut BETTIN (Marburg), Christoph FRIEDRICH (Marburg) und Wolfgang GÖTZ (Wildeshausen), unter Mitarbeit von Henriette BETTIN (Greifswald)
(2009, 342 S., 7 Abb., 9 Stammbäume, 20,95 Euro, ISBN: 978-3-8047-2559-1)

TROMMSDORFF gilt als Vater der wissenschaftlichen Pharmazie. Der Begründer des *Journals der Pharmacie* engagierte sich in standes- und sozialpolitischen Fragen. Seine umfangreiche Korrespondenz spiegelt die Entwicklung von Chemie und Pharmazie im beginnenden 19. Jahrhundert, aber auch die Veränderungen des Apothekenwesens seiner Zeit wider. Die Edition (über 250 Briefpartner und 1500 erhaltene Briefe) stellt eine bedeutende wissenschaftshistorische Quelle dar. Mit dieser Lieferung wird der kommentierte Briefwechsel J. B. TROMMSDORFFS abgeschlossen. Schwerpunkte bilden die Briefe des Anilin-Entdeckers und Trommsdorff-Schülers Otto UNVERDORBEN und die Briefe des Chemie- und Pharmazieprofessors Heinrich August VOGEL. Darüber hinaus enthält der Band zahlreiche Briefe von oder an bedeutende Ärzte und Apotheker, berühmte Chemiker und Naturforscher sowie hochgestellte Persönlichkeiten, beispielsweise Herzog CARL AUGUST VON SACHSEN-WEIMAR-EISENACH und nachgetragene Briefe Johann Wolfgang DÖBEREINERS, August Peter Julius DU MÊNILS sowie Johann Friedrich GMELINS. Der Anhang enthält Familien-Stammbäume zur Orientierung sowie Gesamtsachregister und -verzeichnisse über alle 11 Lieferungen.

Wissenschaftliche Verlagsgesellschaft mbH Stuttgart

Mice, Maggots, Moles:
On the Discussion of Vermin in the 18th Century

Steffi WINDELEN (Göttingen)

Abstract

In the 18th century different kinds of insects (such as maggots, bedbugs, louses and flies), reptiles (such as snakes and lizards), amphibians (such as frogs and toads) and mammals (such as mice, rats, moles and weasels) were called 'vermin'. An animal is not a 'vermin' per se, rather it becomes a vermin in the eyes of man: In the 18th century the term vermin is a collective description for animals with particular characteristics, it works as a category to distinguish and to characterize animals. Nevertheless 'vermin' is not defined and characterized as a particular thing, rather there are different ways of describing it: Although most often animals that have negative characteristics or that are harmful are called vermin. However, 'vermin' are also described as useful, but also as beautiful and as repugnant and it is said that these animals have functional and structural functions in *oeconomia naturae*. Thus 'vermin' is not a class of animals in natural science but an object that is described depending on the frame of reference. There are three frames of reference, the economic, the aesthetic and the frame of order. Every framework has its own logic of observing and describing these animals. For example, in the economic framework animals are observed concerning their material effects on human interests in using something. In this frame useful or harmful qualities are attributed to them. This means, 'vermin' is regarded as useful or harmful to a special aim.

Zusammenfassung

Im 18. Jahrhundert werden verschiedene Insekten (wie Maden, Wanzen, Läuse und Fliegen), Reptilien (wie Schlangen und Eidechsen), Amphibien (wie Frösche und Kröten) und Säugetiere (wie Mäuse, Ratten, Maulwürfe und Wiesel) als ‚Ungeziefer' bezeichnet. Bei diesen Tieren handelt es sich aber nicht per se um ‚Ungeziefer'. Der Begriff dient vielmehr der Beschreibung von Tieren mit spezifischen Merkmalen; er wird dazu verwendet, Tiere zu unterscheiden und zu charakterisieren. Der Begriff ‚Ungeziefer' besitzt im 18. Jahrhundert aber keinen feststehenden Bedeutungsinhalt, sondern es können drei verschiedene Beschreibungsweisen unterschieden werden: Hauptsächlich werden Tiere, denen negative Eigenschaften zugeschrieben werden, als ‚Ungeziefer' bezeichnet, doch zugleich gelten die Tiere auch als nützlich, sie werden ebenfalls für schön und hässlich oder ekelig gehalten. Daneben werden ihnen funktionelle und strukturelle Funktionen im Haushalt der Natur zugeordnet. ‚Ungeziefer' stellt folglich keine naturwissenschaftliche Kategorie dar, es handelt sich vielmehr um ein Objekt, das in Abhängigkeit vom Bezugsrahmen des Beobachters/der Beobachterin bestimmt wird. Bei den drei zu unterscheidenden Bezugsrahmen handelt es sich um den ökonomischen, den ästhetischen und den der Ordnungen. Jeder Bezugsrahmen besitzt eine eigene Logik der Unterscheidung und Beschreibung. Beispielsweise werden Tiere im ökonomischen Bezugsrahmen bezüglich ihrer materiellen Auswirkungen auf menschliche Nutzungsinteressen bewertet, weshalb ihnen nützliche oder schädliche Eigenschaften zugeordnet werden.

Steffi WINDELEN
Georg-August-Universität Göttingen
DFG-Graduiertenkolleg Interdisziplinäre Umweltgeschichte
Bürgerstraße 50
37073 Göttingen
Germany

Perception and Control of Cattle Epidemics in the Electorate of Bavaria in the 18th Century

Carsten STÜHRING (Göttingen)

Abstract

The study investigates the perception and control of cattle epidemics in the 18th century. The geographical focus lies on the electorate of Bavaria. Cattle were an important food basis and were used as draught animal. Furthermore, duties for cattle contributed to princely treasures. Hence, cattle epidemics were an enormous risk for agrarian societies in the electorate of Bavaria.

In an economic historical perspective causes and consequences of cattle epidemics are explored. Administrative authorities and veterinary institutions increasingly used scientific strategies, whereas the rural population often still relied on religious practices like prayers and pilgrimages. In an environmental perspective historical perceptions of animals are described.

Thus, the study focuses on princely instructions to control epidemics and veterinary fonts; the relationship between humans and animals is also analyzed; the behavior of the local population is examined in written petitions; religious sources are votive donations.

Zusammenfassung

In dieser Studie wird die Wahrnehmung und Kontrolle der Rinderseuchen im 18. Jahrhundert näher untersucht. Der geographische Schwerpunkt liegt auf dem Kurfürstentum Bayern. Rinder bildeten eine wichtige Ernährungsgrundlage und wurden als Zugtiere genutzt. Außerdem trugen Abgaben für Tiere zu den kurfürstlichen Schätzen bei. Rinderseuchen stellten daher eine ungeheure Gefahr für die agrarische Gesellschaft im Kurfürstentum Bayern dar.

Aus wirtschaftsgeschichtlicher Perspektive werden die Ursachen und Folgen von Rinderseuchen untersucht. Verwaltungsbehörden und Veterinärinstitute bedienten sich immer häufiger wissenschaftlicher Methoden, während die ländliche Bevölkerung oft immer noch auf religiöse Praktiken wie Gebete und Pilgerfahrten vertraute. Aus umwelthistorischer Perspektive werden historische Mensch-Tier-Verhältnisse beschrieben.

In der Studie liegt daher der Schwerpunkt auf den kurfürstlichen Anweisungen zur Seuchenkontrolle und auf Veterinärschriften; die Schriften werden auch im Hinblick auf das Verhältnis von Mensch und Tier analysiert; das Verhalten der lokalen Bevölkerung wird anhand von schriftlichen Petitionen untersucht; religiöse Quellen sind Votivgaben.

>Carsten STÜHRING, M. A.
>Georg-August-Universität Göttingen
>DFG-Graduiertenkolleg Interdisziplinäre Umweltgeschichte
>Bürgerstraße 50
>37073 Göttingen
>Germany
>Phone: +49 551 399672
>Fax: +49 551 393645
>E-Mail: cstuehr@uni-goettingen.de

Schwarze Elster in the Flow of Time – Landscape Change and History Supported by Hydraulic Engineering

Manuela ARMENAT (Göttingen)

Abstract

The title of my PhD thesis is "Schwarze Elster in the flow of time – landscape change and history supported by hydraulic engineering", and it is a part of the research field focus on "Diking of nature". The thesis focuses on GIS analysis and considers political, socio-economical factors and their results for human and nature from the 19th to the mid 20th century.

German hydraulic engineering progressed due to the increasing need of arable crop as well as transit and flood protection. The straightening of small rivers, the building of stream flow controls like dikes and weirs, as well as the constraining of diseases like malaria and anthrax, turned the natural landscape of the floodplains gradually in a culturally shaped landscape.

In the last years and decades several articles have been published about historical hydraulic engineering and straightening of rivers close to great flooding (BRAZDÍL et al. 2006, SCHMIDT 2000), the risk management and the management of natural disasters and their perception (POLOWODA 2007, ROHR 2005, DEUTSCH und PÖRTGE 2002, PÖRTGE und DEUTSCH 2006), as well as for the reconstruction of climatic development (STURM et al. 2001, BÜRGER et al. 2006, JACOBEIT et al. 2006, PFISTER 2001, GLASER et al. 2003) and the changes of great river flows (ROMMEL 2000, URIBELARREA et al. 2003, VISCHER 1986).

These river straightening and land drainage systems are based on, for example, the Prussian reforms (GRUNDMANN 2001, RAKOW 2003) and their importance for the development of the cultural landscape. Therefore was after MAIER (2004) not only the flood protection an important fact for the river straightening and hydraulic engineering projects but also the growth of Prussian agricultural economy and military interests.

The PhD project focuses on the following key issues: (*i*) how did the ecomorphology of the river change through the development of landuse; (*ii*) how did these processes progress in the major time cuts over the centuries; and (*iii*) which driving forces made the river change possible and (*iv*) which consequences result for the humans and the nature due to the landscape change. With this PhD project I want to give another perspective of hydraulic engineering, flood protection, and landuse change with aspects of political and socio-economical forces.

The resesrach area will be the lower and the middle river stream of the river Schwarze Elster. The Schwarze Elster flows today from the German federal state Saxony through Brandenburg and Saxony-Anhalt and discharges in the river Elbe near Elster.

The methods used in the project include historical-geographical analysis with topographic maps from the 19th to the 20th century. The landuse data are one of the basics for the interpretation and quantification of the landuse change and their connection to natural and anthropogen factors in the research area. Additional the GIS analyzed data will be correlated and interpreted with the help of historical documents in archives.

Zusammenfassung

Das Thema des Dissertationsprojektes lautet „Schwarze Elster im Fluss der Zeit – Landschaftswandel durch wasserbauliche Maßnahmen". Es ist Teil des Forschungsschwerpunktes B3 „Eindämmung von Natur" im Graduiertenkolleg. Im Projekt werden mit Hilfe einer Graphischen Informationssystem (GIS)-gestützten Analyse der Landschaftswandel nachgezeichnet und die politischen, sozioökonomischen Faktoren sowie die Folgen der Maßnahmen für Mensch und Natur anhand von Archivmaterial vom 18. bis zur Mitte des 20. Jahrhunderts untersucht.

Die wasserbaulichen Maßnahmen in Deutschland wurden stark durch den ansteigenden Bedarf an Agrarprodukten sowie durch Hochwasserschutz und den Ausbau der Wassertransportwege beeinflusst. Mit der Begradigung auch kleiner Flüsse, den Bau von Anlagen für die Abflussregelung, wie z. B. von Deichen und Wehranlagen, sowie durch die Maßnahmen zur Eindämmung von Krankheiten, wie Malaria und Milzbrand (Anthrax), wurden die „natürlichen" Landschaften der Flussebenen allmählich in Kulturlandschaften umgewandelt.

Aus den vergangenen Jahren und Jahrzehnten finden sich zahlreiche Publikationen zum Wasserbau und zu Flussregulierungen im Zusammenhang mit Hochwasserereignissen (BRAZDÍL et al. 2006, SCHMIDT 2000), zur Abschätzung von Risiken dieser Naturgefahr, zur Wahrnehmung und zum Umgang mit dem Ereignis, z. B. in Form eines Katastrophenmanagements (POLIWODA 2007, ROHR 2005, DEUTSCH und PÖRTGE 2002, PÖRTGE und DEUTSCH 2006), sowie zur Rekonstruktion der klimatischen Entwicklungen (STURM et al. 2001, BÜRGER et al. 2006, JACOBEIT et al. 2006, PFISTER 2001, GLASER et al. 2003) und zur Veränderung, vor allem der großen Flussläufe (ROMMEL 2000, URIBELARREA et al. 2003, VISCHER 1986).

Getragen wurden die durchgeführten Flussregulierungen und Binnenentwässerungen aber auch z. B. durch die Preußischen Reformen (GRUNDMANN 2001, RAKOW 2003) und deren Bedeutung und Voraussetzung für die landeskulturelle Entwicklung. Damit war nach MAIER (2004) für eine Regulierung nicht nur der Hochwasserschutz ausschlaggebend, sondern spielten auch die in Preußen zunehmende, marktorientierte Landwirtschaft sowie militärische Interessen eine Rolle.

Ziel des Dissertationsprojektes ist es zu klären: (*a.*) wie sich der ökomorphologische Zustand des ausgewählten Fließgewässers durch den Landnutzungswandel geändert hat; (*b.*) wie diese Prozesse in ausgewählten Zeitabschnitten über die Jahrhunderte verliefen, verfolgt und umgesetzt wurden; (*c.*) welche treibenden Faktoren zu dieser Veränderung der Flusslandschaft führten und (*d.*) welche Folgen dieser Landschaftswandel für Mensch und Natur hatte. Damit soll, neben der derzeit weit verbreiteten historischen Hochwasserforschung, ein weiterer Beitrag zu den treibenden sozio-ökonomischen Kräften von Ausbau und Veränderung von Fließgewässern und Landschaft geleistet werden.

Die Teiluntersuchungsgebiete liegen im Unteren und Mittleren Flusslauf der Schwarzen Elster. Die Schwarze Elster durchfließt heute die deutschen Bundesländer Sachsen, Brandenburg und Sachsen-Anhalt, wo sie in der Nähe von Elster in die Elbe mündet.

Zu den im Projekt angewandten Methoden gehört u. a. eine historisch-geographische Analyse mit Hilfe von topographischen Karten aus dem 19. und 20. Jahrhundert. Die Landnutzungsdaten bilden eine der Grundlagen, mit deren Hilfe die Veränderung der Landnutzung und ihr Zusammenhang mit den natürlichen und anthropogenen Faktoren im Untersuchungsgebiet interpretiert und quantifiziert werden. Zusätzlich werden die Daten aus der Analyse in einem Geographischen Informationssystem mit historischen Informationen aus Archivalien korreliert und interpretiert.

References

Brazdíl, R., Kundzewicz, Z. W., and Benito, G.: Historical hydrology for studying flood risk in Europe. Hydrol. Science Journal (Special Issue: Historical Hydrology) *51*/5, 739–764 (2006)

Bürger, K., Dostal, P., Seidel, J., Imbery, F., Barriendos, M., Mayer, H., and Glaser, R.: Hydrometeorological reconstruction of the 1824 flood event in the Neckar River basin (southwest Germany). Hydrol. Science Journal (Special Issue: Historical Hydrology) *51*/5, 864–877 (2006)

Deutsch, M., und Pörtge, K.-H.: Die Hochwassermeldeordnung von 1889 – ein Beitrag zur Geschichte des Hochwasserwarn- und Meldedienstes in Mitteldeutschland. In: Tetzlaff, G., Trautmann, T., und Radtke, K. S. (Eds.): Extreme Naturereignisse – Folgen, Vorsorge, Werkzeuge. Zweites Forum Katastrophenvorsorge, Leipzig, 24.–26. September 2001. S. 396–405. Bonn, Leipzig: Deutsches Komitee für Katastrophenvorsorge e. V. (DKKV) 2002

Glaser, R., Beck, C., und Stangl, H.: Zur Temperatur- und Hochwasserentwicklung der letzten 1000 Jahre in Deutschland. *DWD* (Eds.): Klimastatusbericht 2003. S. 55–67 (2003)

Grundmann, L.: Der Schraden. Eine landeskundliche Bestandsaufnahme im Raum Elsterwerda, Lauchhammer, Hirschfeld und Ortrand. Im Auftrag des Instituts für Länderkunde Leipzig und der Sächsischen Akademie der Wissenschaften zu Leipzig. Köln, Weimar, Wien: Böhlau Verlag GmbH & Cie. 2001

Jacobeit, J., Phillip, A., and Nonnenmacher, M.: Atmospheric circulation dynamics linked with prominent discharge events in Central Europe. Hydrol. Science Journal (Special Issue: Historical Hydrology) *51*/5, 946–965 (2006)

Maier, D.: Die Schwarze Elster. Zur Karriere eines industrialisierten Flusses 1852–1945. Blätter für Technikgeschichte (Technisches Museum Wien) Bd. *65* [2003], 165–179 (2004)

Pfister, C.: Klimawandel in der Geschichte Europas: zur Entwicklung und zum Potential der Historischen Klimatologie. Österreichische Zeitschrift für Geschichtswissenschaften *12*/2, 7–43 (2001)

Polowoda, G. N.: Aus Katastrophen lernen – Sachsen im Kampf gegen die Fluten der Elbe 1784 bis 1845. Köln, Weimar, Wien: Böhlau 2007

Pörtge, K.-H., und Deutsch, M.: „Wir sollten daraus lernen!" Vorschläge und Maßnahmen zur Risikominderung angesichts schwerer Hochwasser (1891–1929). In: Disse, M., Guckenberger, K., Pakosch, S., Yörük, A., und Zimmermann, A. (Eds.): Risikomanagement extremer hydrologischer Ereignisse (Tagungsband „Tag der Hydrologie 2006"). Forum für Hydrologie und Wasserbewirtschaftung Bd. *1*/15, 243–253 (2006)

Rakow, H.: Die Separation in der preußischen Provinz Sachsen und in Anhalt. In: Wollkop, H.-F., und Diemann, R. (Eds.): Historische Landnutzung im thüringisch-sächsisch-anhaltinischen Raum. Tagungsband. S. 14–26. Frankfurt (Main), Berlin, Bern, Bruxelles, New York, Oxford, Wien: Peter Lang GmbH, Europäischer Verlag der Wissenschaften 2003

Rohr, C.: The Danube floods and their human response and perception (14th to 17th C). History of Meteorology *2*, 71–86 (2005)

Rommel, J.: Studie zur: Laufentwicklung der deutschen Elbe bis Geesthacht seit ca. 1600. Auftrag der Bundesanstalt für Gewässerkunde, Stand 2001. Elbe-Informationssystem ELISE. Koblenz, Berlin 2001

Schmidt, M.: Hochwasser und Hochwasserschutz in Deutschland vor 1850. Eine Auswertung alter Quellen und Karten. Martin Schmidt, Harzwasserwerke GmbH, Hildesheim. München: Oldenbourg-Industrieverlag 2000

Sturm, K., Glaser, R., Jacobeit, J., Deutsch, M., Brazdíl, R., Pfister, C., Luterbacher, J., und Wanner, H.: Hochwasser in Mitteleuropa und ihre Beziehungen zur atmosphärischen Zirkulation. Petermanns Geographische Mitteilungen *145*/6, 14–23 (2001)

URIBILARREA, D., PÉREZ-GONZÁLES, and BENITO, G.: Channel changes in the Jarama and Tagus rivers (central Spain) over the past 500 years. Quarternary Sci. Rev. *22*, 2209–2221 (2003)
VISCHER, D.: Schweizerische Flusskorrektionen im 18. und 19. Jahrhundert. Mitteilungen der Versuchsanstalt für Wasserbau, Hydrologie und Glaziologie Nr. *84*, ETH Zürich (1986)

        Manuela ARMENAT, Dipl. Geol.
        Georg-August-Universität Göttingen
        DFG-Graduiertenkolleg Interdisziplinäre Umweltgeschichte
        Bürgerstraße 50
        37073 Göttingen
        Germany
        Phone: +49 551 3913407
        E-Mail: marmena1@gwdg.de

Silva Nervus Belli

Axel Bader (Göttingen)

Abstract

Do the forests sustain damage in wartime? What are the mechanisms responsible for this? Traditional forestry history only addresses this issue marginally. Usually, accounts relate to the days of the Thirty-Years' War. Timber serves as an export commodity, as Kremser writes: "During the Thirty-Years' War, numerous wooded areas in the North of the Empire were exploited, with the timber then being sold to Bremen and Holland."[1] It was also used for reconstruction: "These enormous logging campaigns were joined by extensive forest fires, and what 'was spared by the war was taken by the Empire's own subjects and owners of woodlands in order to support their bare lives with the meagre income or to reconstruct the villages devastated by arson that the bands of soldiers had committed."[2]

Whether the great population losses in the 1618–1648 resulted in reforestation or not is a controversial issue. Kremser maintains that "one would be mistaken to assume that the shrunken population and the little livestock it had gave the forest a period of recovery from excessive grazing and logging. For in the last decade of the great war, there was nothing left to exploit other than the forest."[3] Whereas Dévèze (1976) holds the opposite: "While the forest did initially suffer from the enormous requirements of the warring people, the period following turned out to be more favourable thanks to the decline in population and livestock."[4]

However, the post-war period that Dévèze refers to was not characterized by a recovery of the forest everywhere. Jördens (1931) attributes the timber trade a considerable role in the continuing excessive exploitation of the forests: "When the timber trade began to flourish again after the Thirty-Years' War and a great yard was created to stockpile timber in Hamburg and in Harburg, this excessive logging assumed very threatening proportions."[5]

Häffner (1934) maintains that the frequent requisitions bore the potential to prevent any major deforestation: "From 1620–1650, we can continuously read about horses, wagons,

---

1 Cf. Kremser 1990, p. 254.
2 Langerfeldt 1858, quoted from Kremser 1990, p. 255.
3 Kremser 1990, p. 255.
4 Dévèze 1976, p. 47.
5 Jördens 1931, p. 132. Jördens above all draws his conclusions from contemporary timber and forest regulations that already justify their very existence with the ruining of the forests "in the difficult wartimes" or through "the terrible state of affairs brought about by the war" in their forewords. Cf. Jördens 1931, p. 135.

artisans and fieldwork being requisitioned after Nördlingen, and especially frequently before the Battle of Nördlingen, about 'gifts' of game and fish to pacify this or that party and to Imperial Commanders, and about timber to the Imperial Field Artillery."[6] And the collapse of order in wars was a topic, too: "One bad consequence of the war turmoil was gathering of wood (wood carrying)."[7]

Regarding the situation in France, Devèze (1976) writes: "At least the horrors of the Hundred-Years' War in the 15th century postponed the forestry crisis for a certain period that had already threatened to occur in the 14th century."[8] In spite of this, "[…] unlike in the 15th century, the religious wars did not result in the forest reclaiming ground. On the contrary, they led to a lack of order that was very detrimental to the forests."[9] "The years of peace following the death of the great king [Louis XIV, Author's note] were a considerable blessing for the forests."[10]

This short list of research results could be randomly continued, but with each finding stated being refuted by a precisely opposite one. Even the impacts of war on the forest are controversial: Was timber consumption higher in the war – or wasn't it, were the forests used more intensively, perhaps even excessively – or not? Was timber consumption perhaps even at its highest level in the immediate post-war period? The causal links are just as unclear: Did a war lead to an increase in forest via a depopulation, or did a collapse of order result in an greater exploitation of the forests? Was the resumption of regular timber trading in the post-war period the main reason for excessive exploitation, or was it reconstruction? But then, in terms of forestry, the post-war years would definitely still belong to the war period and not to the peace period, which was generally such a "blessing" for the forest. Obviously, the questions raised have so far only been answered at local level. The big picture of the relationship between forest and war has not been drawn yet.

However, even if the issue is formulated in this manner, as comprehensive as it may appear to be, it remains incomplete. For in addition to nature as such, with all its organisms, the cultural aspect with the projection of human ideas and emotions needs to be integrated. The two sides of the issue become particularly clear in the case of the German forest: First, there is the "German Forest" as such, representing an important source of raw materials (topic of the forestry history literature being discussed) the using of which has always been highly controversial. Then there is another forest: the imaginary, symbolically laden place, in a nutshell, the object of romantic longing. It is understandable that these two notions should have also yielded different histories.

As a place of symbolic projections, the German forest was an invention of romanticism. All forms of artistic expression turned the notion of the forest that had been valid up to then upside down. What had still appeared as uncanny, untamed and wild was now regarded as sacred and sublime. Grimm's Fairy Tales, written precisely at the beginning of this new period, portrayed both views. One strongly expressive feature of this "forest myth" was the alleged affinity between the German Forest and the German People. Setting out from the constant repetition of the well-known passage by Tacitus on the different collections of fairy tales and the pictures by Caspar David Friedrich, the German forest had already developed an impressive impact by the mid-nineteenth century. And thus Wilhelm Heinrich Riehl (1855)

---

6 Häffner 1934, p. 57.
7 Ibid, p. 58.
8 Devèze 1976, p. 32.
9 Ibid, p. 34.
10 Ibid, p. 40.

was able to already draw on a wealth of material to substantiate his proposition on the character of the peoples and the landscape they were surrounded by. The German forest, representing a mawkish, fanciful embodiment of the German people's character, was enthusiastically adopted by all strata of the population. A "forest attitude" suggested itself to support, and even replace, in real life the "national attitude" that had suffered so severely in the Napoleonic Wars. If there were no prospects of a national state, at least one could dream of what were said to have been better times. This feeling of a loss, branded "Waldeinsamkeit" (loneliness in the forest) by Ludwig TIECK (1842), was, however, of a far more fanciful nature than a true political program. With their gnarled trees withstanding storms, the crucifixes and ruins of monasteries, lonely wayfarers and fields of mist, C. D. FRIEDRICH's romantic forests develop an image of vanitas, melancholy and longing.

However, FRIEDRICH's "French Chasseur" represents an unquestionably political painting of these times. The picture of a French huntsman who had lost his way in the forest clearly brought across the message of who – which "cultural nation" – had now taken possession of the forest. It was then not long before Ernst Moritz ARNDT was calling for a forest as a means of protection against the French.

So any examination of the forest in times of war has to bear both aspects in mind: its cultural as well as its natural history. Thus a history of nature being ideologized and culture being naturalized by people emerges.

Why is a survey of the war periods the focus of attention? The example of existential crises – wars – offers the opportunity to highlight the dichotomy of nature and culture in a particularly clear manner, for it is especially then that the immediate dependence of humans on their environment becomes apparent. Such times are characterized by *recollecting* one's own resources. In addition to a direct material reference, the German word for recollection, "*Rückbesinnung*", already integrates the cultural, ideological reference to the natural resource in its meaning. The "sudden" dependence on local natural resources literally forces people to also comprehend this at a cultural level.

Even so, the notion of a peaceful, calm forest has prevailed up to this day. As is so often the case, reality looked very different. Forests were far from being as eternal, peaceful and protective as National Socialist propaganda would have it. Forests were strongly shaped by the war. They were exploited and turned into battle theatres and places of ideological and political appropriation. No accounts have so far been given of this cultural history approach to the forest in wartime.

In order to bridge this gap in research, various North German forests have been selected as focal points for case studies: forests in today's Federal States of Hesse and Lower Saxony that belonged to the Prussian state territory up to 1947. This region is particularly suitable for a case study thanks to its ecological and political variability. For it comprises very different forest areas, and politically, too, it saw numerous changes of rule between 1850 and 1950.

The considerable stretch of time from 1850 to 1950 comprises both the Franco-German War and the two World Wars as well as the post-war period. Thus a short campaign can be compared with long wars of attrition and mobilization at short notice with years of preparation for war.

This case study is aimed at describing forests as sources of substitute products in times of a great paucity of material. In view of the latter, the governments pressed for greater timber production. But the forests were not mere "material stockspiles". The Imperial Government, and even more so the National Socialist Government, attributed meanings to the forest that went way beyond the provision of raw materials. However, this material and cultural mobilization of the forests in National Socialism came last in a long tradition of state control of the forests. In addition, as the traditional executive, the forestry commission had a vested institutional interest in the forests. This mix of interests, demands and claims regarding the forest left back a deep ecological footprint the repercussions of which could not be undone by the reforestation measures of the post-war period, either.

Zusammenfassung

Nehmen die Wälder in Kriegszeiten Schaden? Welche Mechanismen sind dafür verantwortlich? Die traditionelle Forstgeschichte behandelt dieses Thema nur am Rande. Meist beziehen sich die Ausführungen auf die Zeit des Dreißigjährigen Krieges. Holz diente als Exportware, wie KREMSER schreibt: „Im Dreißigjährigen Krieg wurden zahlreiche Waldgebiete im Norden des Reiches ausgeschlachtet, um das Holz nach Bremen und Holland zu verkaufen."[11] Ebenso wurde es zum Wiederaufbau verwendet: „Zu diesen enormen Hieben kamen noch ausgedehnte Waldbrände und was ‚der Krieg verschonte, nahmen die eigenen Unterthanen und Besitzer der Waldungen, um dem spärlichen Erlöse das nackte Leben zu fristen, oder die durch Mordbrennereien der Soldateska verödeten Dörfer wieder aufzubauen.'"[12]

Ob die großen Bevölkerungsverluste im Krieg 1618–1648 zu einer Wiederbewaldung führten oder nicht, ist umstritten. KREMSER meint, „man täuscht sich, wenn man annimmt, dass die geschrumpfte Bevölkerung und ihr weniges Vieh dem Walde eine Zeit der Erholung von exzessiver Weide und Holzhieb verschafft hätten. Denn ‚auszuplündern' gab es im letzten Jahrzehnt des großen Krieges außer dem Walde nichts mehr."[13] Währenddessen DÉVEZE (1976) das Gegenteil annimmt: „Der Wald hat zunächst gelitten unter den enormen Anforderungen der Kriegsleute, aber die Folgezeit gestaltete sich günstiger wegen des Rückgangs der Bevölkerung und des Viehs."[14]

Die von DÉVEZE angesprochene Nachkriegszeit war jedoch nicht überall von einer Walderholung geprägt. JÖRDENS (1931) weist dem Holzhandel eine große Rolle bei der weitergehenden Ausplünderung der Wälder zu: „Als nach dem Dreißigjährigen Krieg der Holzhandel wieder aufblühte und in Hamburg und in Harburg ein großer Holzstapelplatz angelegt wurde, nahm dieser Waldraubbau äußerst bedrohliche Formen an."[15]

Für HÄFFNER (1934) besaßen die häufigen Requisitionen das Potential, größere Kahlschläge zu verhindern: „Von 1620–1650 lesen wir unaufhörlich über Requisitionen von Pferden, Wagen, Handwerkern, Schanzen nach Nördlingen, besonders häufig vor der Schlacht bei Nördlingen, von ‚Geschenken' an Wild und Fisch zur Besänftigung bald dieser, bald jener Partei, an kaiserliche Kommandeure, Holz an die kaiserliche Feldartillerie."[16] Und auch der Zusammenbruch der Ordnung in Kriegen war ein Thema: „Eine üble Folge der Kriegswirren war das Holzsammeln (Holztragen)."[17]

---

11 Vgl. KREMSER 1990, S. 254f.
12 LANGERFELDT 1858, zitiert nach KREMSER 1990, S. 255.
13 KREMSER 1990, S. 255.
14 DEVÈZE 1976, S. 47.
15 JÖRDENS 1931, S. 132. JÖRDENS leitet seine Schlussfolgerungen vor allem aus zeitgenössischen Holz- und Forstordnungen ab, die ihre Existenz mit dem Ruin der Wälder „in den beschwerlichen Kriegszeiten" oder durch das „Kriegsunwesen" schon im Vorwort begründen. Vgl. JÖRDENS 1931, S. 135.
16 HÄFFNER 1934, S. 57.
17 Ebenda, S. 58.

Zur Situation in Frankreich schreibt Devèze (1976): „Die Schrecken des 100-jährigen Krieges im 15. Jahrhundert haben wenigstens eine gewisse Zeit lang die Krise des Forstwesens aufgehalten, die schon im 14. Jahrhundert drohte."[18] Trotzdem haben „[d]ie Religionskriege […] nicht wie im 15. Jahrhundert zu einem Vordringen des Waldes geführt. Im Gegenteil, sie führten zu einem Mangel an Ordnung, der für die Wälder sehr nachteilig war."[19] „Die Friedensjahre, die dem Tod des großen Königs [Ludwig XIV., d. Verf.] folgten, waren für die Wälder sehr segensreich."[20]

Diese kurze Liste der Forschungsergebnisse ließe sich beliebig fortführen, und doch würde stets ein einmal geäußerter Befund von einem genau gegenteiligen kassiert werden. Schon die Auswirkungen des Krieges auf den Wald sind umstritten: War der Holzverbrauch im Krieg höher – oder nicht, wurden die Wälder intensiver genutzt, sogar übernutzt – oder nicht? War der Holzverbrauch eventuell sogar in der unmittelbaren Nachkriegszeit am höchsten? Ebenso unklar sind die kausalen Zusammenhänge: Führte ein Krieg über die Entvölkerung zu einer Waldzunahme oder über einen Zusammenbruch der Ordnung zu einer verstärkten Ausbeutung der Wälder? War in der Nachkriegszeit der Wiederbeginn eines regulären Holzhandels oder der Wiederaufbau der Hauptgrund der Übernutzung? Dann würden waldbaulich gesprochen die Nachkriegsjahre auf jeden Fall aber noch zur Kriegszeit zählen und nicht zur Friedenszeit, die ja allgemein für den Wald so „segensreich" war. Offensichtlich wurden die aufgeworfenen Fragen bisher nur lokal beantwortet. Ein *big picture* des Verhältnisses von Wald und Krieg wurde noch nicht gezeichnet.

Und doch ist selbst diese Form der Fragestellung – so umfassend sie erscheinen mag – unvollständig, denn neben der Natur an sich mit allen ihren Lebensformen gilt es noch, die kulturelle Seite mit der Projektion menschlicher Ideen und Gefühle einzubeziehen. Besonders beim deutschen Wald wird diese Zweiseitigkeit deutlich: Einmal den „deutschen Forst", eine wichtige wirtschaftliche Rohstoffquelle (Thema der besprochenen forsthistorischen Literatur), deren Nutzung stets hoch umstritten war. Daneben gibt es noch einen anderen Wald: Den imaginären, symbolisch aufgeladenen Ort, kurzum das Objekt romantischer Sehnsucht. Es ist verständlich, dass diese beiden Auffassungen auch verschiedene Geschichten hervorbrachten.

Der deutsche Wald als Ort symbolischer Projektionen war eine Erfindung der Romantik. Sämtliche Ausdruckformen der Kunst stellten das bis dahin gültige Bild des Waldes auf den Kopf. Was in barocken Darstellungen noch unheimlich, ungezähmt und wild erschien, wurde nun als weihevoll und erhaben angesehen. Die Grimmschen Märchen, an eben dieser Zeitenwende aufgeschrieben, überliefern beide Sichtweisen. Ausdrucksstarkes Merkmal dieses „Waldmythos" war die untergeschobene Affinität zwischen deutschem Wald und deutschem Volk. Ausgehend von der gebetsmühlenartigen Wiederholung der bekannten Tacitusstelle über die verschiedenen Märchensammlungen und die Bilder von Caspar David Friedrich hatte der deutsche Wald bis zur Mitte des 19. Jahrhunderts bereits eine beachtliche Wirkung entfaltet. Und so konnte Wilhelm Heinrich Riehl (1855) zur Fundierung seiner These über den Charakter der Völker und die sie umgebende Landschaft bereits auf einen großen Materialfundus zurückgreifen. Der deutsche Wald als gefühlsdusselige, schwärmerische Verkörperung des deutschen Volkscharakters wurde von allen Bevölkerungsschichten begeistert aufgenommen. „Waldgesinnung" bot sich an, die in den Napoleonischen Kriegen schwer

---

18 Devèze 1976, S. 32.
19 Ebenda, S. 34.
20 Ebenda, S. 40.

geschundene „Nationalgesinnung" realiter zu unterfüttern, gar zu ersetzen. Wenn schon kein Nationalstaat in Sicht war, so ließ sich doch im Wald in eine vermeintlich bessere Zeit zurückträumen. Dieses Verlustgefühl, von Ludwig Tieck (1842) „Waldeinsamkeit" getauft, war jedoch weit schwärmerischer Natur, als ein tatsächliches politisches Programm. C. D. Friedrichs romantische Wälder entfalten mit ihren knorrigen, sturmtrotzenden Bäumen, den Kruzifixen und Klosterruinen, einsamen Wanderern und Nebelfeldern ein Bild der Vanitas, Melancholie und Sehnsucht.

Ein unzweifelhaft politisches Gemälde dieser Zeit liegt allerdings mit Friedrichs „französischem Chasseur" vor. Das Bild eines französischen Jägers, der sich im Wald verirrte, vermittelte deutlich die Botschaft, wer – welche „Kulturnation" – sich inzwischen den Wald angeeignet hatte. Bis zur Forderung Ernst Moritz Arndts, einen Bannwald gegen die Franzosen zu errichten, war es dann gar nicht mehr weit.

Eine Arbeit über den Wald zu Kriegszeiten muss also beide Seiten beachten: Die Kultur- wie die Naturgeschichte desselben. So entsteht eine Geschichte der Ideologisierung der Natur und der Naturalisierung der Kultur durch den Menschen.

Warum steht die Betrachtung der kriegerischen Perioden im Mittelpunkt? Das Beispiel existenzieller Krisen – von Kriegen – eröffnet die Möglichkeit, die Dichotomie von Natur und Kultur besonders deutlich aufzuzeigen, denn gerade dann wird eine unmittelbare Abhängigkeit der Menschen von ihrer Umwelt offensichtlich. Diese Zeit ist durch eine *Rückbesinnung* auf eigene Ressourcen geprägt. Der Begriff *Rückbesinnung* bezieht neben einer direkten materiellen auch die kulturelle, ideologische Bezugnahme auf die Naturressource schon im Wortsinne mit ein. Die „plötzliche" Abhängigkeit von heimischen Naturschätzen zwingt den Menschen förmlich dazu, diese auch kulturell nachzuvollziehen.

Trotzdem herrscht bis heute das Bild eines friedlichen, ruhigen Waldes vor. Die Realität sah – wie so oft – ganz anders aus. Wälder waren keinesfalls so ewig, friedlich und schützend, wie die nationalsozialistische Propaganda glauben machen wollte. Wälder wurden durch den Krieg stark geprägt, sie wurden ausgebeutet und zu Orten des Kampfes und der ideologisch-politischen Aneignung. Diese kulturgeschichtliche Annäherung an den Wald im Krieg blieb bisher ungeschrieben.

Um diese Forschungslücke zu füllen, wurden verschiedene norddeutsche Wälder als Fallstudienschwerpunkte ausgewählt. Wälder in den heutigen Bundesländern Hessen und Niedersachsen, die bis 1947 zum preußischen Staatsgebiet gehörten. Diese Region eignet sich wegen ihrer ökologischen und politischen Variabilität besonders für eine Fallstudie. Denn sie umfasst sehr unterschiedliche Waldgebiete, und auch politisch erfuhr sie zwischen 1850 und 1950 zahlreiche Herrschaftswechsel.

Die betrachtete Zeitspanne von 1850 bis 1950 umfasst sowohl den deutsch-französischen Krieg als auch die beiden Weltkriege und die Nachkriegszeit. Damit lassen sich ein kurzer Feldzug mit langen Abnutzungskriegen und kurzfristige Mobilmachungen mit langjährigen Kriegsvorbereitungen vergleichen.

Diese Fallstudie dient dazu, Wälder als Quellen der Ersatzprodukte in Zeiten großer Materialknappheit zu beschreiben. Angesicht dessen drangen die Regierungen auf eine Holzproduktionssteigerung. Aber Wälder waren eben nicht nur „Materiallager". Gerade die kaiserliche und besonders die nationalsozialistische Regierung luden den Wald mit Bedeutungen auf, die weit über die Rohstoffbereitstellung hinausgingen. Diese materielle und kulturelle Mobilisierung der Wälder im Nationalsozialismus stand aber nur an letzter Stelle einer langen Tradition staatlicher Kontrolle über die Forste. Darüber hinaus besaß die Forstverwaltung als traditionelle Exekutive ein institutionelles Eigeninteresse an den Wäldern. Diese Gemengelage an Interessen, Forderungen und Ansprüchen an den Wald hinterließ einen tiefen ökologischen Fußabdruck, dessen Nachwirkungen auch die Wiederaufforstungen der Nachkriegszeit nicht ungeschehen machen konnten.

Literatur

Devèze, M.: Histoire des Forêts. (Geschichte der Wälder, übersetzt von K. Hasel.) Freiburg 1976
Häffner, A.: Forst- und Jagdgeschichte der fürstlichen Standesherrschaft Oettingen-Wallerstein. Nördlingen 1934
Jördens, C.: Wirtschaftsgeschichte der Forsten in der Lüneburger Heide vom Ausgang des Mittelalters bis zum Beginn des neunzehnten Jahrhunderts. Braunschweig 1931
Kremser, W.: Niedersächsische Forstgeschichte. Eine integrierte Kulturgeschichte des nordwestdeutschen Forstwesens. Rotenburg (Wümme) 1990
Langerfeldt, C. H.: Geschichte des Forstwesens im Herzogthum Braunschweig. In: Festgabe für die Mitglieder der XX. Versammlung deutscher Land- und Forstwirthe. Braunschweig 1858
Riehl, W. H.: Die Naturgeschichte des Volkes als Grundlage einer deutschen Social-Politik. Stuttgart 1855
Tieck, L.: Liebeswerben: Waldeinsamkeit. Breslau 1842

Axel Bader
Georg-August-Universität Göttingen
DFG-Graduiertenkolleg Interdisziplinäre Umweltgeschichte
Bürgerstraße 50
37073 Göttingen
Germany
Phone: +49 551 392124
E-Mail: axel_p_bader@yahoo.de

Die Gründung der Leopoldina
(Academia Naturae Curiosorum) im historischen Kontext
Johann Laurentius Bausch zum 400. Geburtstag

Leopoldina-Symposium

vom 29. September bis 1. Oktober 2005 in Schweinfurt (Bibliothek Otto Schäfer)

> Acta Historica Leopoldina Nr. 49
> Herausgegeben von Richard TOELLNER (Kloster Amelungsborn), Uwe MÜLLER (Schweinfurt), Benno PARTHIER und Wieland BERG (Halle/Saale)
> (2008, 336 Seiten, 42 Abbildungen, 22,95 Euro, ISBN: 978-3-8047-2471-6)

Ziel dieser interdisziplinären, internationalen Tagung war es, die Gestalt des Johann Laurentius Bausch (1605–1665) in ihren biographischen, sozialen und wissenschaftsgeschichtlichen Bedingungen darzustellen sowie die Gründung der Leopoldina in den Rahmen der internationalen Akademiengeschichte des 17. Jahrhunderts einzuordnen. Es wurde der über die bisherige Literatur hinausgehende aktuelle Forschungsstand in neun Vorträgen präsentiert, die der vorliegende Band in erweiterter und aktualisierter Form dokumentiert und vertieft durch Anhänge mit der Edition der Leges der Akademie und Bibliographien der im frühen Akademieprogramm veröffentlichten Monographien und ihrer Vorgänger aus anderthalb Jahrhunderten sowie einer Analyse der Selbstdarstellung der Leopoldina in ihrer Korrespondenz mit der Royal Society von 1670 bis 1677.

Wissenschaftliche Verlagsgesellschaft mbH Stuttgart

Replanting the World.
Colonial Forestry in the German "Kaiserreich" 1884–1918

Lars KREYE (Göttingen)

With 1 Figure

Abstract

During the second half of the 19th century, colonial powers tried to establish direct state control over the world's tropical woodland resources. Their foremost goal was the exploitation of valuable timber species and other raw material for industrial usage. To be able to reach this goal, a new science was developed: colonial forestry.

My project investigates the role of colonial forestry in the developing colonial system of the German Empire from 1884–1918. At the core of my analysis are questions about the transfer of seeds, plants and forest technology in hitherto nearly unknown bio-physical and social environments. Thereby one goal is to reconstruct the worldwide network of institutions and actors, which constituted the colonial system of the German Empire in regard to the exploitation of renewable resources (Fig. 1).

The investigation is divided into two parts. The first part reconstructs the national network of colonial forestry in the German Empire. Beginning at the top the system was centered in the political realm at the Colonial Department of the German Foreign Office, which was called from 1907 on "Reichskolonialamt". The scientific element of the system was the Botanical Central Department for the Colonies at the Royal Botanical Garden and Museum in Berlin – "Botanische Zentralstelle für die Kolonien am Königlich Botanischen Garten und Museum".[1] The German Colonial Economic Committee – "Kolonialwirtschaftliches Kommitee" – was the third main element of the system and responsible for economic relations. These three institutions collaborated, and important decisions concerning the plantation of forest products in the colonies were made by them. The "Reichskolonialamt" was the corridor

---

[1] The task of the "Botanische Zentralstelle" was not only to collect and store specimen and knowledge, but to organize the transfer of seeds and plants from all over the world to the German colonies. It was in regard of plant expertise the most important knot of knowledge in the colonial network of the transnational German Empire. In comparison to the "Botanische Zentralstelle" schools of scientific forestry in the provinces of the German Empire, like the academies of Hann. Münden and Tharandt, played at this point of time only an inferior role. They were more onlookers of the processes in developing a colonial system to control renewable resources. Therefore famous contemporary teachers of scientific forestry, like JENTSCH or BÜSGEN, complained that there was a lack of expertise about tropical forests, because direct experience from the German colonies was only poorly available for teaching. The gap between theory and practice concerning tropical forestry was broad. And, if one believes in the laments, the forest academies and colonial schools of the Empire were not able to prepare their students adequately for the colonial forestry service (BArch, R 1001/7659).

Fig. 1 Caption: Forester and forest worker in the Cameroons: "Sawing of trees". Source: Colonial Picture Archive, Library of the University of Frankfurt on the Main, pict. no. 025-0284-085

of power, the "Botanische Zentralstelle" was the think tank and responsible for the transfer of plants and seeds to the colonies. The "Kolonialwirtschaftliche Komitee" functioned as the system's interface to the German domestic economy. Its main task was to raise money for colonial transactions.

While the first part of the work deals with these three institutions at the center of the system, the second part will stress the meaning of the colonies. It will be investigated how the system dealt with ecological knowledge about tropical plants and forests, which was generated in the colonies. Here, new seeds and plants were introduced in small scale experiments by the colonial scientific stations, but also in large scale enterprises by planters and foresters. The small scale experiments were performed to gain information about the possibilities to plant specific biological populations. If small scale experiments were successful, the possibility of large scale plantation had to be calculated by foresters with regard to the broader ecological conditions as well as e.g. working and transportation costs. The calculations and experiments generated knowledge about the ecological conditions of the colonies, which was necessary to make conclusions about the possibility to plant specific species.

This colonial knowledge did not always endorse imperial interests of the mother country to gain cheap raw materials for industrial production. Therefore, there was an inherent tension in the colonial system of the "Kaiserreich" in regard of planting renewable resources: It was not always possible to act ecologically to fulfill the economic requirements. My project will investigate how the colonial system dealt with the tensions and in which cases ecological concerns were subdued to economic interests.

Zusammenfassung

In der zweiten Hälfte des langen 19. Jahrhunderts versuchten imperiale Großmächte, die direkte Kontrolle über die Waldressourcen der tropischen Welt zu erlangen. Ziel war die Ausbeutung wertvoller Nutz- und Bauholzreserven sowie anderer industriell verwertbarer forstlicher Nebenprodukte.

Meine Arbeit untersucht beispielhaft für das Deutsche Kaiserreich, wie sich ein koloniales System zur Ausbeutung von nachwachsenden Rohstoffen entwickelte. Besonderes Augenmerk wird auf die weltweiten, netzwerkartigen Beziehungen gerichtet, die für die Herausbildung dieses Systems konstitutiv waren.

Im politischen Zentrum des Systems stand die Kolonialabteilung des Auswärtigen Amtes, die ab 1907 im neu gegründeten Reichskolonialamt aufging. Hier wurden Entscheidungen über Pflanzungsversuche, aber auch über die Verteilung von Wissen innerhalb des kolonialen Systems getroffen. Während in der Sphäre der Macht die Schaltzentrale des Systems lag, war der „think tank" bei der Botanischen Zentralstelle für die deutschen Kolonien in Berlin zu suchen. Hier wurden nicht nur Pflanzenproben aus den Kolonien zentral gesammelt und ausgewertet, die Zentralstelle war auch der Knotenpunkt eines weltweit agierenden Netzwerks zum Transfer von Samen und Pflanzenmaterial in die deutschen Kolonien.[2] Das Kolonialwirtschaftliche Komitee war als drittes Hauptelement des Systems die Schnittstelle zur deutschen Wirtschaft und verantwortlich, die monetären Mittel für die Transfers zu beschaffen.

Während die oben genannten Institutionen im Zentrum des Systems den Handel und den Transfer von Pflanzenmaterial organisierten, wurden in den Kolonien die Anbauversuche mit dem bereitgestellten, neuen Pflanzmaterial unternommen. Hier wurde auch das Wissen über die ökologischen Verhältnisse vor Ort generiert. Diese waren nicht unbedingt förderlich für die kolonialen Interessen des Mutterlandes. Und so wurden dem imperialen Interesse zur Ausbeutung erneuerbarer Ressourcen die ökologischen Bedenken aus den Kolonien oftmals untergeordnet. Hieraus ergaben sich grundsätzliche Spannungen im kolonialen System des Kaiserreichs, denen im Forschungsprojekt nachgegangen werden soll.

Lars KREYE
Georg-August-Universität Göttingen
DFG-Graduiertenkolleg Interdisziplinäre Umweltgeschichte
Bürgerstraße 50
37073 Göttingen
Germany

Phone: +49 551 3922389
Fax: +49 551 393645
E-Mail: lkreye@biologie.uni-goettingen.de

---

2 Hingegen spielten die Forsthochschulen und auch die Kolonialschulen im Reich zu diesem Zeitpunkt eine eher untergeordnete Rolle. Hier fehlte das Erfahrungswissen zum Umgang mit tropischen Wäldern fast gänzlich, wie aus den zeitgenössischen Äußerungen bekannter Forstwissenschaftler zu entnehmen ist. Die forstwissenschaftlichen Hochschulen sahen sich kaum in der Lage, eine adäquate forstliche Ausbildung für den Kolonialdienst zu gewährleisten.

350 Jahre Leopoldina – Anspruch und Wirklichkeit
Festschrift der Deutschen Akademie der Naturforscher Leopoldina 1652–2002

Herausgegeben von
Benno PARTHIER (Halle/Saale) und Dietrich VON ENGELHARDT (Lübeck)
(2002, 816 Seiten, 130 Abbildungen, 8 Tabellen, 54,90 Euro, ISBN: 3-928466-45-3)

Die älteste deutschsprachige Akademie prüft „Anspruch und Wirklichkeit" ihrer Vergangenheit und lässt 350 Jahre wechselvoller Geschichte in ihren naturwissenschaftlichen und medizinischen Rahmenbedingungen Revue passieren. Die Festschrift wendet sich an eine interessierte Öffentlichkeit, die allmählich diese besondere Akademie in der deutschen und internationalen Akademienlandschaft mit ihrer spezifischen wissenschaftlich-kulturellen Bedeutung wahrnimmt, nachdem die Wirkungen von 40 Jahren defizitärer Existenz „hinter dem eisernen Vorhang" überwunden werden konnten. (Klappentext)

Inhalt:
Teil I: Geschichte der Leopoldina in Schwerpunkten
Teil II: Die Leopoldina im Spiegel einzelner Wissenschaftsdisziplinen
Teil III: Querschnittsthemen
Teil IV: Anhänge

Mit Beiträgen von:
Gunnar BERG (Halle/Saale), Johanna BOHLEY (Halle/Saale), Dietrich VON ENGELHARDT (Lübeck), Menso FOLKERTS (München), Bernhard FRITSCHER (München), Sybille GERSTENGARBE (Halle/Saale), Fritz HARTMANN (Hannover), Lothar JAENICKE (Köln), Ilse JAHN (Berlin), Joachim KAASCH (Halle/Saale), Michael KAASCH (Halle/Saale), Kai Torsten KANZ (Lübeck), Andreas KLEINERT (Halle/Saale), Eberhard KNOBLOCH (Berlin), Dorothea KUHN (Marbach), Irmgard MÜLLER (Bochum), Uwe MÜLLER (Schweinfurt), Gisela NICKEL (Ober-Olm), Thomas NICKOL (Halle/Saale), Benno PARTHIER (Halle/Saale), Horst REMANE (Halle/Saale), Hermann-J. RUPIEPER (Halle/Saale), Klaus SANDER (Freiburg i. Br.), Thomas SCHNALKE (Berlin), Werner SCHROTH (Halle/Saale), Eugen SEIBOLD (Freiburg i. Br.), Eduard SEIDLER (Freiburg i. Br.), Richard TOELLNER (Rottenburg-Bieringen) und Gudrun WOLFSCHMIDT (Hamburg).

Druck-Zuck GmbH, Seebener Straße 4, 06114 Halle/Saale
Buchbestellung on-line: www.druck-zuck.net

Flooded. Social Perspectives on Natural Disaster in 19th Century Germany

Patrick Masius (Göttingen)

Abstract

Floods have affected human societies since ages. Today, they are a major concern to people and politics on Germany. Studying disasters of the past can help to understand modes of reaction and precaution that might be still relevant. This project will analyze catastrophes in terms of social perception, political organization and economic mechanisms. Furthermore, it will be examined how the environment and the social impact on resources has contributed to flooding. The general focus, therefore, lies on the interactive relation between environment and humans. In one direction questions of coping (short term) and how natural disaster has shaped social organization (long term) will be addressed. Complementary, it is of interest how humans have generated hazards themselves by changing their environment. The theoretical background is given by theories of vulnerability and resilience. Additionally, a time dimension is induced into the vulnerability discourse which enables specific environmental history perspective on disasters. It is this connection of historical environmental change with socio-economic short term reaction that offers a promising approach to catastrophes. In this respect, the "Gründerzeit" at the end of the 19th century, a key period of industrialization and modernization, will be reviewed in terms of anthropogenic impacts on water environments and newly developed political, social and economic coping mechanisms. Case studies will be drawn from the floods at the Rhine (1882/83) and the Baltic Sea (1872). In conclusion today's strategies of disaster management will be set in historical context and be evaluated.

Zusammenfassung

Von Überschwemmungen war die menschliche Gesellschaft schon in vielen Epochen betroffen. Heutzutage sind sie ein sehr wichtiges Anliegen der Menschen und der Politik in Deutschland. Die Analyse von Katastrophen in der Vergangenheit soll dazu beitragen, Reaktionsmuster und Vorsichtsmaßnahmen zu identifizieren, die auch heute noch relevant sein könnten. Mein Projekt wird Katastrophen der Vergangenheit unter dem Gesichtspunkt der sozialen Wahrnehmung, der politischen Organisation und wirtschaftlicher Mechanismen analysieren. Des Weiteren soll untersucht werden, wie die Umwelt ebenso wie soziale Auswirkungen auf Ressourcen zu Überschwemmungen beigetragen haben. Der allgemeine Schwerpunkt liegt daher auf dem interaktiven Verhältnis zwischen Mensch und Umwelt. Einerseits werden Fragen der Bewältigung (kurzfristig) und wie die Naturkatastrophe die soziale Organisation geprägt hat (langfristig) behandelt. Ergänzend dazu ist es von Interesse, wie der Mensch selbst Risiken herbeigeführt hat, indem er seine Umwelt verändert hat. Den theoretischen Hintergrund bilden Theorien über Vulnerabilität und Resilienz. Zusätzlich wird in die Vulnerabilitätsdebatte eine zeitliche Dimension mit aufgenommen, durch die eine spezifische umweltgeschichtliche Perspektive im Hinblick auf Katastrophen ermöglicht wird. In dieser Verbindung von umwelthistorischer Veränderung mit der kurzfristigen sozioökonomischen Reaktion liegt eine vielversprechende Herangehensweise an Katastrophen. In dieser Hinsicht wird die „Gründerzeit" Ende des 19. Jahrhunderts, eine entscheidende Periode der Industrialisierung und Modernisierung, im Hinblick auf anthropogene Auswirkungen auf Wassereinzugsgebiete und neu entwickelte politische, soziale und wirtschaftliche Bewältigungsmechanismen untersucht. Als Fallstudien dienen die Rheinüberschwemmungen von 1882/83 und die Ostseesturmflut von 1872. Anschließend werden die heutigen Strategien des Katastrophenmanagements historisch in Bezug gesetzt und bewertet.

Patrick MASIUS
Georg-August-Universität Göttingen
DFG-Graduiertenkolleg Interdisziplinäre Umweltgeschichte
Bürgerstraße 50
37073 Göttingen
Germany
Phone: +49 551 394813
Fax: +49 551 393645
E-Mail: pmasius@gwdg.de

The Influence of Medical Topographies on Urban Development and Residents' Health in Urban Environments in the 18th/19th Century

Tanja Zwingelberg (Göttingen)

Abstract

The project deals with the findings of what is known as medical topographies in the late 18th, early 19th century. At the centre of the research is the question of what advice the topographies hold for urban development. This includes the question of health sustainment, health improvement and the containment of diseases in terms of physical, mental and social well-being as determined by the WHO. Using this as a starting point, it will be analyzed whether such pieces of advice were considered at all by local and societal authorities, and if so, to what extent. The aim is to discuss possible links between advice in medical topographies, urban development and reactions within local actions.

If no evidence for advice-based urban development can be found, the further aim will be to find out about the reasons which prevented the implementation of this advice.

The existence or non-existence of a health-oriented influence in urban development will be analyzed on the basis of two exemplary cities, Berlin and Hamburg. The selected areas will be examined according to urban- and street-layout in general, to arrangement and design of housing space, to drinking water supply and sewage as well as to site quality and hygienic quality of building materials.

Within those research areas, emphasis is to be put on potential changes in action which were motivated by medical advice. The aim of the doctoral thesis is it to reduce the research deficit in the area of health orientation with reference to urban development in the historical context of the 18th and early 19th century.

Zusammenfassung

Das Projekt befasst sich mit dem zeitgenössischen medizinischen Diskurs, also auch mit den sogenannten medizinischen Topographien, des späten 18. und frühen 19. Jahrhunderts. Im Zentrum der Forschung steht die Frage, welche Ratschläge sich aus den Topographien und dem allgemeinen medizinischen Diskurs für die Stadtentwicklung ableiten lassen. Hier geht es beispielsweise um Fragen wie Aufrechterhaltung der Gesundheit, Verbesserung der Gesundheit und die Eindämmung von Krankheiten im Sinne von physischem, mentalem und sozialem Wohlergehen, wie es heute durch die WHO definiert wird. Davon ausgehend, soll analysiert werden, ob solche Ratschläge überhaupt von den kommunalen und gesellschaftlichen Instanzen in Betracht gezogen wurden, und wenn ja, in welchem Umfang. Das Ziel besteht darin, mögliche Verbindungen zwischen den Ratschlägen in den medizinischen Topographien, städtischer Entwicklung und den Reaktionen im Rahmen lokaler Aktionen zu diskutieren.

Wenn sich keine Hinweise auf eine auf diesen Ratschlägen basierende Stadtentwicklung finden lassen, wird unser nächstes Anliegen sein, herauszufinden, weshalb die Umsetzung solcher Ratschläge nicht erfolgte.

Die Existenz bzw. Nichtexistenz eines gesundheitsorientierten Einflusses bei der Stadtentwicklung wird auf Grundlage zweier Beispielstädte, Berlin und Hamburg, analysiert. Die ausgewählten Gebiete werden entsprechend der Stadtphysiognomie im Allgemeinen, der Trinkwasser- und sanitären Versorgung ebenso wie der Qualität des Standorts und der Hygienequalität der Baumaterialien untersucht.

Im Rahmen dieser Untersuchung sollen mögliche Veränderungen bei den Maßnahmen, die auf medizinische Ratschläge zurückzuführen sind, besonders berücksichtigt werden. Es ist das Ziel meiner Dissertation, zu einer Verringerung des Forschungsdefizits beizutragen, das auf dem Gebiet der Gesundheitsorientierung in Bezug auf die Stadtentwicklung im historischen Kontext des 18. und frühen 19. Jahrhunderts besteht.

    Tanja Zwingelberg
    Georg-August-Universität Göttingen
    DFG-Graduiertenkolleg Interdisziplinäre Umweltgeschichte
    Bürgerstraße 50
    37073 Göttingen
    Germany
    Phone: +49 551 399789
    Fax: +49 551 393645
    E-Mail: tzwinge@gwdg.de

Conceptions and Ways of Dealing with Post-Mining Landscapes.
A Cultural Analysis of Planning Discourses

Markus Schwarzer (Göttingen)

Abstract

Since the enormous structural transformation following German reunification took place, there has been an intensive debate about the reclamation, recultivation and design of the former lignite mining districts in Lusatia and around Leipzig. In the process, there have been concerted efforts to create a new, positive image for the landscape that has for a long time been considered to be destroyed. Next to concepts of agricultural and forest recultivation, conceptions of recreation and tourism, preservation of industrial monuments, and nature conservation are priorities when dealing with post-mining landscapes. The ideas, values, and symbolism of this landscape which are generated in these creative planning discourses will be at the center of my study. The main point is to depict the positions and concepts involved in the formation of post-mining landscapes, which are in part very hotly discussed within the subject areas mentioned. In doing so, the following questions must be considered: how are these landscapes respectively perceived, described and evaluated? What role does the pre-mining landscape play in these concepts? Which typical formation concepts characterize the subject areas? When analyzing these concepts, the institutional context should be reflected, and concrete formation processes should be compared selective in a framework which must be tightly delineated. The goal of this study is to examine the creative positions and concepts in this very diverse discourse, to organize them in a typified way, and to illustrate them exemplarily.

Zusammenfassung

Seit dem enormen Strukturwandel nach der Wiedervereinigung wird intensiv über die Sanierung, Rekultivierung und Gestaltung der stillgelegten Braunkohletagebaugebiete in der Lausitz und im Mitteldeutschen Revier um Leipzig debattiert. Dabei wird für die lange als zerstört geltende Landschaft gezielt um ein neues, positives Image gerungen. Erholungs- und Tourismuskonzepte sowie Industriedenkmalpflege und Naturschutz bilden neben Konzepten zur land- und forstwirtschaftlichen Rekultivierung Schwerpunkte bei der Gestaltung der Bergbaufolgelandschaften. Im Mittelpunkt meiner Untersuchung werden die in diesen planerisch-gestalterischen Diskursen generierten Vorstellungen, Werte und Symboliken jener Landschaften stehen. Es geht im Kern darum, die Positionen und Konzepte zur Gestaltung der Bergbaufolgelandschaften, welche auch innerhalb der genannten Themenfelder zum Teil sehr

Markus Schwarzer

kontrovers diskutiert werden, darzustellen. Dabei sind folgende Fragen leitend: Wie werden diese Landschaften jeweils wahrgenommen, beschrieben und bewertet? Welche Rolle spielt die Landschaft vor dem Bergbau in den Konzepten? Welche typischen Gestaltungskonzepte zeichnen die Themenfelder aus? Bei der Analyse der Konzepte soll der institutionelle Kontext reflektiert und punktuell auch konkrete Gestaltungsprozesse in einem eng abzugrenzenden Rahmen verglichen werden. Das Ziel der Arbeit ist, die gestalterischen Positionen und Konzepte in diesem sehr facettenreichen Diskurs zu untersuchen, in einer typisierenden Weise zu ordnen und exemplarisch zu veranschaulichen.

Markus Schwarzer
Georg-August-Universität Göttingen
DFG-Graduiertenkolleg Interdisziplinäre Umweltgeschichte
Bürgerstraße 50
37073 Göttingen
Germany
Phone: +49 551 399798
Fax: +49 551 393645
E-Mail: mschwar@uni-goettingen.de